AF324617

The Paraoxonases: Their Role in Disease Development and Xenobiotic Metabolism

PROTEINS AND CELL REGULATION

Volume 6

Series Editors:

Professor Anne Ridley
*Ludwig Institute for Cancer Research
and Department of Biochemistry and Molecular Biology
University College London
London
United Kingdom*

Professor Jon Frampton
*Professor of Stem Cell Biology
Institute for Biomedical Research,
Birmingham University Medical School,
Division of Immunity and Infection
Birmingham
United Kingdom*

Aims and Scope

Our knowledge of the ways in which a cell communicates with its environment and how it responds to information received has reached a level of almost bewildering complexity. The large diagrams of cells to be found on the walls of many a biologist's office are usually adorned with parallel and interconnecting pathways linking the multitude of components and suggest a clear logic and understanding of the role played by each protein. Of course this two-dimensional, albeit often colourful representation takes no account of the three-dimensional structure of a cell, the nature of the external and internal milieu, the dynamics of changes in protein levels and interactions, or the variations between cells in different tissues.

Each book in this series, entitled *"Proteins and Cell Regulation"*, will seek to explore specific protein families or categories of proteins from the viewpoint of the general and specific functions they provide and their involvement in the dynamic behaviour of a cell. Content will range from basic protein structure and function to consideration of cell type-specific features and the consequences of disease-associated changes and potential therapeutic intervention. So that the books represent the most up-to-date understanding, contributors will be prominent researchers in each particular area. Although aimed at graduate, postgraduate and principle investigators, the books will also be of use to science and medical undergraduates and to those wishing to understand basic cellular processes and develop novel therapeutic interventions for specific diseases.

The Paraoxonases: Their Role in Disease Development and Xenobiotic Metabolism

Edited by

Bharti Mackness
University of Manchester, U.K.

Mike Mackness
University of Manchester, U.K.

Michael Aviram
Technion Faculty of Medicine, Haifa, Israel

and

György Paragh
University of Debrecen, Hungary

A C.I.P. Catalogue record for this book is available from the Library of Congress.

ISBN 978-1-4020-6560-6 (HB)
ISBN 978-1-4020-6561-3 (e-book)

Published by Springer,
P.O. Box 17, 3300 AA Dordrecht, The Netherlands.

www.springer.com

Printed on acid-free paper

Cover Illustration by Dr. D. I. Draganov

CONTENTS

CONTRIBUTORS

A. Alipour   Department of Internal Medicine, Sint Franciscus Gasthuis, Rotterdam, The Netherlands

C. Alonso-Villaverde   Centre de Recerca Biomèdica and Department of Internal Medicine, Hospital Universitari de Sant Joan, Reus, Spain

M. Aviram   The Lipid Research Laboratory, Technion Faculty of Medicine, The Rappaport Family Institute for Research in the Medical Sciences, Rambam Medical Center, Haifa, Israel

N. Bourquard   Atherosclerosis Research Unit, Department of Medicine, David Geffen School of Medicine at UCLA, Los Angeles, CA, USA

V.H. Brophy   Departments of Medicine (Div. Medical Genetics) and Genome Sciences, University of Washington, Seattle, WA 98195, USA; Roche Diagnostics, Alameda, CA, USA

M.C. Cabezas   Department of Internal Medicine, Sint Franciscus Gasthuis, Rotterdam, The Netherlands

J. Camps   Centre de Recerca Biomèdica and Department of Internal Medicine, Hospital Universitari de Sant Joan, Reus, Spain

C. Carlson   Department of Genome Sciences, University of Washington, Seattle, WA 98195, USA; The Fred Hutchinson Cancer Research Center, Seattle, WA, USA

D.M. Cerasoli   Research Division, U.S. Army Medical Research Institute of Chemical Defense, 3100 Ricketts Point Rd., Aberdeen Proving Ground, MD 21010-5400, USA

E. Chabriere   Laboratoire de Cristallographie et Modélisation des Matériaux Minéraux et Biologiques, CNRS-Université Henri Poincaré, BP 239, 54506 Vandoeuvre-lès-Nancy, France

C. Clery-Barraud   Département de Toxicologie, Centre de Recherches du Service de Santé des Armées, BP 87, 38702 La Tronche cedex, France

T.B. Cole   Department of Environmental and Occupational Health Sciences, University of Washington, Seattle, WA, USA; Departments of Genome Sciences and Medicine (Div. Medical Genetics), University of Washington, Seattle, WA, USA

B. Coll   Centre de Recerca Biomèdica and Department of Internal Medicine, Hospital Universitari de Sant Joan, Reus, Spain

L.G. Costa   Department of Environmental and Occupational Health Sciences, University of Washington, Seattle, WA, USA; Department of Human Anatomy, Pharmacology and Forensic Medicine, University of Parma, Italy

C. Dalgård   Institute of Public Health – Environmental Medicine. University of Southern Denmark, Odense, Denmark

S.P. Deakin   Division of Endocrinology, Diabetes and Nutrition, University Hospital, Geneva, Switzerland

D.I. Draganov   Department of Metabolism, WIL Research Laboratories, LLC, Ashland, OH 44 805, USA

M. Elias   Laboratoire de Cristallographie et Modélisation des Matériaux Minéraux et Biologiques, CNRS-Université Henri Poincaré, BP 239, 54506 Vandoeuvre-lès-Nancy, France

J.W.F. Elte   Department of Internal Medicine, Sint Franciscus Gasthuis, Rotterdam, The Netherlands

N. Ferré   Department of Clinical Biochemistry and Molecular Genetics, Hospital Clínic Universitari, Barcelona, Spain

A.M. Fogelman   Atherosclerosis Research Unit, Department of Medicine, David Geffen School of Medicine at UCLA, Los Angeles, CA, USA

T. Fulop   First Department of Medicine, Medical and Health Science Center, University of Debrecen, Hungary

C.E. Furlong   Departments of Genome Sciences and Medicine (Div. Medical Genetics), University of Washington, Seattle, WA 98195, USA

F. Gil   Department of Legal Medicine and Toxicology, University of Granada Medical School, Avda, Madrid, 11 18071 Granada, Spain

V. Grijalva Atherosclerosis Research Unit, Department of Medicine, David Geffen School of Medicine at UCLA, Los Angeles, CA, USA

S.Y. Hama David Geffen School of Medicine at UCLA, Los Angeles, CA, USA

M. Harangi Department of Medicine, Medical and Health Science Center, University of Debrecen, Hungary

A.F. Hernández Department of Legal Medicine and Toxicology, University of Granada Medical School, Avda, Madrid, 11 18071 Granada, Spain

G. Hough David Geffen School of Medicine at UCLA Los Angeles, CA, USA

H. Jakubowski Department of Microbiology and Molecular Genetics, UMDNJ-New Jersey Medical School, International Center for Public Health, 225 Warren Street, Newark, NJ 07101, USA; Institute of Bioorganic Chemistry, Polish Academy of Sciences, 61-704 Poznań, Poland

R.W. James Division of Endocrinology, Diabetes and Nutrition, University Hospital, Geneva, Switzerland

K.L. Jansen Department of Environmental and Occupational Health Sciences, University of Washington, Seattle, WA, USA

G.P. Jarvik Departments of Medicine (Div. Medical Genetics) and Genome Sciences, University of Washington, Seattle, WA 98195, USA

J. Joven Centre de Recerca Biomèdica and Department of Internal Medicine, Hospital Universitari de Sant Joan, Reus, Spain

M. Kayikcioglu Ege University Faculty of Medicine, Department of Cardiology, Ege, Turkey

A. Khalil First Department of Medicine, Medical and Health Science Center, University of Debrecen, Hungary

D.E. Lenz Research Division, U.S. Army Medical Research Institute of Chemical Defense, 3100 Ricketts Point Rd., Aberdeen Proving Ground, MD 21010-5400, USA

W.-F. Li Division of Environmental Health and Occupational Medicine, National Health Research Institutes, Zhunan, Taiwan

O. López   Department of Legal Medicine and Toxicology, University of Granada Medical School, Avda, Madrid, 11 18071 Granada, Spain

A.J. Lusis   Department of Medicine, UCLA, Los Angeles, CA 90095-1697, USA

B. Mackness   Division of Cardiovascular and Endocrine Sciences, University of Manchester, Department of Medicine, Manchester Royal Infirmary, Oxford Road, Manchester M13 9WL, UK

M.I. Mackness   Division of Cardiovascular and Endocrine Sciences, University of Manchester, Department of Medicine, Manchester Royal Infirmary, Oxford Road, Manchester M13 9WL, UK

J. Marsillach   Centre de Recerca Biomèdica and Department of Internal Medicine, Hospital Universitari de Sant Joan, Reus, Spain

P. Masson   Département de Toxicologie, Centre de Recherches du Service de Santé des Armées, BP 87, 38702 La Tronche cedex, France

K. Nakamura   School of Medicine, UCSF, San Francisco, CA, USA

M. Navab   Atherosclerosis Research Unit, Department of Medicine, David Geffen School of Medicine at UCLA, Los Angeles, CA, USA

C.J. Ng   Atherosclerosis Research Unit, Department of Medicine, David Geffen School of Medicine at UCLA, Los Angeles, CA, USA

D. Nickerson   Department of Genome Sciences, University of Washington, Seattle, WA 98195, USA

E.A. Ozer   Department of Internal Medicine, Roy J. and Lucille A. Carver College of Medicine, University of Iowa, Iowa City, IA 52242, USA

G. Paragh   Department of Medicine, Medical and Health Science Center, University of Debrecen, Hungary

S. Parra   Centre de Recerca Biomèdica and Department of Internal Medicine, Hospital Universitari de Sant Joan, Reus, Spain

T. Parrón   Delegación Provincial de Salud, Almería, Spain

G. Pena   Centro de Salud Motril Este, SAS, Granada, Spain

A. Pla Department of Legal Medicine and Toxicology, University of Granada Medical School, Avda, Madrid, 11 18071 Granada, Spain

J. Ranchalis Departments of Medicine (Div. Medical Genetics) and Genome Sciences, University of Washington, Seattle, WA 98195, USA

S.T. Reddy Atherosclerosis Research Unit, Department of Medicine, David Geffen School of Medicine at UCLA, Los Angeles, CA, USA

F. Renault Département de Toxicologie, Centre de Recherches du Service de Santé des Armées, BP 87, 38702 La Tronche cedex, France

R.J. Richter Departments of Medicine (Div. Medical Genetics) and Genome Sciences, University of Washington, Seattle, WA 98195, USA

M. Rieder Department of Genome Sciences, University of Washington, Seattle, WA 98195, USA

A.P. Rietveld Department of Internal Medicine, Sint Franciscus Gasthuis, Rotterdam, The Netherlands

D. Rochu Département de Toxicologie, Centre de Recherches du Service de Santé des Armées, BP 87, 38702 La Tronche cedex, France; Bundeswehr Institute of Pharmacology and Toxicology, 80937 Munich, Germany

L. Rodrigo Department of Legal Medicine and Toxicology, University of Granada Medical School, Avda, Madrid, 11 18071 Granada, Spain

M. Roest Laboratory for clinical Chemistry and Haematology, UMC Utrecht, The Netherlands

M. Rosenblat The Lipid Research Laboratory, Technion Faculty of Medicine, The Rappaport Family Institute for Research in the Medical Sciences, Rambam Medical Center, Haifa, Israel

N. Rozengurt Department of Pathology and Laboratory Medicine, UCLA, Los Angeles, CA, USA

F.G. Sagin Ege University Faculty of Medicine, Department of Biochemistry, Ege, Turkey

O. Sapir The Lipid Research Laboratory, Technion Faculty of Medicine, The Rappaport Family Institute for Research in the Medical Sciences, Rambam Medical Center, Haifa, Israel

I. Seres   Department of Medicine, Medical and Health Science Center, University of Debrecen, Hungary

J.L. Serrano   Delegación Provincial de Salud, Almería, Spain

D.M. Shih   Atherosclerosis Research Unit, Department of Medicine, David geffen School of Medicine at UCLA, Los Angeles, CA, USA

B. Sozmen   Atatürk Research and Trainig Hospital, Department of Internal Medicine, Izmir/Republic of Turkiye

E.Y. Sozmen   Ege University Faculty of Medicine, Department of Biochemistry, Egen, Turkey

D.A. Stoltz   Department of Internal Medicine, Roy J. and Lucille A. Carver College of Medicine, University of Iowa, Iowa City, IA 52242, USA

J.F. Teiber   Department of Internal Medicine, Division of Epidemiology, The University of Texas Southwestern Medical Center, Dallas, TX 75390, USA

A. Tward   Department of Medicine, UCLA, Los Angeles, CA 90095-1697; University of California, San Francisco, CA, USA

N. Villa-Garcia   Atherosclerosis Research Unit, Department of Medicine, David Geffen School of Medicine at UCLA, Los Angeles, CA, USA

H.A.M. Voorbij   Laboratory for clinical Chemistry and Haematolgy, UMC Utrecht, The Netherlands

D.T. Yeung   Research Division, U.S. Army Medical Research Institute of Chemical Defense, 3100 Ricketts Point Rd., Aberdeen Proving Ground, MD 21010-5400, USA

J. Zabner   Department of Internal Medicine, Roy J. and Lucille A. Carver College of Medicine, University of Iowa, Iowa City, IA 52242, USA

PREFACE

Research into the paraoxonase (PON) multigene family has blossomed in the last 10 years. Before this time only PON1 was known and research was restricted to toxicologists investigating the metabolism of organophosphate insecticides and nerve gases and a few scientists searching for "natural" substrates.

Since this time two new members PON2 and PON3 have been discovered. All 3 PONs have been shown to act as antioxidants and the PON family has taken centre stage as major players in the development of a wide variety of diseases such as cardiovascular disease, diabetes mellitus, rheumatism, Alzheimers and many others while remaining important in determining organophosphate toxicity.

In September 2006 the 2nd International Conference on Paraoxonases took place in Hajdúszoboszló, Hungary, bringing together the worlds foremost experts in the field. The current book is a distillation of the plenary lectures which took place at the meeting, resulting in a comprehensive up to date, state of the art review of the current and the historic paraoxonase research.

PART 1

THE BERT LA DU MEMORIAL LECTURE

PARAOXONASES: AN HISTORICAL PERSPECTIVE

C.E. FURLONG

*Departments of Medicine (Div. Medical Genetics) and Genome Sciences
and, University of Washington, Seattle, WA 98195, USA*

Abstract: This chapter provides a brief overview of the history of studies on human paraoxonases. It honors the memory of the late Dr. Bert La Du (1920–2005), who with his graduate students, postdoctoral fellows and collaborators made many contributions to our knowledge of this family of enzymes and the genes that encode them. Dr. La Du was honored for these contributions at the First International Conference on Paraoxonases (PONs) – *"Paraoxonases: Basic and Clinical Directions of Current Research"* held in Ann Arbor, Michigan in 2004. Many of the scientists who trained with and/or collaborated with the late Dr. La Du were present at this Second International Conference on Paraoxonases and have contributed to this volume. This chapter begins with a review of some of the early esterase enzymology and the discovery of plasma paraoxonase activity. The pioneering work of Dr. Norman Aldridge who differentiated the A- and B-esterases is described.

The studies that defined the polymorphic distribution of PON1 in human populations are discussed along with the many different biochemical assays that were developed to explore this interesting polymorphism. The experiments that led to the purification and cloning of human and rabbit PON1s are described along with the properties of this first enzyme know to retain its signal sequences for use in anchoring it into the HDL particle are discussed.

Recent advances by Tawfik and co-workers which include the generation of a PON1 sequence that could be expressed, crystallized and characterized are presented along with the characterization of the many different substrates of this promiscuous enzyme including physiological lactone and xenobiotic lactone substrates. The lactonase activities were characterized by both Tawfik's team and Dr. La Du's research group.

The expression and characterization of PON1, PON2 and PON3 by Dr. La Du's research team is also discussed. This effort along with related work by other research groups has greatly expanded our knowledge of the many different activities of the PON family of enzymes. It is probably appropriate to include these proteins in the antioxidant family of proteins.

The history of the role of the PONs in lipid metabolism and the association of the genetic variability in the PON family of enzymes is discussed. The important take home lesson from understanding the relationship of genetic variability of PON1 and risk for vascular disease was often stressed by Dr. La Du as well as other leaders in PON1 research is that is both the quantity (plasma PON1 level) as well as the quality of PON1 (position 192 genotype) that need to be considered when evaluating risk of disease.

Experiments on the relationship of the genetic variability of PON1 and risk of exposure to organophosphorus compounds are also discussed. The take home message is

*B. Mackness et al. (eds.), The Paraoxonases: Their Role in Disease Development
and Xenobiotic Metabolism*, 3–32.

the same, in some cases the quality of PON1 (Q192R) is important, but in all cases, the quantity of plasma PON1 is important. This consideration holds for all epidemiological studies that examine the relationship of PON genetic variability and disease

Keywords: paraoxonase, PON1, organophosphate, organophosphorus compounds (OPs), chlorpyrifos, chlorpyrifos oxon, diazinon, diazoxon, regulation of gene expression, nerve agents, PON1 status, vascular disease, carotid artery disease, developmental regulation of PON1 level, quorum sensing, quorum sensing factors, anti-oxidants

This introductory chapter provides a brief overview of the history of studies on human paraoxonases. It honors the memory of the late Dr. Bert La Du (1920–2005), who made many contributions to our knowledge of this interesting family of enzymes, the genes that encode them, and who was honored for these contributions at the First International Conference on Paraoxonases (PONs) – *"Paraoxonases: Basic and Clinical Directions of Current Research"* held in Ann Arbor, Michigan in 2004. If we had to pick a single individual who contributed the most to our understanding of the paraoxonase family of enzymes and genes that encode them, it would be Dr. La Du. The reader is referred to excellent reviews for more extensive details of early PON1 research (Draganov and La Du, 2004; Geldmacher-von Mallinckrodt and Diepgen, 1988; La Du, 1992; Mackness et al., 1998) and to specialized reviews for the relationship of paraoxonase genetic variability and disease risk (e.g., Durrington et al., 2001; van Himbergen et al., 2006; Ng et al., 2005). The first book on paraoxonase (PON1) appeared in 2002 (Costa and Furlong, 2002). So many reviews on paraoxonase have appeared that it would be useful to have a "review of the paraoxonase reviews", including reviews of genetic factors involved in cardiovascular and other diseases that include sections on PON genetic variability.

1.   EARLY STUDIES

Dr Abraham Mazur, working at the Edgewood Arsenal, Maryland, is credited with the discovery that organophosphorus compounds can be hydrolyzed by enzymes (Table 1). Using manometric assays, he found that diisopropyl fluorophosphate was hydrolyzed by extracts of various tissues from human and rabbit, with the liver, kidney, small intestine and plasma having the highest activities (Mazur, 1946). Mazur's studies were followed by an insightful series of studies by Norman Aldridge (1953a, b) who examined rates of paraoxon (E600) hydrolysis in different tissues of rats and rabbits (Figs. 1, 2). One of his interesting observations that proved to be useful for purifying paraoxonase 1 was that rabbits have very high PON1 levels in their plasma (Figs. 1, 3). Aldridge divided esterases into two categories, those that were inhibited by interaction with substrates (B-esterases) and those that could catalytically hydrolyze substrates (A-esterases). Paraoxonase falls into the A-esterase category (Aldridge, 1953b).

Table 1. Table II from Mazur (1946). Hydrolysis of Diisopropyl Fluorophosphate by Tissue Extracts. The velocity of hydrolysis of diisopropyl fluorophosphate was determined as c.mm. of CO_2 liberated in 30 minutes from a bicarbonate buffer, pH 7.4, by 0.5 cc. of a 1:10 tissue extract. The plasma and red blood cell activities were calculated from the activities determined experimentally on more concentrated solutions and also expressed as those of a 1:10 dilution. All data are corrected for the hydrolysis of the diisopropyl fluorophosphate in water.

Tissue	Velocity of hydrolysis by	
	Rabbit	Human
	c.mm. CO_2	*c.mm. CO_2*
Liver	274	457
Kidney	187	319
Small intestine	129	
Plasma	127	19
Lung	55	50
Heart	30	
Brain	20	29
Muscle	12	40
Red cells	9	13

Aldridge examined the substrate and inhibitor specificity of the A- and B-esterases in several species and tissues. He noted that the substrate specificity of the A-esterases and B-esterases from different species varied considerably. Following Aldridge's publications, the serum A-esterase was referred to as paraoxonase, based on its ability to hydrolyze the toxic oxon metabolite of parathion, paraoxon (Fig 4). Following the discovery of two closely related enzymes, paraoxonase has been referred to since as PON1, and the other two related enzymes as PON2 and PON3 (Primo-Parmo et al., 1996). This terminology will be used here to represent the three related proteins and the genes that encode these proteins (*PON1, PON2 and PON3*).

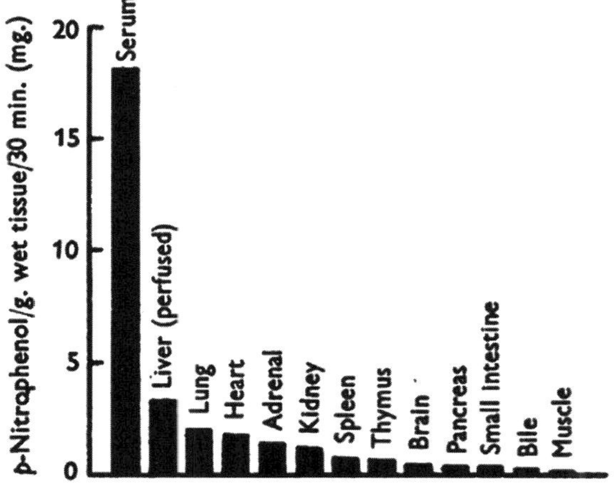

Figure 1. E600-esterase activity of tissues of rabbit. Reproduced from Aldridge, 1953b with permission

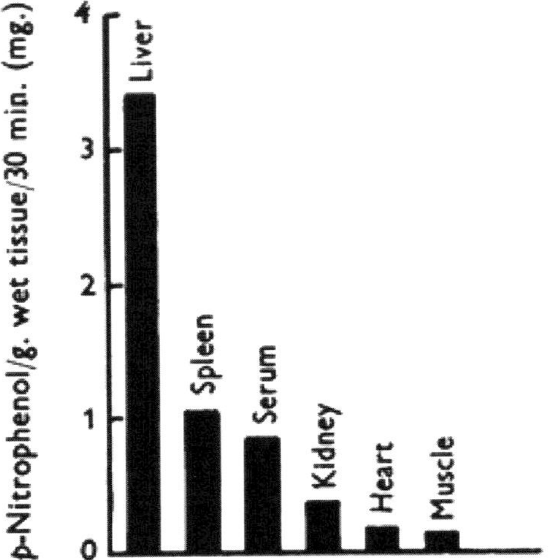

Figure 2. E-600 Esterase activity of rat tissues. Reproduced from Aldridge, 1953b with permission

Studies carried out in the 1960s and 1970s examined the polymorphic distribution of plasma PON1 activity in human populations. Two major findings came out of these early studies. The first was that there was a large variability of plasma/serum PON1 activity among individuals. The second was that the gene frequency for the low activity allele varied considerably among populations of different geographic origin with about 50% of individuals of Northern European origin homozygous for

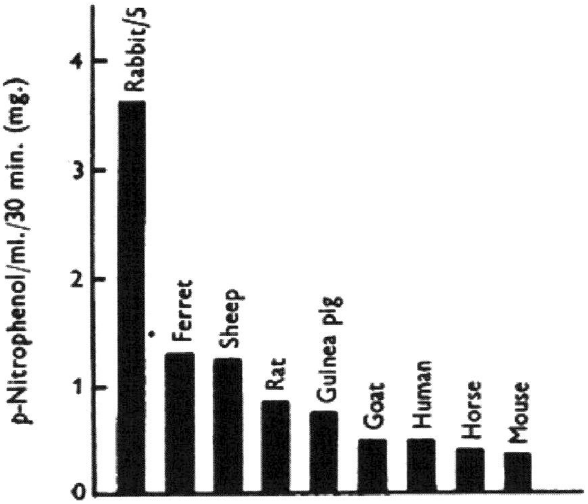

Figure 3. E600-esterase activity of sera of various species. Reproduced from Aldridge, 1953b with permission. Note that the rabbit PON1 activity is 5X that shown

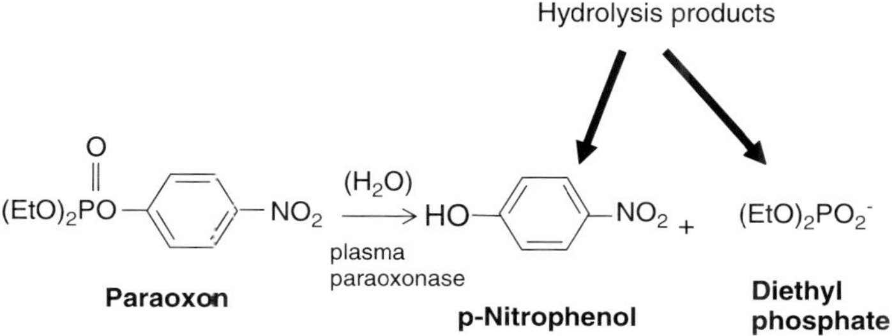

Figure 4. Reaction for which paraoxonase (PON1) was named

the low activity allele, the molecular basis of which was unknown at that time. Some populations such as the Maoris of New Zealand and those of African or Asian origin had low percentages of individuals homozygous for the low activity allele (Diepgen and Geldmacher v. Mallinkrodt, 1986; Fig. 5).

Following these early studies, many different assays were developed to examine the PON1 polymorphism. They varied in pH values, salt concentration and presence and absence of EDTA (Table 2). EDTA inhibition of hydrolysis of paraoxon at high pH values was shown later to affect the polymorphic PON1, but not the albumin esterase hydrolysis of paraoxon (reviewed in Geldmacher-von Mallinckrodt and Diepgen, 1988). The EDTA resistant hydrolysis of paraoxon was not observed in plasma from an analbuminemic individual at high pH values. Thus, it was clearly an activity of albumin that could be observed at high pH values. The albumin activity measured at high pH values represents a significant percentage of the observed paraoxonase activity in individuals with low paraoxonase activity (Ortigoza-Ferado et al., 1984).

In ensuing years, Dr. La Du's research team developed several different assays for examining the PON1 polymorphism (Eckerson et al., 1983). The cumulative distribution of paraoxonase to arylesterase activity ratio in 1M NaCl provided a reasonable resolution of three phenotypes, low, intermediate and high ratios of activity (Fig. 6). Histograms of these ratios also provided a good separation of three activity phenotypes. A number of laboratories have used this protocol to assign phenotypes.

Plotting rates of hydrolysis of one substrate (phenyl acetate) against a second substrate (paraoxon) provided a clear resolution of the individuals with low activities from individuals with intermediate and high activities. However, there was incomplete resolution of the phenotypes (A/B and B/B) of individuals with intermediate and high activities (Fig. 7). This approach did, however, provide the all important measure of plasma PON1 level for each individual in the population and served as the basis for the development of similar assays that did resolve all three PON1-192 phenotypes (see below). Another approach involved measuring paraoxonase activity (with and without 1M NaCl) and phenylacetate hydrolytic activity (Eckerson et al., 1983). It was clear from all of these analyses that there was a large variability of activity observed within each PON1 phenotype (A/A, A/B and

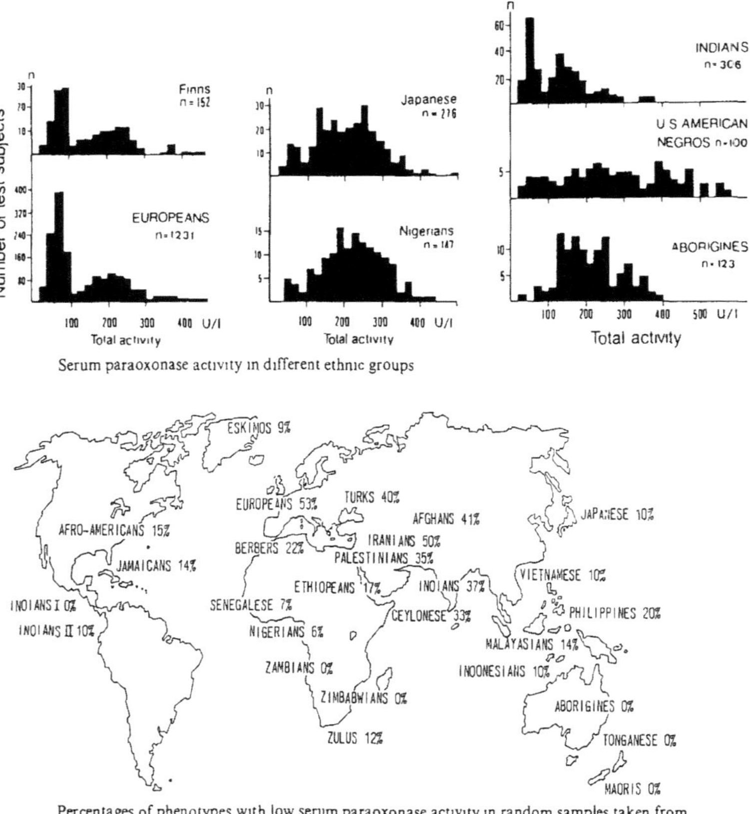

Figure 5. PON1 activity distribution in different ethnic groups (top); percentages of populations having low PON1 activity (bottom). Modified from Diepgen and Geldmacher v. Mallinkrodt, 1986. Reproduced with permission

B/B), the designation at that time for what was later shown to be due to an amino acid substitution (glutamine/arginine) at position 192 of the 354 amino acid PON1 protein (355 before cleavage of the initiator methionine residue).

Following these observations, we used the same two-substrate approach to examine the PON1 activities for hydrolysis of chlorpyrifos oxon, diazoxon, phenylacetate, soman and sarin (Davies et al., 1996, Fig. 8). The later two were included following the releases of sarin in Japan (Ohbu et al., 1997; Suzuki et al., 1995).

These analyses (Fig. 8), and more that followed (see e.g., Fig. 9), showed that plotting the rates of diazoxon hydrolysis against rates of paraoxon hydrolysis at 2M NaCl, pH 8.5 provided a complete separation of the three phenotypes, without any overlap of the three sets of Q192R phenotype data points (Jarvik et al., 2000, 2003b; Richter and Furlong, 1999; Richter et al., 2004).

Table 2. (Table 1 from Ortigoza-Ferado et al., 1984, reproduced with permission) PROCEDURES AND CONDITIONS USED TO INVESTIGATE THE POLYMORPHISM OF HUMAN SERUM PARAOXONASE

LABORATORY	PROCEDURE USED	ASSAY CONDITIONS		REPORTED FREQUENCY DISTRIBUTION
		Buffer used	Final pH value*	
Krisch [1]	Paraoxon hydrolysis (2.4 mM)	0.1 M Na glycinate, pH 11.2	10.5	Bimodal
Geldmacher-v. Mallinckrodt et al. [2, 12]	(A) Cholinesterase protection (32–160 nM paraoxone)	30 mM Na-bicarbonate, pH 7.7	...	Trimodal
[3]	(B) Paraoxon hydrolysis (0.1 mM)	0.1 M Na glycinate, pH 11.2	10.5	Trimodal
Zech and Zurcher [5]	Paraoxon hydrolysis (1 mM)	Triethanolamine 100 mM, pH 7.4, 10 mM $CaCl_2$	...	Bimodal
Playfer et al. [6]	Paraoxon hydrolysis (0.65 mM)	Na-glycinate 52.4 mM, pH 10.5	8.4	Bimodal
Eiberg and Mohr [7]	Paraoxon hydrolysis (0.65 mM)	(A) Tris-HCl 200 mM. pH 7.5	7.5	Bimodal
		(B) Tris-glycine 200 mM, pH 9.5	...	
Carro-Ciampi et al [8]	Paraoxon hydrolysis (0.45 mM)	Na-glycinate 25 mM, pH 10, ±500 mM NaCl± 5 mM $CaCl_2$	10.0	Bimodal
Eckerson et al. [10]	Paraoxon hydrolysis (1 mM)	Glycine 50 mM, pH 10.5, 1 mM $CaCl_2$, ±NaCl	10.5	Bimodal
Mueller et al. [11]	Paraoxon hydrolysis (1.2 mM)	Tris 4.5 mM, Na-glycinate 69 mM	10.3	Trimodal

*The final pH values were determined by us by reproducing the published reaction mixtures and measuring the resulting pH values.

[Assays developed after the molecular basis of high vs. low paraoxonase activity had been elucidated (Adkins et al., 1993; Humbert et al., 1993) (described below) are included in this section for the sake of continuity.]

At the time the above experiments were carried out, it was not know whether the dramatic differences in rates of sarin hydrolysis between $PON1_{Q192}$

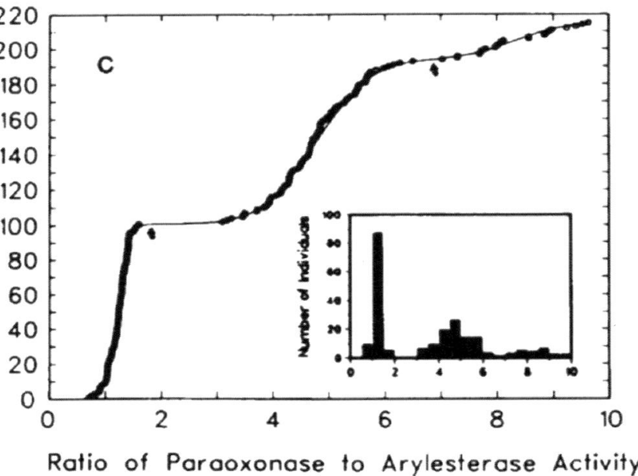

Figure 6. Cumulative distribution of paraoxonase activity in 215 unrelated Caucasians. Inset, histogram of ratio of Paraoxonase to Arylesterase Activity. Reproduced from Eckerson et al., 1983, with permission

homozygotes and $PON1_{R192}$ homozygotes was of physiological significance. More recent observations suggest that the differences in sarin hydrolysis are probably not of physiological consequence (Yamada et al., 2001), much like the differences in paraoxon hydrolysis of the two $PON1_{192}$ alloforms described below (Li et al., 2000).

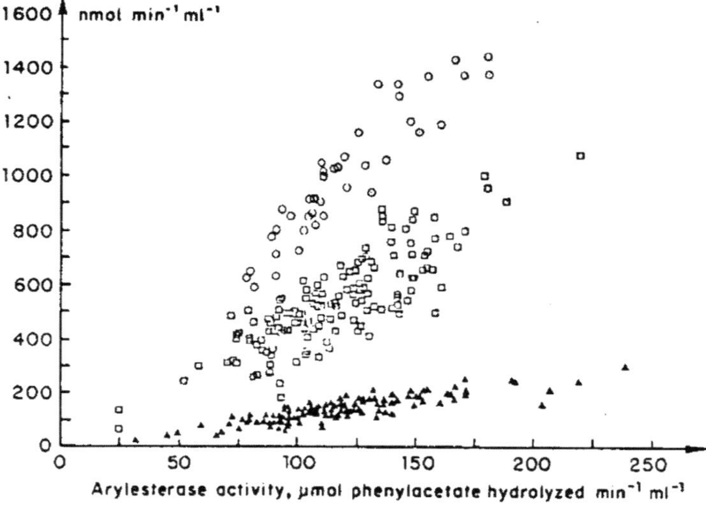

Figure 7. Individual paraoxonase and arylesterase activities. Data collected from 348 people in the United States. Reproduced from La Du et al., 1986, (Fig. 2) with permission

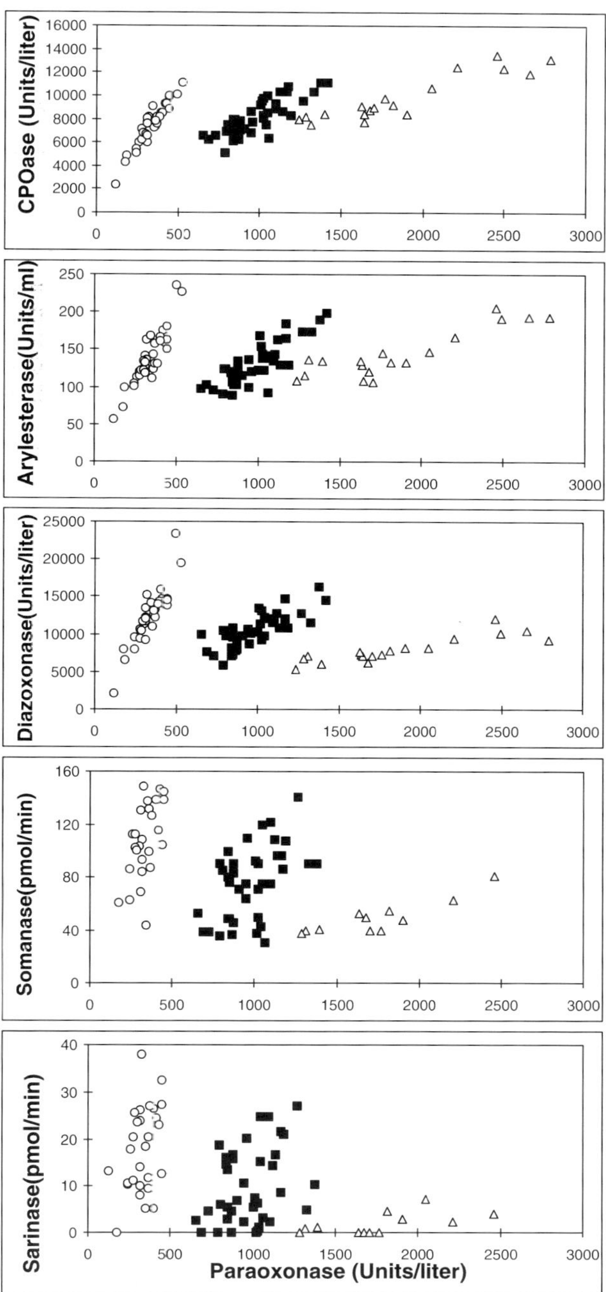

Figure 8. Plots of rates of hydrolysis of the indicated substrates against rates of paraoxon hydrolysis. Genotypes (Q192R) were determined by PCR analysis ($\bigcirc = Q/Q$; $\blacksquare = Q/R$; $\triangle = R/R$). (Modified from Davies et al., 1996, with permission)

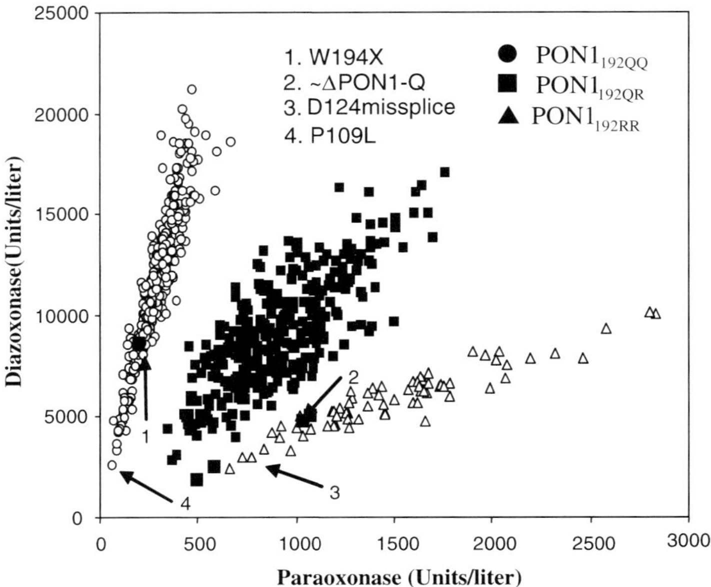

Figure 9. PON1 status plot with identification of new mutations by sequencing PON1 genes from individuals whose Q/R 192 genotype/phenotype was discordant from the functional PON1 2-substrate analysis. Modified from Jarvik et al., 2003b, with permission

Another consequence of the publication of Fig. 8 was that many investigators simply examined the figure and assumed that PON1$_{R192}$ homozygotes would be more sensitive to diazoxon hydrolysis than PON$_{Q192}$ individuals. As noted below, this is not the case. Following a careful characterization of the kinetic properties of the two purified PON1$_{192}$ allofoms, assay conditions were intentionally set to differentially inhibit diazoxon hydrolysis by the PON1$_{R192}$ alloform relative to the PON1$_{Q192}$ alloform so that the two-substrate analysis would provide a clear separation of the three PON1 phenotypes (Q/Q; Q/R and R/R) (Richter and Furlong, 1999). The PON1$_{R192}$ alloform is more sensitive to salt inhibition than the PON1$_{Q192}$ alloform. As noted below, however, PON1$_{R192}$ actually appears to protect a bit better against diazoxon exposure than does PON1$_{Q192}$ (Li et al., 2000).

2.   PURIFICATION OF PON1

Due to its tight association with high density lipoproteins (HDL), PON1 was difficult to purify. Early attempts by a number of research teams used either arylesterase activity (phenylacetate hydrolysis) or paraoxonase activity to follow the purification from several different species (see e.g., Choi and Forster, 1967a, b; Don et al., 1975; Kitchen et al., 1973; Mackness et al., 1988; Main, 1960; Wilde and Kekwick, 1964; Zimmerman et al., 1989). Finally in 1991, the purification of rabbit (Furlong et al., 1991) and human PON1s (Furlong, 1991; Gan et al., 1991) were reported (Figs. 10, 11). Both reports concluded that PON1 hydrolyzed both paraoxon and phenylacetate. These activities co-eluted in all of the columns run by

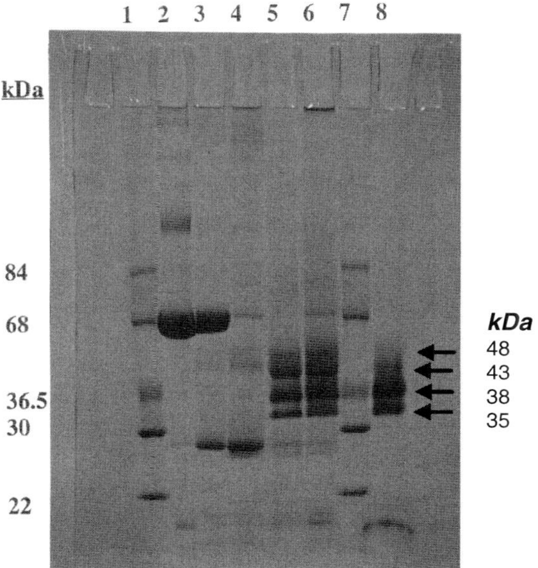

Figure 10. Purification of rabbit PON1 with activity stain. Lanes: 1, 7. MW Stds; 2. serum; 3. Agarose blue; 4. G75; 5. DEAE; 6. G75-2; 8 DEAE-Separate prep. Modified from Furlong et al., 1991, with permission. The bands at 35 and 38 kDa stained bright red

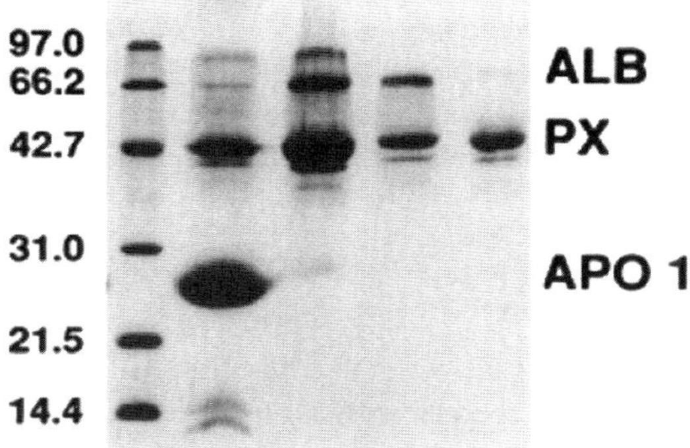

Figure 11. Purification of human PON1. Lanes: 1. Agarose blue; 2. DEAE I; 3. DEAE II; 4. DEAE II — agarose blue recycle. (reproduced from Gan et al., 1991, with permission)

both research groups. Gan et al. (1991) reported a molecular weight of human PON1 of about 43,000 with up to three carbohydrate chains representing about 15.8% of the total weight. Furlong et al. (1991) reported two bands ($\sim$ 35kDa & 38kDa) in their purified rabbit preparation that remarkably activity stained following SDS-gel electrophoresis. Re-running the excised bands generated two additional less active bands of 43 kDa and 48 kDa. Two bands were also observed in their purified human PON1 preparation ($\sim$ 44.7 kDa & 47.9 kDa). It appeared that the higher molecular weight bands represent partially unfolded PON1. As noted below, the rabbit PON1 is much more stable than the human PON1 due to its tighter binding of the Ca^{++} ion required for maintaining the active structure of PON1. The slower mobility of the human PON1 in SDS gels is likely due to a less tightly folded protein structure, possible related to the loss of the Ca^{++} ion required for structural integrity of PON1 (Kuo and La Du, 1998). Protein sequencing data indicated that PON1 retained its signal sequence (Furlong et al., 1991; Hassett et al., 1991).

The availability of gel-purified, active rabbit PON1 allowed for the sequencing of this protein and the design of oligonucleotide probes that were used to isolate rabbit PON1 cDNA clones from a rabbit liver cDNA library. Sequencing of the rabbit cDNA clone confirmed that the signal sequence was maintained by the mature PON1, minus the initiator methionine residue (Hassett et al., 1991). The rabbit PON1 cDNA was used as a probe to isolate human cDNA clones from a human liver cDNA library. The characterized human cDNA clones revealed two amino acid polymorphisms (L55M and Q192R). It was interesting that none of the characterized clones contained a canonical polyadenylation site. The three characterized human cDNA clones utilized three different alternative polyadenylation sites (Hassett et al., 1991). The rabbit and human nucleotide and amino acid sequences were greater than 85% similar (Adkins et al., 1993; Hassett et al., 1991).

3.   MOLECULAR BASIS OF THE PON1 POLYMORPHISM

The molecular basis of the PON1 activity polymorphism was reported by both La Du's research team and ours in 1993 (Adkins et al., 1993; Humbert et al., 1993). It was found that the Q192R polymorphism was responsible for the differences in catalytic efficiency of hydrolysis of paraoxonase. Humbert et al. (1993) also reported the physical mapping of the PON1 gene to human chromosome 7q21–22 where it had been previously mapped by genetic methods (Knowlton et al., 1985; Schmiegelow et al., 1986; Wainwright et al., 1985; White et al., 1985).

4.   STRUCTURAL STUDIES

The first significant information about the structure of PON1 was the finding that PON1 retained its signal sequence following secretion from the liver, the first protein know to do so (Hassett et al., 1991). Definitive proof of the role of the signal sequence in HDL attachment came from Dr. La Du's laboratory. Sorenson et al. (1999) demonstrated that the signal sequence provided a hydrophobic anchor for attachment of

PON1 to HDL. The identification of two PON1 calcium ion binding sites by Dr. La Du's team (Kuo and La Du, 1998), one for structural stability and the other for catalysis, added a bit more information about the structural requirements of PON1. It was interesting that rabbit PON1 has a higher affinity for the structural calcium ion than human PON, explaining the higher stability of rabbit PON1. Site specific mutagenesis experiments by Josse et al. (1999, 2001) identified the following amino acid residues that are essential for PON1's catalytic activity: W280, H114, H154, H242, D53, D168, D182, D268, D278, E52, E194 and the two cysteine residues C41 and C352 that are in disulfide linkage). La Du et al. (1993; modified in La Du, 2002) presented a schematic model of the structural features of PON1 (Fig. 12, Table 3).

A major breakthrough in PON1 structural studies came with the engineering by directed evolution (Aharoni et al., 2004) of a form of PON1 that could be crystallized and subjected to X-ray crystallographic structural determination (Harel et al., 2004).

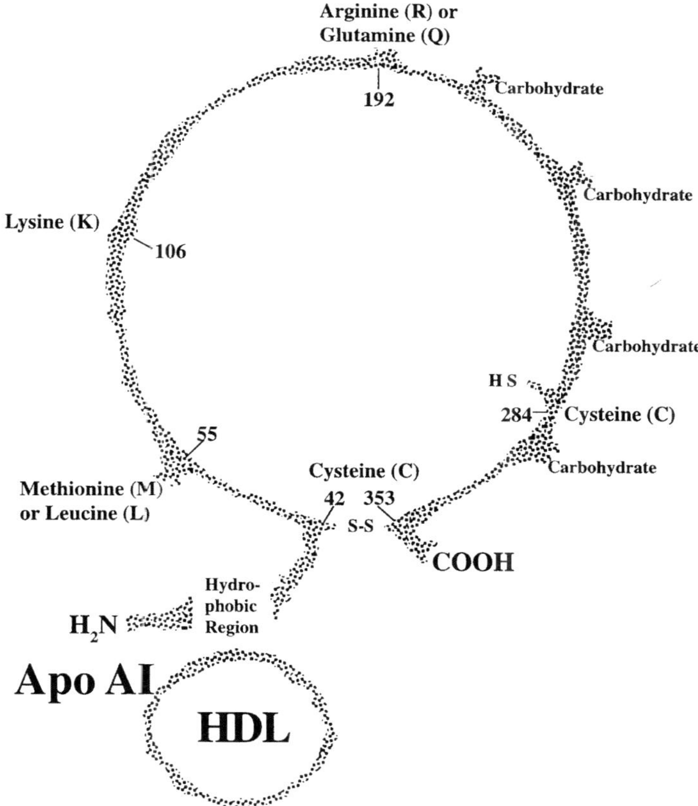

Figure 12. Human PON1 structural characteristics, including the two common polymorphic sites, the disulfide linkage, free sulfhydryl residue and potential locations of carbohydrate groups. Reproduced from La Du, 2002 with permission

Table 3. Calcium binding properties of rabbit and human PON1s

	K_{d1}	K_{d2}
Human	$3.6 \pm 0.9 \times 10^{-7}\,M$	$6.6 \pm 1.2 \times 10^{-6}\,M$
Rabbit	$1.4 \pm .5 \times 10^{-8}\,M$	$5.3 \pm 0.9 \times 10^{-6}\,M$

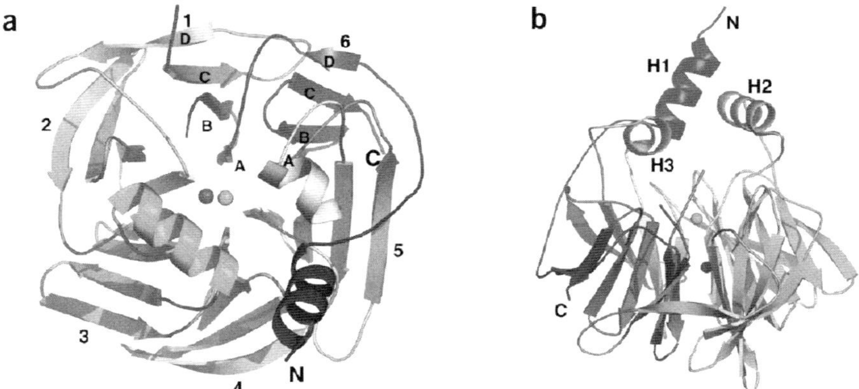

Figure 13. Overall Structure of PON1. (a) View of the six-bladed -propeller from above. The top of the propeller is, by convention, the face carrying the loops connecting the outer -strand of each blade (strand D) with the inner strand (A) of the next blade. Shown are the N and C termini, and the two calcium atoms in the central tunnel of the propeller [Ca1, green (left); Ca2, red (right)]. (b) A side view of the propeller, including the three helices at the top of the propeller (H1–H3). N-terminal residues 1–15 and a surface loop connecting strands 1B and 1C (residues 72–79) are not visible in the structure. Reproduced from Harel et al., 2004 (Fig. 1) with permission

The directed evolution involved rounds of domain shuffling and screening for increased hydrolytic activity and solubility when expressed in *E. coli*. The analysis showed that PON1 has a 6-bladed propeller structure with the hydrophobic leader sequence and an amphipathic helix H2 that appear to participate in binding PON1 to HDL (Fig. 13). Further, site specific mutagenic analysis showed that the catalytic efficiency of PON1 toward specific substrates could be improved by specific amino acid substitutions (Aharoni et al., 2004). Engineered recombinant human PON1 will be the optimal candidate molecule to use in treating OP exposures or for prophylactic prevention of OP poisoning in cases if imminent exposure (Aharoni et al., 2004; Li et al., 1995, 2000; Rochu et al., 2007).

5.   PON1 PROTECTS AGAINST OP EXPOSURE

The research that establishes an important role in PON1 protecting against specific OP exposures is reviewed in Chaps. 13 and 14 of this volume and will be briefly summarized here. Early experiments by Main (1956), Costa et al. (1990)

and Li et al. (1993, 1995) demonstrated that injection of purified rabbit PON1 provided protection against paraoxon and much better protection against chlorpyrifos (CPS)/chlorpyrifos oxon (CPO) exposures, particularly against CPO exposures.

Follow on experiments with genetically modified mice generated by Shih, Lusis and colleagues at UCLA provided convincing evidence that the absence of PON1 in PON1 null mice rendered them dramatically more sensitive to chlorpyrifos oxon exposure with increased sensitivity to high doses of chlorpyrifos (Shih et al., 1998). The PON1 null mice also showed dramatically increased sensitivity to diazoxon and increased sensitivity to diazinon at high doses (Li et al., 2000). The finding that PON1 null mice did not exhibit increased sensitivity to paraoxon led to a series of experiments that examined catalytic efficiency of human PON1s as an explanation for why PON1 protected against some OP compounds and not others (Li et al., 2000). The answer proved to be rather straight forward. PON1 protected against OPs that are hydrolyzed with high catalytic efficiency. Diazoxon was hydrolyzed with nearly equivalent efficiency by both PON1-192 alloforms and injection of either purified human PON1-192 alloform into PON1 null mice restored resistance to diazoxon exposure. On the other hand, $PON1_{R192}$ had better catalytic efficiency for chlorpyrifos oxon hydrolysis than $PON1_{Q192}$ and provided better protection against this oxon. Even though $PON1_{R192}$ hydrolyzed paraoxon with much higher efficiency than $PON1_{Q192}$, the efficiency was still too low to provide protection against dermal exposure.

Taken together, these experiments indicated that it should be possible to engineer recombinant human PON1 to provide enzyme with sufficient catalytic efficiency to protect against or treat cases of OP poisoning. For example, one amino acid change, the naturally occurring Q192 to R192 increases the catalytic efficiency by approximately 9-fold. This is approximately the increase in catalytic efficiency needed to protect against or treat exposure to soman.

6. PON1 AND CADIOVASCULAR DISEASE

Looking back, the road to understanding a role of PON1 in lipid metabolism and protecting against vascular disease began as early as 1961 with the characterization of plasma esterases (Uriel, 1961). A summary time line of characterizing PON1 as a lipoprotein-associated esterase is shown in Table 4. As noted above, these efforts paralleled the characterization of PON1 as a phosphotriesterase with some of the efforts encompassing both arylesterase and paraoxonase characterization.

A key finding in understanding the role of PON1 in preventing vascular disease was the demonstration by Mackness et al. (1991) that PON1 prevented the accumulation of oxidized lipids in low density lipoprotein particles.

Following this observation, Watson et al. (1995) reported that PON1 reduced the biological activity of minimally oxidized LDL. However, Connelly et al. (2005) reported that PON1 did not reduce or modify the oxidation of phospholipids by peroxynitrite. It is clear that more research is required to understand the molecular mechanisms of the role of PON1 in lipid metabolism.

Table 4. PON1: Role in Lipid Metabolism (Biochemical Studies)

α-Lipoprotein Esterase Activities

1961	Uriel. Characterization of cholinesterase and other carboxylic esterases after electrophoresis and immunoelectrophoresis on agar. I. Application to the study of esterases of normal human serum. Ann Inst Pasteur (Paris) 101:104–119
1968	Skarnes RC. In vivo interaction of endotoxin with a plasma lipoprotein having esterase activity. J Bacteriol 95:2031–2034
1975	Don MM, Masters CJ, Winzor DJ. Further evidence for the concept of bovine plasma arylesterase as a lipoprotein. Biochem J 151:625–630

Arylesterase/PON1 Associated with HDL

1973	Kitchen BJ, Masters CJ, Winzor DJ. Effects of lipid removal on the molecular size and kinetic properties of bovine plasma arylesterase. Biochem J 135:93–99
1978	Bog-Hansen TC, Krog HH, Back U. Plasma lipoprotein-associated arylesterase is induced by bacterial lipopolysaccharide. Febs Lett 93:86–90
1983	Mackness MI, Walker CH. Partial purification and properties of sheep serum 'A' esterases. Biochem Pharmacol 32:2291–2296
1985	Mackness MI, Hallam SD, Peard T, Warner S, Walker CH. The separation of sheep and human serum 'A'-esterase activity into the lipoprotein fraction by ultracentrifugation. Comp Biochem Physiol 83B:675–677
1993	Blatter MC, James RW, Messmer S, Barja F, Pometta D. Identification of a distinct human high-density lipoprotein subspecies defined by a lipoprotein-associated protein, K-45. Identity of K-45 with paraoxonase. Eur J Biochem 211:871–9
1994	Kelso GJ, Stuart WD, Richter RJ, Furlong CE, Jordan-Starck TC, Harmony JA Apolipoprotein J is associated with paraoxonase in human plasma. Biochemistry 33:832–839

PON Associated with apoliprotein A-1

1988	Mackness MI, Walker CH. Multiple forms of sheep serum A-esterase activity associated with the high-density lipoprotein. Biochem J 250:539–545
1989	Mackness MI. Commentary. 'A'-esterases. Enzymes looking for a role? Biochem Pharmacol 38:385–390
1989	Mackness MI. Possible medical significance of human serum 'A-esterases. In: Reiner E, Aldridge WN, Hoskin FCG (eds) Enzymes Hydrolysing Organophosphorus Compounds. Ellis Horwood, Chichester, pp 202–213

Modulation of LDL Oxidation

1991	Mackness MI, Arrol S, Durrington PN. Paraoxonase prevents accumulation of lipoperoxides in low-density lipoprotein. FEBS Lett 286:152–154
1995	Watson AD, Berliner JA, Hama SY, La Du BN, Faull KF, Fogelman AM, and Navab, M. Protective effect of high density lipoprotein associated paraoxonase - inhibition of the biological activity of minimally oxidized low-density lipoprotein. J Clin Invest 96:2882–2891
1997	Gordon T, Castelli WP, Hjortland MC, Kannel WB, Dawber TR. High density lipoprotein as a protective factor against coronary heart disease. The Framingham Study. Am J Med 62:707–714.

Questions about Paraoxonase Activities

2005	Connelly PW, Draganov D, Maguire GF. Paraoxonase-1 does not reduce or modify oxidation of phospholipids by peroxynitrite. Free Radic Biol Med 38:164–174

Following these reports, many studies examined PON1 polymorphisms as risk factors for cardiovascular and other diseases. Nearly all of these studies made use of only DNA based assays to characterize the polymorphisms (reviewed in: Brophy et al., 2002; Mackness and Mackness, 2004). Several meta analyses of the genotyping studies were carried out that noted at best a slight increase in risk of the PON1 R/R192 genotypes for CAD. The studies were complicated by the different gene frequencies in different ethnic groups and the presence or absence of diabetes (reviewed in Mackness and Mackness, 2004).

As noted in many studies that have examined the functional variability of PON1 status in populations, there is as much as a 15-fold variability in plasma PON1 protein levels among adult individuals with the same 192 phenotype (QQ; QR or RR). Since simple biochemical and physiological principles dictate that it is the level of protein that determines how rapidly substrates including toxicants are metabolized, it makes no sense to ignore the large inter-individual variability in PON1 levels (Figs. 5, 7, 8 and 9) when examining risk of exposure or disease. In fact, the variability in levels may be much more important than any of the SNP analyses. It appears that less than half of the variability in PON1 levels is accounted for by genetic variability in the PON1 gene (Jarvik et al., 2003). Genotyping all of the known PON1 SNPs (see Chap. 7) will not allow for an accurate estimation of PON1 plasma levels. Thus, the simple high throughput functional assay that provides both the functional PON_{192} genotype as well as accurate plasma PON1 levels (PON1 status) provides the information necessary for examining PON1 genetic variability and risk of exposure or disease (Jarvik et al., 2000, 2003b; Richter et al., 1999, 2004).

A few studies have examined PON1 status, or activities and levels as a risk factor for vascular disease (reviewed in Mackness and Mackness, 2004). The studies of Jarvik et al. (2000) clearly showed that low PON1 levels are a risk factor for carotid artery disease (Fig. 14). It is important to note that, as in many other studies, no differences were observed in gene frequencies of PON1 polymorphisms between patients and controls. Rates of diazoxon hydrolysis were used to compare PON1 activity within each PON1 192 genotype. More recent studies indicate that rates of phenylacetate hydrolysis allow comparison of PON1 levels within and across $PON1_{192}$ genotypes (Fig. 15) as originally shown by Eckerson et al. (Fig. 7).

7.   DEVELOPMENTAL REGULATION OF PON1 EXPRESSION

Low PON1 and esterase levels at birth contribute to the increased sensitivity of young to OP exposures (See for example, Atterberry et al., 1997; Li et al., 1997; Liu et al., 1999; Moser et al., 1998). In humans, it takes between 6 months and 2 years following birth to develop mature levels of PON1 (Augustinsson and Barr, 1963; Cole et al., 2003; Ecobichon and Stephens, 1973). In rats and mice, mature PON1 levels are reached at PND 21 (Li et al., 1997). Interestingly, when human PON1 transgenes are introduced into PON1 knockout mice, including the 5′ human

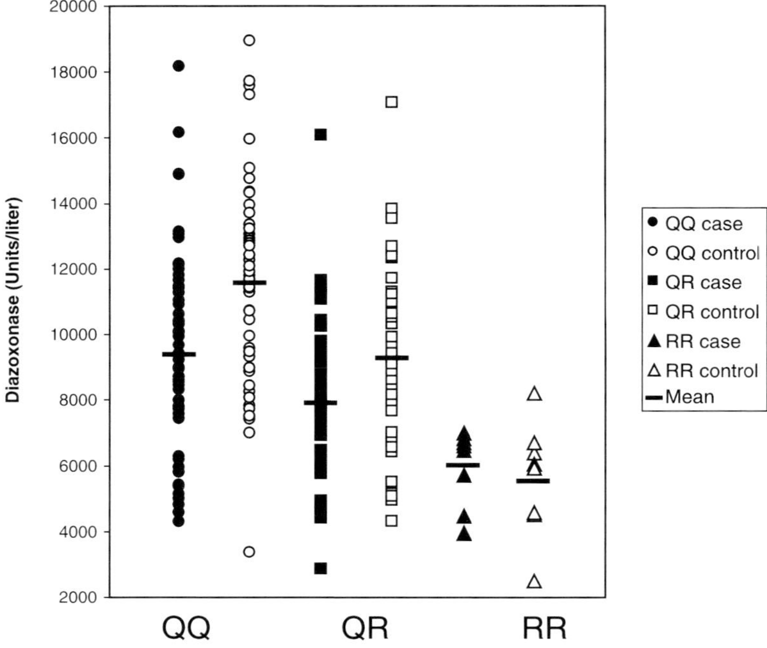

Figure 14. PON1 levels with in genotypes determined by rates of diazoxon hydrolysis at high salt (differentially inhibits *PON1R$_{192}$*). Modified and reproduced from Jarvik et al. 2000 with permission

regulatory sequences, the appearance of hPON1 follows the mouse developmental program, peaking at PND 21 (Cole et al., 2003), indicating a conservation of regulatory elements between these two species. There is also a large variability of PON1 levels among newborn humans (Fig. 15). Surprisingly, some newborns have higher PON1 levels than some mothers. A serious concern would be the exposure of an expectant mother with low PON1 status to chlorpyrifos/chlorpyrifos oxon or diazinon/diazoxon (Furlong et al., 2006). Comparison of chlorpyrifos oxon sensitivity in PON1 knockout mice compared with wild type mice shows that already by 4 days of age, the wild type mice are more resistant to exposure than the PON1$^{-/-}$ mice (Cole et al., unpublished data).

8.   ADDITIONAL MEMBERS OF THE PON FAMILY OF GENES/PROTEINS

In 1996, La Du and colleagues (Primo-Parmo et al., 1996) published a paper describing two new members of the PON family of genes. They isolated cDNAs representing both *PON2* and *PON3* from cDNA libraries. Like *PON1*, *PON2* also mapped to human chromosome 7 and mouse chromosome 6. *PON2* also has 9 exons whose junctions correspond with the exon/intron junctions

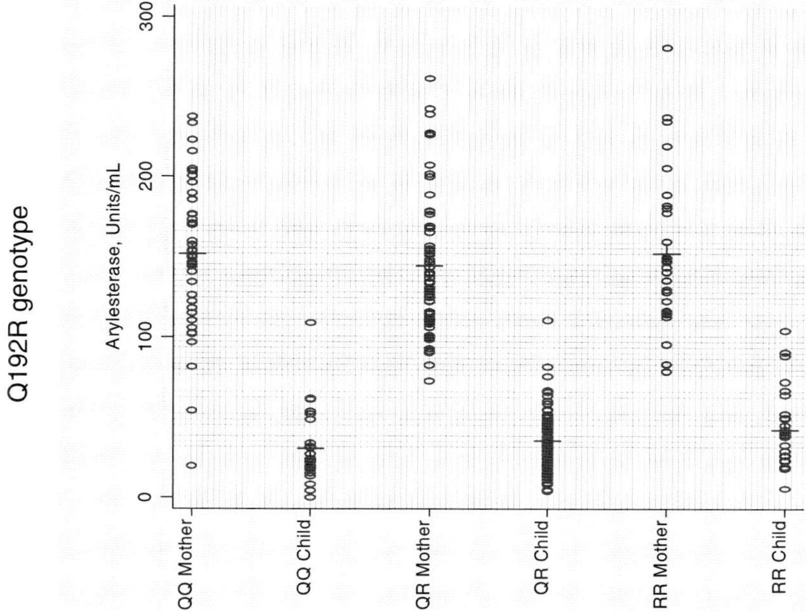

Figure 15. PON1 levels in mothers and newborns with the indicated $PON1_{192}$ genotypes. Individual data points for arylesterase activities (AREase) in mothers (solid circles) and newborns (open circles) for each $PON1_{192}$ genotype as indicated. Means are indicated by the crossbars. Note that the average level of arylesterase was the same across PON1192 genotypes. Modified and reproduced from Furlong et al., 2006 with permission

of *PON1* (Clendenning et al., 1996; Primo-Parmo et al., 1996). Primo-Parma et al. (1996) also noted that all of the characterized *PON2* and *PON3* sequences were missing the codon for K105. Their cloning, sequencing and database mining efforts generated sequences for mouse, rabbit, human (see also Hassett et al., 1991), and dog PON1s; human, mouse, dog, turkey and chicken PON2s and human and mouse PON3s. The PON2 S311C polymorphism was revealed in this effort. They noted an approximate 85% identity between each of the respective human and mouse PONs. Interestingly, the PON3 polyadenylation site for PON3 was only 5 residues downstream from the putative stop codon. The *PON* genes appear to be approximately 27–28 kb (Clendenning et al., 1996; Primo-Parmo et al., 1996).

Mochizuki et al. confirmed the *PON2* discovery and also reported the existence of alternative splice variant RNAs and another coding polymorphism, A147G (Mochizuki et al., 1998). They also used a contig of six overlapping non-chimeric YAC clones and six PAC clones to provide a fine structure map of the 3 *PON* genes. The order was found to be *PON1, PON3* and *PON2* followed by *PDK4*, with *PON1* the most centromeric.

9.   PURIFICATION OF PON2 AND PON3

Ozols (1999) was the first to purify a protein that corresponded to the *PON3* sequence described by Primo-Parmo et al. (1996). The protein was purified from rabbit liver microsomes and sequenced. The protein consisted of 350 residues and was 60% identical with rabbit serum PON1 and 84% similar with the sequence designated *PON3* by Primo-Parma et al. (1996). Like PON1, the PON3 sequence retained the hydrophobic amino-terminus. Unlike PON1, PON3 was not glycosylated. Draganov et al. (2000) in La Du's laboratory were the first to purify and characterize a mammalian (rabbit) plasma PON3. They also isolated the rabbit PON3 cDNA. They expressed the rabbit PON3 in 293T/17 cells and found that the cloned enzyme had the same specificity and molecular mass as the PON3 purified from rabbit serum. Its ability to hydrolyze lactones and not OPs provided the specificity necessary to follow the purification. Rodrigo et al. (2003) purified and characterized rat liver microsomal PON3. The sequence of the rat PON3 protein was 95% identical with the deduced cDNA sequence of the mouse PON3 cDNA. The rat PON3 sequence was 67% identical to rat PON1, similar to homologies observed between PON1s and PON3s of other species.

Draganov et al. (2005) purified all three human PONs from a baculovirus-mediated expression system. Characterization of the substrate specificities of the purified recombinant PONs is described below. Brushia et al. (2001) had earlier reported the expression and purification of $PON1_{Q192}$ from a baculovirus expression system.

Recently Lu et al. (2006) have described the expression of active human PON3 in *E coli*. Most of the protein was expressed as inclusion bodies, which were solubilized with high levels of Triton X-100, then refolded.

Another interesting expression system for PON1 was recently reported by Zhu et al. (2006) who expressed and purified human his-tagged $PON1_{Q192}$ in a silk worm system. The purified protein had nearly identical kinetic properties to serum purified PON1.

10.   PONs' DRUG AND LACTONE METABOLIZING ACTIVITIES

Considerable interest in the ability of PONs to metabolize drugs began in 1998 when Tougou et al. (1998) demonstrated that PON1 could activate the prodrug prulifloxacin (NM441) to its active form NM394. Following this observation, Biggadike et al. (2000) demonstrated that PON1 hydrolyzed (inactivated) pharmaceutically important glucocorticoid γ-lactones and cyclic carbonates, thus confining the site of drug action to the target sites such as lung. Expression of PON1, PON2 and PON3 in a baculovirus system allowed La Du and colleagues to investigate the substrate specificities of each member of the PON family of enzymes (Draganov et al., 2005). They characterized the activity of the three PONs against a number of organophosphates, aromatic esters, lactones and hydroxycarbolylic acid substrates (Table 5). PON1 exclusively hydrolyzed OP substrates, while PON3 hydrolyzed more

Table 5. Specific enzymatic activities of the purified recombinant human PONs (Draganov et al., 2005, reproduced with permission)

Substrate	PON1	PON2	PON3
Organophosphatase activity (U/mg)			
Paraoxon	1.94 ± 0.11	ND	0.205 ± 0.05
Chlorpyrifos oxon (0.32 mM)	40.9 ± 0.9	ND	ND
Diazoxon	113 ± 5	ND	ND
Arylesterase activity (U/mg)			
Phenyl acetate	1.120 ± 50	0.086 ± 0.0013	4.1 ± 0.3
$p\text{-NO}_2$-acetate	15.0 ± 0.03	0.7 ± 0.07	39.0 ± 4.1
$p\text{-NO}_2$-propionate	13.6 ± 0.04	0.96 ± 0.06	20.7 ± 3.2
$p\text{-NO}_2$-butyrate	1.3 ± 0.015	1.4 ± 0.03	11.4 ± 0.7
Lactonase activity (U/mg)			
Dihydrocoumarin	129.9 ± 8.30	3.1 ± 0.2	126.1 ± 12
2-Coumaronone	135.7 ± 10.3	10.9 ± 0.4	40.7 ± 3.8
Homogentisic acid lactone	329.5 ± 13.1	ND	ND
γ-Butyrolactone	32.1 ± 2.73	ND	0.81 ± 0.1
γ-Valerolactone	45.0 ± 3.7	ND	6.2 ± 0.4
γ-Hexalactone	51.7 ± 4.2	ND	23.9 ± 3.2
γ-Heptalactone	57.2 ± 2.3	ND	27.7 ± 2.7
γ-Octalactone	69.2 ± 4.3	ND	25.6 ± 3.2
γ-Nonalactone	144.7 ± 11.3	ND	30.9 ± 2.7
γ-Decanolactone	173.8 ± 14.7	ND	45.6 ± 3.6
γ-Undecanolactone	127.6 ± 10.5	ND	71.4 ± 3.1
α-Angelica lactone	183.0 ± 16	ND	20.7 ± 3.2
γ-Phenyl-γ-butyrolactone (0.5 mM)	63.0 ± 3.1	0.68 ± 0.08	11.4 ± 0.7
δ-Valerolactone	671 ± 14	ND	14.5 ± 0.7
δ-Hexalactone	72 ± 2.3	ND	11.7 ± 1.2
δ-Nonalactone	150 ± 12.3	ND	11.1 ± 0.9
δ-Decanolactone	251 ± 13	ND	44.3 ± 3.2
δ-Undecanolactone	287 ± 17	ND	84.4 ± 2.7
δ-Tetradecanolactone (0.5 mM)	154 ± 24	ND	22.7 ± 2.2
5-HETEL (10 μM)	75.4 ± 8.36	1.83 ± 0.08	27.5 ± 3.6
DL-3-Oxo-hexanoyl-HSL (250 μM)	0.0334 ± 0.0031	0.2683 ± 0.0384	ND
L-3-Oxo-hexanoyl-HSL (250 μM)		0.5080 ± 0.0661	
DL-Heptanoyl-HSL (25 μM)	0.0036 ± 0.0004	0.0311 ± 0.0026	0.0049 ± 0.0023
DL-Dodecanoyl-HSL (25 μM)	0.0167 ± 0.0005	0.4588 ± 0.0371	0.0877 ± 0.0014
DL-Tetradecanoyl-HSL (25 μM)	0.0035 ± 0.0013	0.4239 ± 0.0204	0.0255 ± 0.0003
Lovastatin (25 μM)	ND	ND	0.266 ± 0.022

(Continued)

Table 5. (Continued)

Substrate	PON1	PON2	PON3
Spironolactone (25 μM)	ND	ND	0.011 ± 0.001
Canrenone (25 μM)	ND	ND	0.013 ± 0.001
Lactonizing activity (U/mg)			
Counaric acid (100 μM)	0.047 ± 0.004	ND	0.013 ± 0.0007
4-HDoHE (10 μM)	1.51 ± 0.16	0.52 ± 0.03	13.7 ± 2.0

5-HETEL, ($\pm$)5-hydroxy-6E,8Z,11Z,14Z-eicosatetraenoic acid 1,5-lactone; HSL, homoserine lactone; ND, not detectable under these assay conditions. Data are averages from two to four measurements $\pm$ SD or range. One unit $= 1\,\mu$mol of substrate metabolized per minute. All substrates were at 1 mM final concentration unless indicated otherwise.

bulky substrates such as lovastatin and spironolactone. All three PONs metabolized 5-hydroxy-eicosatetraneoic acid 1,5-lactone and 4-hydroxy-docosahezaenoic acid, oxidation products of arachidonic acid and docosahexaenoic acid respectively. They concluded that the PONs are lactonases with overlapping and distinct substrate specificities. Khersonsky and Tawfik (2005) carried out extensive studies on more than 50 different PON1 substrates aimed at understanding the mechanism of hydrolysis and probable physiological-type substrates. The substrates included esters, phosphotriesters and lactones. Their results suggested that PON1 is primarily a lactonase in agreement with La Du and coworkers. While the primary role of PONs may be as lactonases, non-the-less, they play a physiologically important role in protecting against specific OP exposures as described above. From the few studies carried out on the metabolism of various drugs by PONs, it is clear that much research is left to do on the physiological importance of the pharmacokinetics of drug metabolism by this interesting family of enzymes. The genetically modified mouse strains generated by Shih and colleagues should provide important insights into these questions, as should studies examining the relationship of PON status and drug activation/inactivation.

It should be emphasized here that the contributions of Lusis, Shih and colleagues in generating genetically modified mice with the PON genes knocked out and knocked in have been invaluable, as evident throughout this Second International Conference on PONs. These strains of mice will continue to provide valuable insights into the physiological function of this interesting family of enzymes well into the future.

11.   QUORUM SENSING FACTOR HYDROLYSIS BY PONs

One of the interesting observations in the survey of substrate specificities of the PONs by Draganov et al. (2005) was the finding that all three PONs hydrolyzed the bacterial quorum sensing acyl homoserine lactones, with an apparent specificity for the L-isomers. Of the three PONs, PON2 appeared to have the highest activity against the quorum sensing factors. These observations were confirmed by Ozer et al. (2005; Fig. 16). The question of the physiological significance has been addressed using the genetically modified PON mouse model. Experiments using airway epithelial cells cultured from PON2 knockout mice and a quorum-sensing reporter strain of *P. aeruginosa*, showed quorum sensing to be enhanced in cells cultured from PON2 deficient mice (Stolz et al., 2006).

It is likely that inactivation of quorum sensing factors may have been one of the earliest physiological functions of the PON family of enzymes. Related lactonase activities are found in microbes (Kanagasundaram and Scopes, 1992), nematodes (Chelur et al., 2002) and fungi (Kobayashi et al., 1998) and are likely involved in protection against pathogenic microbes, and microbial competition in the environment. PON2 is thought to be the earliest member of the PON family of proteins to evolve (Draganov and La Du, 2004) and, as noted above, is the most efficient at hydrolyzing the *Pseudomonas* quorum sensing factors. The second member to evolve is proposed to be PON3, followed by PON1.

12. SUMMARY

It is probable that sorting out of the physiological significance of the many activities of the three human PONs will continue for years to come. It is clear from the animal model studies described above that several of the known activities of PON1 do have physiological importance. Hydrolysis of specific OP compounds by PON1 (diazoxon and chlorpyrifos oxon) is physiologically important while the hydrolysis of others, such as paraoxon, the compound for which PON1 was named, is much less physiologically important (Li et al., 2000). Whatever the roles of PON1 in lipid metabolism, low levels of PON1 appear to be a risk factor for vascular disease (reviewed in Deakin and James 2004; Jarvik et al., 2000, 2003a; Mackness and Mackness, 2004; Mackness et al., 2001). The lactonase activity against quorum sensing factors (Ozer et al., 2005, Stolz et al., 2006; Tiber et al., 2003) also appears

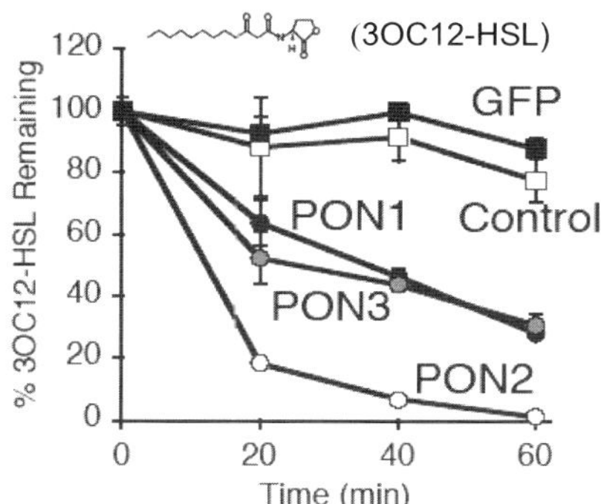

Figure 16. Human PONs 1, 2, and 3 degrade 3OC12-HSL. CHO cells were transfected with adenovirus expressing hPON1, hPON2, hPON3, or GFP. After 48 h, 10 lM 3OC12-HSL was added to the medium and remaining 3OC12-HSL was measured at 20, 40, and 60 min. Data are means ±*SEM*; *n* = 5. Modified from Ozer et al., 2005 (Fig. 5) with permission

to have physiological significance. The role of PONs in pharmacokinetics will obviously be important and merits much more research effort.

Comments are in order regarding the many studies that have examined only SNPs in searches for the association of the genetic variability of PON1 to risk of exposure or disease. It should be clear from the above discussion that in all cases, PON1 levels are important and in some cases where the catalytic efficiencies of the two 192 alloforms (Q/R) differ, the position 192 genotype is also important. In no case are PON1 levels unimportant. If it is of interest to determine how a specific SNP may influence PON1 levels, then a study should be designed to answer that question. If not, determination of genotype alone may well be a waste of time and resources. The high throughput, two-substrate assays provide both functional position 192 genotype as well as PON1 levels. These are the two factors that are important for analyzing risk. It is recommended that arylesterase activity also be determined to provide an accurate measure of plasma PON1 across genotypes (Furlong et al., 2006). This point has been emphasized in a number of publications by La Du's research group (reviewed in Draganov and La Du, 2004), James and co-workers (reviewed in Deakin and James, 2004) our research group (Jarvik et al., 2000, 2003a, b; Li et al., 1993; Richter and Furlong, 1999) and the Mackness research team (reviewed in Mackness and Mackness, 2004). We introduced the term PON1 status in 1993 (Li et al., 1993) with the hope that researchers would realize that both quantity (activity levels) and quality (catalytic efficiency specified by the Q192R polymorphism) of PON1 are important in determining risk of exposure or disease.

To date, means of obtaining measures of PON2 and PON3 status have not been developed. While PON1 plasma activity for a given individual is generally quite stable over time (Zech and Zurcher, 1974; for a review of factors affecting PON1 levels see Costa et al., 2005), environmental factors may influence PON2 levels (Shiner et al., 2004, 2007; Rosenblat et al., 2003) so that measures of PON2 expression over time may vary more than PON1. None-the-less, it will be worthwhile to develop equivalent measures of activities for PON2 and PON3.

Dr. Bert La Du, with his research team and collaborators, made many significant contributions to our understanding the structure and function of the PON family of enzymes and the identification and characterization of the genes encoding this family of enzymes. Many of his former associates attended and contributed to this Second International Paraoxonase Conference. Dr. La Du's contributions will continue far into the future through all of the students, postdoctoral fellows and collaborators that trained in his laboratory. It is fitting that this conference was dedicated to the memory of Dr. Bert La Du.

ACKNOWLEDGEMENTS

The author expresses his appreciation for the guidance and encouragement of Dr. Arno Motulsky who initiated our studies on PON1 in the late 1970s. He also expresses his gratitude for the advice and productive interactions with Dr. Bert La Du and co-workers throughout the years of research on the paraoxonases. None of the research carried out at the University of Washington would have been possible without the collaboration of many graduate students, postdoctoral fellows and colleagues whose names appear on the cited references. The following grants provided support for this effort: ES09883, ES-09601/EPA-R826886, ES04696 and P30ES07033.

REFERENCES

Adkins, S., Gan, K.N., Mody, M., and La Du, B.N., 1993, Molecular basis for the polymorphic forms of human serum paraoxonase/arylesterase: Glutamine or arginine at position 191 for the respective A or B allozymes. Am. J. Hum. Genet. 52, 598–608

Aharoni, A., Gaidukov, L., Yagur, S., Toker, L., Silman, I., and Tawfik, D.S., 2004, Directed evolution of mammalian paraoxonases PON1 and PON3 for bacterial expression and catalytic specialization. Proc. Natl. Acad. Sci., 101, 482–487

Aldridge, W.N., 1953a, Serum esterases I. Two types of esterase (A and B) hydrolysing p-nitrophenyl acetate, propionate and butyrate and a method for their determination. Biochem J, 53, 110–117

Aldridge, W.N., 1953b, Serum esterases II. An enzyme hydrolysing diethyl p-nitrophenyl acetate (E600) and its identity with the A-esterase of mammalian sera. Biochem. J., 53, 117–124

Atterberry, T.T., Burnett, W.T., and Chambers, J.E., 1997, Age-related differences in parathion and chlorpyrifos toxicity in male rats: target and nontarget esterase sensitivity and cytochrome P450–mediated metabolism. Toxicol. Appl. Pharmacol., 147, 411–418

Augustinsson, K.B., and Barr, M., 1963, Age variation in plasma arylesterase activity in children. Clin Chim Acta, 8, 568–573

Biggadike, K., Angell, R.M., Burgess, C.M., Farrell, R.M., Hancock, A.P., Harker, A.J., Irving, W.R., Ioannou, C., Procopiou, P.A., Shaw, R.E., Solanke, Y.E., Singh, O.M., Snowden, M.A., Stubbs, R.J., Walton, S., and Weston, H.E., 2000, Selective plasma hydrolysis of glucocorticoid gamma-lactones and cyclic carbonates by the enzyme paraoxonase: an ideal plasma inactivation mechanism. J. Med. Chem. 43,19–21

Blatter, M.C., James, R.W., Messmer, S., Barja, F., and Pometta, D., 1993, Identification of a distinct human high-density lipoprotein subspecies defined by a lipoprotein-associated protein: K-45. Identity of K-45 with paraoxonase. Eur. J. Biochem. 211, 871–9

Bog-Hansen, T.C., Krog, H.H., and Back, U., 1978, Plasma lipoprotein-associated arylesterase is induced by bacterial lipopolysaccharide. Febs. Lett. 93, 86–90

Brophy, V., Jarvik, G.P., and Furlong, C.E., 2002, PON1 Polymorphisms. *In: Paraoxonase (PON1) in Health and Disease: Basic and Clinical Aspects* (L.G. Costa and C.E. Furlong, eds.) Kluwer Academic Press, Boston, USA, pp. 53–77

Brushia, R.J., Forte, T.M., Oda, M.N., La Du, B.N., and Bielicki, J.K., 2001, Baculovirus-mediated expression and purification of human serum paraoxonase 1A. J. Lipid Res., 42, 951–958

Chelur, D.S., Ernstrom, G.G., Goodman, M.B., Yao, C.A., Chen, L., Hagan, R.O., and Chalfie, M., 2002, The mechanosensory protein MEC-6 is a subunit of the C. elegans touch-cell degenerin channel, Nature, 420, 669–673

Choi, S.S., and Forster, T.L., 1967a, Triton X-155 as a stabilizer of bovine plasma arylesterase activity. Dairy Sci,, 50, 837–39

Choi, S.S., and Forster, T.L., 1967b, Purification of bovine. plasma arylesterase. J. Dairy Sci., 50, 1088–1091

Clendenning, J.B., Humbert, R., Green, E.D., Wood, C., Traver, D., and Furlong, C.E.,1996, Structural organization of the human PON1 gene. Genomics 35, 586–589

Cole, T.B., Jampsa, R.L., Walter, B.,J., Arndt, T.L., Richter, R.J., Shih, D.M., Tward, A., Lusis, A.J., Jack, R.M., Costa, L.G., and Furlong, C.E., 2003, Expression of human paraoxonase (PON1) during development. Pharmacogenetics, 13, 357–364

Connelly, P.W., Draganov, D., and Maguire, G.F., 2005, Paraoxonase-1 does not reduce or modify oxidation of phospholipids by peroxynitrite. Free Radic. Biol. Med., 38, 164–74

Costa, L.G., and Furlong, C.E. eds., 2002, *Paraoxonase (PON1) in Health and Disease: Basic and Clinical Aspects.* Kluwer Academic Publishers, Boston, USA

Costa, L.G., McDonald, B.E., Murphy, S.D., Omenn, G.S., Richter, R.J., Motulsky, A.G., and Furlong, C.E., 1990, Serum paraoxonase and its influence on paraoxon and chlorpyrifos-oxon toxicity in rats. Toxicol. Appl. Pharmacol., 103, 66–76

Costa, L.G., Vitalone, A., Cole, T.B., and Furlong, C.E., 2005, Modulation of paraoxonase (PON1) activity. Biochem. Pharmacol., 69, 541–550

Davies, H., Richter, R.J., Keifer, M., Broomfield, C., Sowalla, J., and Furlong, C.E., 1996, The effect of the human serum paraoxonase polymorphism is reversed with diazoxon, soman and sarin. Nat. Genet. 14, 334–336

Deakin, S.P., and James, R.W., 2004, Genetic and environmental factors modulating serum concentrations and activities of the antioxidant enzyme paraoxonase-1. Clin Sci (Lond), 107, 435–47

Diepgen, T.L., and Geldmacher, V., Mallinkrodt, M., 1986, The human serum paraoxonase polymorphism. Arch. Toxicol. Suppl., 9, 154–158

Don, M.M., Masters, C.J., Winzor, D.J., 1975, Further evidence for the concept of bovine plasma arylesterase as a lipoprotein. Biochem. J., 151, 625–630

Draganov, D.I., and La Du, B.N., 2004, Pharmacogenetics of paraoxonses, a brief review. Naunyn-Schmiedeberg's Arch. Pharmacol., 369, 78–88

Draganov, D.I., Stetson, P.L., Watson, C.E., Billecke, S.S., and La Du, B.N., 2000, Rabbit serum paraoxonase 3 (PON3) is a high density lipoprotein-associated lactonase and protects low density lipoprotein against oxidation. J., Biol., Chem., 275, 33435–33442

Draganov, D.I., Teiber, J.F., Speelman, A., Osawa, Y., Sunahara, R., La Du, B.N., 2005, Human paraoxonases (PON1, PON2, and PON3) are lactonases with overlapping and distinct substrate specificities. J. Lipid Res., 46, 1239–1247

Durrington, P.N., Mackness, B., and Mackness, M.I., 2001, Paraoxonase and Atherosclerosis. Atheroscler. Thromb. Vasc. Biol., 21, 473–480

Eckerson, H.W., Wyte, C.M., and La Du, B.N., 1983, The human serum paraoxonase/arylesterase polymorphism. Am. J. Hum. Genet. 35, 1126–1138

Ecobichon, D.J., and Stephens, D.S., 1973, Perinatal development of human blood esterases, Clin Pharmacol. Ther., 14, 41–47

Furlong, C.E., Richter, R.J., Chapline, C., Crabb, J.W., 1991, Purification of rabbit and human serum paraoxonase. Biochemistry, 30,10133–10140

Furlong, C., Holland, N., Richter, R., Bradman, A., Ho. A., and Eskenazi, B., 2006, PON1 status of farmworker mothers and children as a predictor of organophosphate sensitivity, Pharmacogenet. Genomics, 16, 183–190

Gan, K., Smolen, A., Eckerson, H.W., La Du, B.N., 1991, Purification of human serum paraoxonase/arylesterase: Evidence for one esterase catalyzing both activities. Drug Metab. Dispos. 19, 100–106

Geldmacher-von Mallinckrodt, M., and Diepgen, T.L., 1988, The human serum paraoxonase polymorphism and specificity. Toxicol. Environ. Chem., 18, 79–196

Harel, M., Aharoni, A., Gaidukov, L., Brumshtein, B., Khersonsky, O., Meged, R., Dvir, H., Ravelli, R.B.G., McCarthy, A., Toker, L., Silman, I., Sussman, J.L, and Tawfik, D.S., 2004, Structure and evolution of the serum paraoxonase family of detoxifying and anti-atherosclerotic enzymes. Nat. Struct. Mol. Biol., 11, 412–419

Hassett, C., Richter, R.J., Humbert, R., Chapline, C., Crabb, J.W., Omiecinski, C.J., and Furlong, C.E., 1991, Characterization of cDNA clones encoding rabbit and human serum paraoxonase, the mature protein retains its signal sequence. Biochemistry, 30, 10141–10149

Humbert, R., Adler, D.A., Disteche, C.M., Hassett, C., Omiecinski, C.J., and Furlong, C.E. 1993, The molecular basis of the human serum paraoxonase activity polymorphism. Nat. Genet., 3, 73–76

Jarvik, G.P., Hatsukami, T.S., Carlson, C.S., Richter, R.J., Jampsa, R., Brophy, V.H., Margolin, S., Rieder, M.J., Nickerson D.A., Schellenberg G.D., Heagerty P.J., and Furlong C.E., 2003a, Paraoxonase activity, but not haplotype utilizing the linkage disequilibrium structure predicts vascular disease. Arterioscler. Thromb. Vasc. Biol., 23, 1465–1471

Jarvik, G.P., Jampsa, R., Richter, R.J, Carlson, C., Rieder, M., Nickerson, D., and Furlong, C.E., 2003b, Novel Paraoxonase (PON1) nonsense and missense mutations predicted by functional genomic assay of PON1 status. Pharmacogenetics, 13, 291–295

Jarvik, G.P., Rozek, L.S., Brophy, V.H., Hatsukami, T.S., Richter, R.J., Schellenberg, G.D., and Furlong, C.E., 2000, Paraoxonase phenotype is a better predictor of vascular disease than PON1192 or PON155 genotype. Atheroscler. Thromb. Vas.c Biol., 20, 2442–2447

Josse, D., Lockridge, O., Xie, W., Bartels, C.F., Schopfer, L.M., and Masson, P., 2001, The active site of human paraoxonase (PON1). J. Appl. Toxicol. Suppl. 1, S7–11

Josse, D., Xie, W., Renault, F., Rochu, D., Schopfer, L.M., Masson, P., and Lockridge, O., 1999, Identification of residues essential for human paraoxonase (PON1) arylesterase/organophosphatase activities. Biochemistry, 38, 2816–2825

Kanagasundaram, V., and Scopes, R., 1992, Isolation and characterization of the gene encoding gluconolactonase from *Zymononas mobilis*. Biochim. Biophys. Acta, 1171, 198–200

Kelso, G.J., Stuart, W.D., Richter, R.J., Furlong, C.E., Jordan-Starck, T.C., and Harmony, J.A., 1994, Apolipoprotein J is associated with paraoxonase in human plasma. Biochemistry, 33, 832–839

Khersonsky, O., and Tawfik, D.S., 2005, Structure-reactivity studies of serum paraoxonase PON1 suggest that its native activity is lactonase. Biochemistry, 44, 6371–6382

Kitchen, B.J., Masters, C.J., and Winzor, D.J., 1973, Effects of lipid removal on the molecular size and kinetic properties of bovine plasma arylesterase. Biochem. J., 135, 93–99

Knowlton, R.G., Cohen-Haguenauer, O., Van Cong, N., Frezal, J., Brown, V.A., Barker, D., Braman, J.C., Schumm, J.W., Tsui, L.C., Buchwald, M., et al., 1985, A polymorphic DNA marker linked to cystic fibrosis is located on chromosome 7. Nature, 318, 380–382

Kobayashi, M., Shinohara, M., Sakoh, C., Kataoka, M., and Shimizu, S., 1998, Lactone-ring-cleaving enzyme: Genetic analysis, novel RNA editing and evolutionary implications. Proc. Natl. Acad. Sci. USA, 95, 12787–12792

Kuo, C.-L., La Du, B.N., 1998, Calcium binding by human and rabbit serum paraoxonases. Structural stability and enzymatic activity. Drug. Metab. Dispos. 26, 653–660

La Du, B.N., 1992, Human serum paraoxonase/arylesterase. In, *Pharmacogenetics of Drug Metabolism* (W. Kalow, ed.) Pergamon Press, New York USA pp. 51–91

La Du, B.N., 2002. Historical considerations. In, *Paraoxonase (PON1) in Health and Disease, Basic and Clinical Aspects*. (L.G. Costa and C.E. Furlong, eds.) Kluwer Academic Press, Boston, USA,. Pp. 53–77

La Du, B.N., Adkins, S., Kuo, C.-L., and Lipsig, D., 1993, Studies on human serum paraoxonase/arylesterase, Chem.-Biol. Interact. 87, 25–34

Li, W.-F., Costa, L.G., and Furlong, C.E., 1993, Serum paraoxonase status: a major factor in determining resistance to organophosphates. J. Toxicol. Environ. Health, 40, 337–346

Li, W.-F., Costa, L.G., and Furlong, C.E., 1997, Paraoxonase (Pon1) gene in mice: sequencing, chromosomal location, and developmental expression. Pharmacogenetics, 7, 137–144

Li, W.-F., Costa, L.G., Richter, R.J., Hagen, T., Shih, D.M., Tward, A., Lusis A.J., and Furlong, C.E., 2000, Catalytic efficiency determines the in vivo efficacy of PON1 for detoxifying organophosphates, Pharmacogenetics, 10, 767–780

Li, W.-F., Furlong, C.E., and Costa, L.G., 1995, Paraoxonase protects against chlorpyrifos toxicity in mice. Toxicol. Lett., 76, 29–226

Liu, J., Oliver, K, and Pope, C.N., 1999, Comparative neurochemical effects of repeated methyl parathion or chlorpyrifos exposures in neonatal and adult rats. Toxicol. Appl. Pharmacol. 158, 186–196

Lu, H., Zhu, J., Zang, Y., Ze, Y., and Qin, J. 2006, Cloning, purification, and refolding of human paraoxonase-3 expressed in *Escherichia coli* and its characterization. Protein Expr. Purif., 46, 92–99

Mackness, B., Davies, G.K., Turkie, W., Lee, E., Roberts, D.H., Hill, E., Roberts, C., Durrington, P.N., and Mackness, M.I., 2001, Paraoxonase status in coronary heart disease: are activity and concentration more important than genotype? Arterioscler. Thromb. Vasc. Biol.,21, 1451–1457

Mackness, B., Durrington, P.N., and Mackness, M.I., 1998, Human Paraoxonase. Gen Pharmac, 31, 329–336

Mackness, M., Arrol, S., Durrington, P.N., 1991, Paraoxonase prevents accumulation of lipoperoxides in low-density lipoprotein. FEBS Lett., 286, 152–154

Mackness, M., and Mackness, B., 2004, Paraoxonase 1 and atherosclerosis: is the gene or the protein more important? Free Radic. Biol. Med., 37, 1317–1323

Mackness, M.I., and Walker, C.H., 1983, Partial purification and properties of sheep serum 'A' esterases. Biochem. Pharmacol., 32, 2291–2296

Mackness, M.I., and Walker, C.H., 1988, Multiple forms of sheep serum A-esterase activity associated with the high-density lipoprotein. Biochem. J., 250, 539–545

Main, A.R., 1956, The role of A-esterase in the acute toxicity of paraoxon, TEEP and parathion. Can. J. Biochem. Physiol., 34, 197–216

Main, A.R., 1960, The purification of the enzyme hydrolysing diethyl-p-nitrophenyl phosphate (paraoxon) in sheep serum. Biochem. J., 74, 10–20

Mazur, A., 1946, An enzyme in animal tissue capable of hydrolyzing the phosphorus-fluorine bond of alkyl fluorophosphates. J. Biol. Chem., 164, 271–289

Mochizuki, H., Scherer, S.W., Xi, T., Nickle, D.C., Majer, M., Huizenga, J.J., Tsui, L.C., and Prochazka, M., 1998, Human PON2 gene at 7q21.3: cloning, multiple mRNA forms, and missense polymorphisms in the coding sequence. Gene, 213(1–2), 149–157

Moser, V.C., Chanda, S.M., Mortensen, S., and Padilla, S., 1998, Age and gender-related differences in sensitivity to chlorpyrifos in the rat reflect developmental profiles of esterase activity, Toxicol. Sci., 46, 211–222

Ng, C.J., Shih, D.M., Hama, S.Y., Villa, B., Navab, M., and Reddy, S.T., 2005, The paraoxonase gene family and atherosclerosis, Free Radic. Biol. Med., 38, 153–163

Ohbu, S., Yamashina, A., Takasu, N., Yamaguchi, T., Murai, T., Nakano, K., Matsui, Y., Mikami, R., Sakurai, K., and Hinohara, S., 1997, Sarin poisoning on Tokyo subway. South. Med. J., 90, 587–593

Ortigoza-Ferado, J., Richter, R., Hornung, S.K., Motulsky, A.G., and Furlong, C.E., 1984, Paraoxon hydrolysis in human serum mediated by a genetically variable arylesterase and albumin. Am. J. Hum. Genet. 36, 295–305

Ozer, E.A., Pezzulo, A., Shih, D.M., Chun, C., Furlong, C., Lusis, A.J., Greenberg, E.P., and Zabner, J., 2005, Human and murine paraoxonase 1 are host modulators of Pseudomonas aeruginosa quorum-sensing, FEMS Microbiol. Lett., 253, 29–37

Ozols, J., 1999, Isolation and complete covalent structure of liver microsomal paraoxonase. Biochem. J. 338, 265–272

Primo-Parmo, S.L., Sorenson, R.C., Teiber, J., and La Du, B.N., 1996, The human serum paraoxonase/arylesterase gene (PON1) is one member of a multigene family. Genomics, 33, 498–507

Richter, R.J., and Furlong, C.E., 1999, Determination of paraoxonase (PON1) status requires more than genotyping. Pharmacogenetics, 9, 745–753

Richter, R.J., Jampsa, R.L., Jarvik, G.P., Costa, L.G., and Furlong, C.E, 2004, Determination of paraoxonase 1 (PON1) status and genotypes at specific polymorphic sites, In *Current Protocols in Toxicology*, MD Mains, LG Costa, DJ Reed, E Hodgson, eds., John Wiley and Sons, NY, NY. Pp. 4.12.1–4.12.a19.

Rochu, D., Chabriere, E., and Masson, P., 2007, Human paraoxonase: A promising approach for pre-treatment and therapy of organophosphorus poisoning, Toxicol., 233, 47–59

Rodrigo, L., Gil, F., Hernandez, A.F., Lopez, O., and Pla, A., 2003, Identification of paraoxonase 3 in rat liver microsomes: purification and biochemical properties, Biochem. J., 376(Pt 1), 261–268

Rosenblat, M., Draganov, D., Watson, C.E., Bisgaier, C.L., La Du, B.N., Aviram, M., 2003, Mouse macrophage paraoxonase 2 activity is increased whereas cellular paraoxonase 3 activity is decreased under oxidative stress. Arterioscler. Thromb. Vasc. Biol., 23, 468–474

Schmiegelow, K., Eiberg, H., Tsui, L.C., Buchwald, M., Phelan, P.D., Williamson, R., Warwick, W., Niebuhr, E., Mohr, J., Schwartz, M., et al., 1986, Linkage between the loci for cystic fibrosis and paraoxonase. Clin. Genet. 29, 374–377

Shih, D.M., Gu, L., Xia, Y.-R., Navab, M., Li, W.-F., Hama, S., Castellani, L.W., Furlong, C.E., Costa, L.G., Fogelman, A.M., and Lusis, A.J., 1998, Mice lacking serum paraoxonase are susceptible to organophosphate toxicity and atherosclerosis. Nature 394, 284–287

Shiner, M., Fuhrman, B., and Aviram, M., 2004, Paraoxonase 2 (PON2) expression is upregulated via a reduced-nicotinamide-adenine-dinucleotide-phosphate (NADPH)-oxidase-dependent mechanism during monocytes differentiation into macrophages. Free Radic. Biol. Med., 37, 2052–2063

Shiner, M., Fuhrman, B., and Aviram, M., 2007, Macrophage paraoxonase 2 (PON2) expression is up-regulated by pomegranate juice phenolic anti-oxidants via PPARgamma and AP-1 pathway activation, Atherosclerosis, 2007 Feb 9; [Epub ahead of print]

Skarnes R.C., 1968, In vivo interaction of endotoxin with a plasma lipoprotein having esterase activity. J Bacteriol. 95, 2031–2034

Sorenson, R.C., Bisgaier, C.L., Aviram, M., Hsu, C., Billecke, S., and La Du, B.N. 1999, Human serum Paraoxonase/Arylesterase's retained hydrophobic N-terminal leader sequence associates with HDLs by binding phospholipids : apolipoprotein A-I stabilizes activity. Arterioscler. Thromb. Vasc. Biol. 19, 2214–2225

Stoltz, D.A., Ozer, E.A., Ng, C.J., Yu, J., Reddy, S.T., Lusis, A.J., Bourquard, N., Parsek, M.R., Zabner, J., and Shih, D.M., 2006, Paraoxonase-2 Deficiency Enhances *Pseudomonas aeruginosa* Quorum Sensing in Murine Tracheal Epithelia, Am. J Physiol. Lung. Cell Mol.Physiol., November 22 doi:10.1152/ajplung.00370.2006, 1–43

Suzuki, T., Morita, H., Ono, K., Maekawa, K., Nagai, R., and Yazaki, Y., 1995, Sarin poisoning in Tokyo subway. Lancet, 345, 980

Teiber, J.F., Draganov, D.I., and La Du, B.N., 2003, Lactonase and lactonizing activities of human serum paraoxonase (PON1) and rabbit serum PON3. Biochem. Pharmacol., 66, 887–896

Tougou, K., Nakamura, A., Watanabe, S. Okuyama, Y., and Morino, A., 1998, Paraoxonase has a major role in the hydrolysis of prulifloxacin (NM441): a prodrug of a new antibacterial agent Drug Metab. Dispos., 26, 355–359

Uriel, J., 1961, Characterization of cholinesterase and other carboxylic esterases after electrophoresis and immunoelectrophoresis on agar. I. Application to the study of esterases of normal human serum. Ann Inst Pasteur (Paris), 101, 104–119

van Himbergen, T.M., van Tits, L.J.H., Roest, M., and Stalenhoef, A.F.H., 2006, The story of PON1: how an organophosphate-hydrolyzing enzyme is becoming a player in cardiovascular medicine. Neth. J. Med., 64, 34–38

Wainwright, B.J, Scambler, P.J., Schmidtke, J., Watson, E.A., Law, H.Y., Farrall, M., Cooke, H.J., Eiberg, H., and Williamson, R., 1985, Localization of cystic fibrosis to human chromosome 7cen-q22. Nature, 318, 384–385

Watson, A., Berliner, J.A., Hama, S.Y., La Du, B.N., Faull, K.F., Fogelman, A.M., and Navab, M., 1995, Protective effect of high density lipoprotein associated paraoxonase – inhibition of the biological activity of minimally oxidized low-density lipoprotein. J. Clin. Invest., 96, 2882–2891

Wheeler, J.G., Keavney, B.D., Watkins, H., Collins, R., and Danesh, J., 2004, Four paraoxonase gene polymorphisms in 11212 cases of coronary heart disease and 12786 controls: meta-analysis of 43 studies, Lancet, 363, 689–9

White, R., Woodward, S., Leppert, M., O'Connell, P., Hoff, M., Herbst, J., Lalouel, J.M., Dean, M., and Vande Woude, G., 1985, A closely linked genetic marker for cystic fibrosis. Nature, 318, 382–384

Wilde, C.E., and Kekwick, R.G.O., 1964, The arylesterases of human serum. Biochem. J., 91, 297–307

Yamada, Y., Takatori, T., Nagao, M., Iwase, H., Kuroda, N., Yanagida, J., and Shinozuka, T., 2001, Expression of paraoxonase isoform did not confer protection from acute sarin poisoning in the Tokyo subway terrorist attack. Int. J. Legal Med., 115, 82–84

Zech, R., and Zurcher, K., 1974, Organophosphate splitting serum enzymes in different mammals. Comp. Biochem. Physiol., 48B, 427–433

Zhu, J., Ze, Y., Zhang, C., Zang, Y., Lu, H., Chu, P., Sun, M., and Qin, J., 2006, High-level expression of recombinant human paraoxonase 1 Q in silkworm larvae (Bombyx mori). Appl. Microbiol. Biotechnol., 72,103–108

Zimmerman, J.K., Grothusen, J.R., Bryson, P.K., and Brown, T.M., 1989, Partial purification and characterization of paraoxonase from rabbit serum, In: *Enzymes Hydrolysing Organophosphorus Compounds*. E.Reiner, W.N.Aldridge, F.C.G. Haskin, eds., Ellis Horwood Ltd., Chichester: pp. 128–142

PART 2

PONs AND ATHEROSCLEROSIS

GLUCOSE INACTIVATES PARAOXONASE 1 (PON1) AND DISPLACES IT FROM HIGH DENSITY LIPOPROTEIN (HDL) TO A FREE PON1 FORM

MIRA ROSENBLAT, ORLY SAPIR AND MICHAEL AVIRAM

The Lipid Research Laboratory, Technion Faculty of Medicine, The Rappaport Family Institute for Research in the Medical Sciences, Rambam Medical Center, Haifa, Israel

Abstract: The aim of the present study was to analyze the direct effect of glucose *in vitro* on HDL-associated paraoxonase 1 (PON1) activity and stability, and on HDL-mediated macrophage cholesterol efflux. Furthermore, the above parameters were determined in HDL from diabetic patients in comparison to healthy subjects.

Incubation of serum from healthy subject (glucose levels lower than 100 mg%) with increasing concentrations of glucose (200–600 mg%) for 24 hours at 37 °C, resulted in a significant glucose dose-dependent reduction in paraoxonase activity in the serum and also in the HDL fraction (by up to 49). Enrichment of serum with glucose, up to 400 mg% significantly increased the inactivation rate of the free PON1 but not that of the HDL-bound PON1, by 7 fold.

Direct incubation of HDL with increasing glucose concentrations (0–400 mg%) for 24 hours at 37 °C resulted in a glucose dose-dependent decrement in HDL PON1 lactonase activity by up to 38%. In parallel, the HDLs ability to induce macrophage cholesterol efflux was reduced by up to 44%. Similarly, the lactonase activity of recombinant PON1 that was incubated with 400 mg% glucose was decreased by 31%, and this glucose-treated PON1 was not able to stimulate HDL-mediated macrophage cholesterol efflux, compared with the stimulatory effect of recombinant PON1 that was incubated without glucose. Furthermore, we were able to demonstrate by Western Blot analysis that glucose caused the dissociation of PON1 from the HDL, and this effect was glucose dose-dependent.

Finally, in diabetic patients, serum PON1 lactonase activity was significantly lower, by 60% vs. control healthy subjects. Analysis of serum PON1 stability, revealed that in diabetic patients the free PON1, as well as the HDL-bound PON1 inactivation rates were both significantly higher than those observed in the controls. Similarly, PON1 lactonase activity in HDL isolated from the diabetic patients vs. controls HDL was significantly lower by 77%. In addition, cholesterol efflux rate from J774 A.1 macrophages that was induced by diabetic patients HDL was significantly decreased by 53%, compared

Address correspondence to: Prof. Michael Aviram, D.Sc. The Lipid Research Laboratory, Rambam Medical Center, Haifa, Israel, 31096.TS
Fax: 972-4-8542130; Tel: 972-4-8542970; E mail: aviram@tx.technion.ac.il

B. Mackness et cl. (eds.), The Paraoxonases: Their Role in Disease Development and Xenobiotic Metabolism, 35–49.
© 2008 *Springer.*

to the effect of HDL from healthy subjects. We thus conclude that PON1 inactivation
and dissociation from HDL by high glucose concentrations may be responsible for the
accelerated atherosclerosis in diabetic patients

Keywords: Diabetes, glucose, paraoxonase 1, HDL, macrophages, lactonase activity, cholesterol
efflux

1.   INTRODUCTION

Paraoxonase 1 (PON1) is an HDL-associated esterase/lactonase (Sorenson et al.,
1999; Teiber et al., 2003; Gaidukov and Tawfik, 2005; Khersonsky and Tawfik,
2005), and its activity is inversely related to the risk of cardiovascular diseases
(Getz and Reardon, 2004; Mackness et al.,[1] 2004). The crystal structure elucidation
of PON1 obtained by directed evolution showed that PON1 consists of a six-bladed
ß propeller with a unique active site (Harel et al., 2004). ApoA-I the major protein
in HDL, stabilizes PON1, binds it with very high affinity, and selectively stimulates
its lactonase activity (Sorenson et al., 1999; James and Deakin, 2004; Gaidukov
and Tawfik, 2005). PON1 was shown to have a role in atherosclerosis development.
PON1 anti-atherogenic properties include protection of LDL, HDL and macrophages
against oxidative stress (Rozenberg et al.,[2] 2005; Aviram et al.,[2] 1998; Mackness
et al.,[2] 1993; Aviram and Rosenblat,[1] 2004), as shown by attenuation of oxidized-
LDL uptake by macrophages (Fuhrman et al., 2002), inhibition of macrophage
cholesterol biosynthesis (Rozenberg et al.,[1] 2003), and stimulation of HDL-mediated
cholesterol efflux from macrophages (Rosenblat et al.,[3] 2005). Recently, we have
demonstrated that these anti-atherogenic properties of PON1 are all related to its
lipo-lactonase activity (Rosenblat et al.,[1] 2006).
   Diabetic patients are at high risk to develop atherosclerosis (Mackness et al.,
2006; Li et al., 2005). Most studies have found that PON1 activity is reduced in
type 1 and type 2 diabetic patients, independently of the PON1 genotype (Boemi
et al., 2001; Letellier et al., 2002; Juretic et al., 2006). The low serum paraoxonase
activity in type 2 diabetes was recently shown to be correlated with the levels of
the patients' plasma oxidized LDL and with vascular complications (Tsuzura et al.,
2004; Karabina et al., 2005). A recent study (Ferretti et al.,[1] 2004) demonstrated
that HDL from type 1 diabetic patients showed higher levels of lipid hydroperoxides
and lower activity of HDL-PON1 than healthy subjects. Moreover, HDL from these
patients protected less efficiently erythrocyte membranes against oxidative damage
compared with HDL from control subjects. We (Rosenblat et al.,[2] 2006) have
recently found that in diabetic patients, a significant PON1 paraoxonase activity is
found in the lipoprotein deficient serum (LPDS) fraction in comparison to healthy
subjects, where most of PON1 paraoxonase activity is in HDL. PON1 in LPDS,
unlike PON1 in HDL, can no longer protect against lipid peroxidation or stimulate
macrophage cholesterol efflux.
   The dissociation of PON1 from the HDL particles to LPDS in diabetic patients
may be related to the reduced amount of apoA-I observed in the HDL from the

diabetic patients, or to the fact that apoA-I in diabetic patients is nitrated (Gugliucci et al., 2006). Another possibility is that this phenomenon may be associated with the increased blood glucose levels in diabetic patients. Oxidation or glycation of HDL could cause structural modifications, which affect the activities of HDL-associated enzymes and HDL anti-atherogenic properties (Ferretti et al.,[3] 2006). HDL that was incubated for three days with increasing glucose concentrations (0–100 mM) showed significant increase in HDL lipid peroxidation, modified apolipoprotein conformation, and a significant decreased PON1 activity (Ferretti et al.,[2] 2001). Glycation of purified paraoxonase protein (by incubation with 25 mM glucose) caused a 40% reduction in its enzymatic activity, and this glycated PON1 did not inhibit monocytes adhesion to human aortic endothelial cells (Hedrick et al., 2000). Furthermore, it was shown that glycation of HDL impaired the binding of HDL to fibroblasts and the efflux of cholesterol from the cells (Duell et al.,[1] 1990; Duell et al.,[2] 1991). Other studies however demonstrated that HDL glycation did not increase its susceptibility to oxidation or diminish its capacity to induce cholesterol efflux from macrophages (Rashdani et al., 1999; Duell et al.,[3] 1991).

However, in vivo, small dense HDL_3 particles that were separated from type 2 diabetic patients, in comparison to HDL_3 from healthy subjects, were significantly less able to protect LDL from oxidation, and to induce macrophage cholesterol efflux (Nodecourt et al., 2005). The goal of the present study was to analyze the direct effect of glucose on HDL-associated PON1 activities, and on HDL ability to induce macrophage cholesterol efflux.

Furthermore, we questioned whether glucose interaction with HDL could result in the dissociation of PON1 from the HDL.

2.   RESULTS

2.1.   Inactivation of PON1 in Serum by Glucose

Incubation of serum from control healthy subject (containing 100 mg% glucose) with increasing glucose concentrations for 48 hours at 37 °C, resulted in a significant reduction in serum paraoxonase activity, which was glucose dose-dependent. Upon using 600 mg% of glucose, PON1 activity decreased by up to 49% of the activity that was measured prior to the incubation ("0 time", Fig. 1A). Next we performed a time study, where 200 mg% of glucose was added to serum from control healthy subject (containing 87 mg% of glucose). Both serum samples (87 mg% or 287 mg%) were than incubated at 37 °C for 24, 48 or 78 hours. Serum paraoxonase activity in the control serum was reduced with incubation time by 17%, 35% and 42%, respectively (Fig. 1B). In serum containing 287 mg% of glucose however, a more significant reduction in serum paraoxonase activity with time was noted, as shown, by 28%, 43% and 53%, decrease after 24, 48 or 72 hours, respectively (Fig. 1B). Since the maximal effect of glucose was observed already after 24 hours of incubation, this time period was used in further experiments. The above results suggest that high glucose concentrations inactivate serum PON1 activity.

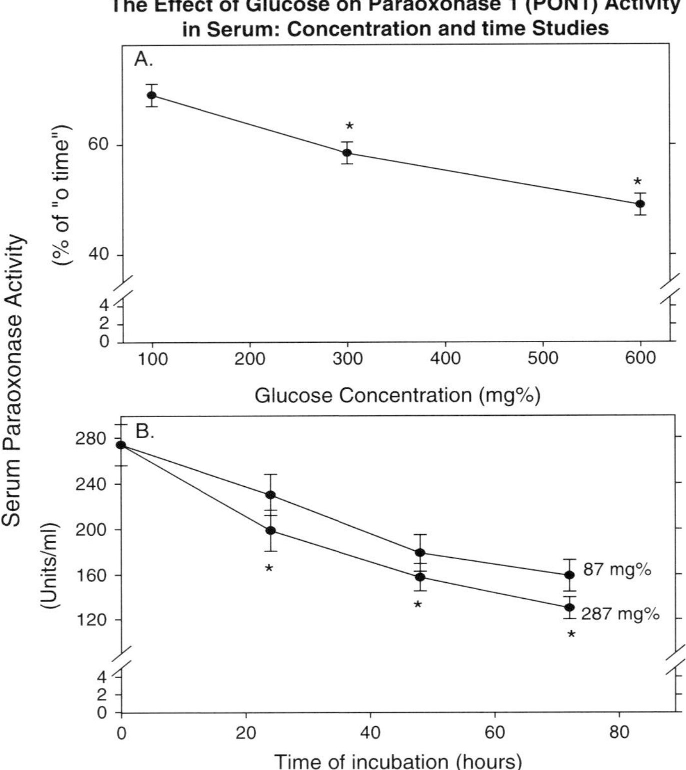

Figure 1. The effect of glucose on serum paraoxonase 1 (PON1) activity: concentration and time study.(A) Serum from control healthy subject containing 100 mg% glucose, was incubated for 48 hours at 37 °C with no addition, or with two glucose concentration (up to 300 and 600 mg%). Paraoxonase activity was measured before the incubation ("o time") and at the end of the incubation period. (B) Serum from control healthy subject containing 87 mg% glucose was incubated for 24, 48 and 72 hours at 37 °C with no addition, or with the addition of 200 mg% glucose. Paraoxonase activity was measured at the different time points. Results are given as mean ±SD of three different experiments. *p < 0.01 vs. o time

As in healthy subjects most of serum PON1 is associated with HDL, we next examined whether high serum glucose concentrations inactivate also HDL-associated PON1. For this purpose, HDL was isolated by ultracentrifugation from serum that was preincubated with increasing glucose concentrations (72, 272 or 472 mg% of glucose) for 24 hours at 37 °C. Indeed, paraoxonase activity in HDL that was isolated from serum with high glucose levels, (272 mg%, or 472 mg%) was significantly reduced, by 18% or by 23%, respectively, as compared to paraoxonase activity measured in HDL isolated from control serum containing only 72 mg% of glucose (Fig. 2).

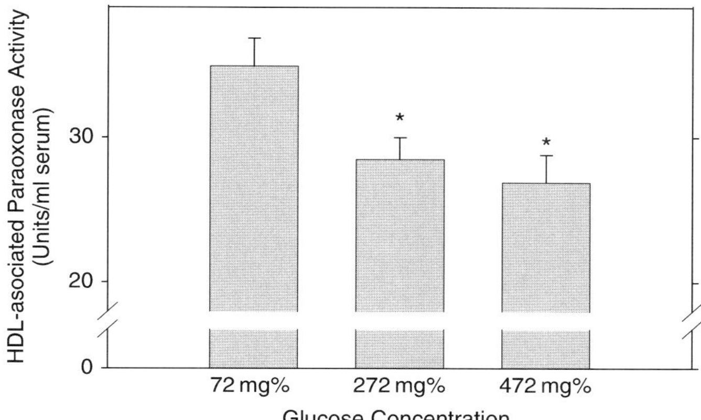

Figure 2. The effect of glucose on serum HDL-associated paraoxonase 1 (PON1) activity. HDL was isolated by ultracentrifugation from serum that was incubated for 24 hours at 37 °C with no addition (72 mg% of glucose) or with increasing glucose concentrations (272, or 472 mg% of glucose). PON1 paraoxonase activity was determined in the HDL samples at the end of the incubation period. Results are given as mean ±SD of three different experiments. *p < 0.01 vs. glucose concentration of 72 mg%

2.2. PON1 Stability in Glucose-enriched Serum

In order to determine the effect of glucose on the stability of PON1 in serum, serum from control healthy subject (100 mg% of glucose) was preincubated for 30 minutes at 25 °C with increasing glucose concentrations. Then, the low affinity calcium chelator nitrilotriacetic acid (NTA) was added to the serum samples, and serum arylesterase activity (against phenyl acetate) was monitored along increasing time period (0–10 hours). PON1 kinetic curves were plotted and inactivation rates were calculated (Fig. 3).

The kinetic curves demonstrate a two-phase pattern: the first, fast declining phase is attributed to the free PON1, and the second, slow declining phase is related to the HDL-bound PON1. Incubation for 30 minutes at 25 °C of serum from control subject (83 mg% of glucose) with increasing glucose concentrations, revealed that enrichment of the serum with glucose, up to a concentration of 200 mg% or 400 mg% significantly increased the inactivation rate of the free PON1 form by 4 or 7 fold (0.5%/min in the control serum, and 2.1%/min or 3.6%/min in serum that was enriched with 200 mg% or 400 mg%, respectively, Fig. 3). In contrast, the inactivation rate of the HDL-bound PON1 was not significantly affected by glucose enrichment (Fig. 3). These results suggest that high glucose concentration in serum inactivates mostly the free PON1, but not the more stable HDL-associated PON1.

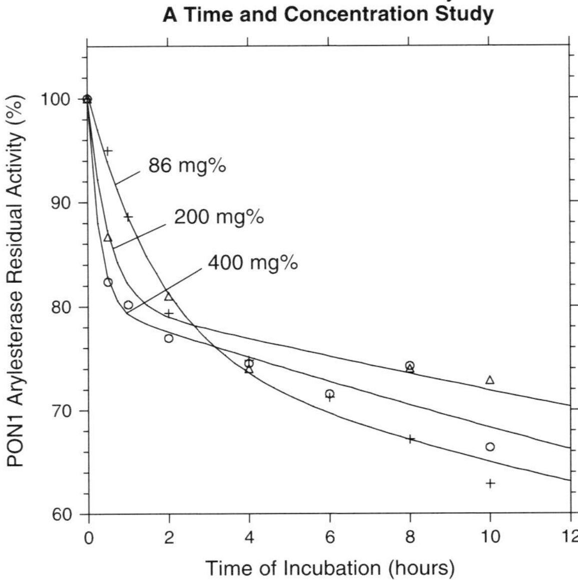

Figure 3. Effect of glucose on serum Paraoxonase 1 (PON1) stability: concentration study. Serum from control healthy subject (containing 86 mg% of glucose) was preincubated for 30 minutes at 25 °C with increasing glucose concentrations (final glucose concentrations of 200 or 400 mg%). Then, the low affinity calcium chelator nitrilotriacetic acid was added to the serum samples, and serum arylesterase activity (against phenyl acetate) was monitored along increasing time period (0–24 hours). Arylesterase activity is given as % of residual activity. PON1 kinetic curves are shown. The results of one representative experiment are shown

Next we questioned whether glucose incubation with HDL could result in the dissociation of PON1 from the HDL particles as a free form. Indeed, by Western Blot analysis, using specific antibody to human PON1, we were able to demonstrate that the amount of free PON1 in the HDL samples that were incubated for 24 hours at 37 °C with glucose (100 mg% or 200 mg%), was increased (by 10% or by 15%), respectively, as compared to the levels of free PON1 in the HDL that was incubated with no glucose (Fig. 4).

2.3.   Effect of Glucose on HDL-associated PON1 Activity, and on HDL-mediated Macrophage Cholesterol Efflux

HDL-mediated cholesterol efflux from macrophage foam cells is considered to be a most important process in the attenuation of atherosclerosis development (Von Eckardstein et al., 2001). We have previously shown that HDL-associated PON1 stimulates the lipoprotein ability to induce cholesterol efflux from macrophages (Rosenblat et al.,[2] 2006; Rosenblat et al.,[1] 2006). Furthermore, by using specific PON1 inhibitor and recombinant mutants (in the PON1 active site histidine dyad),

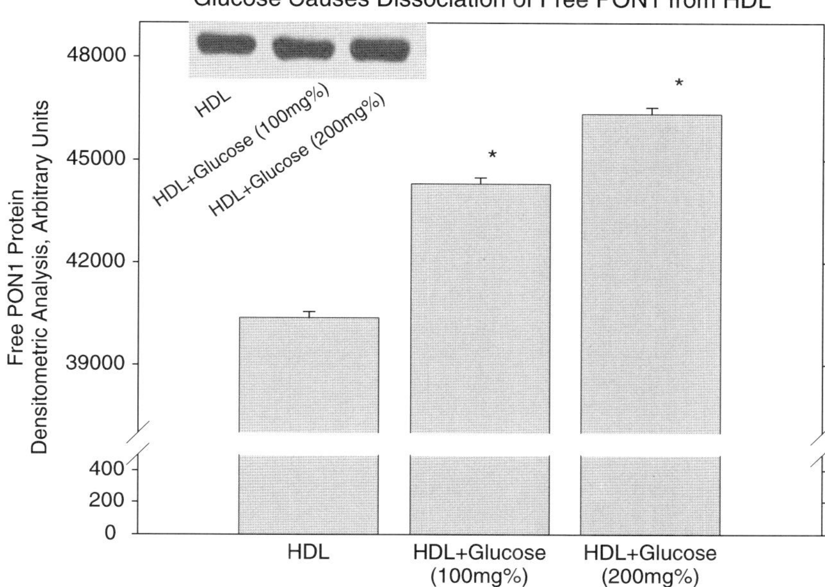

Figure 4. Glucose incubation with HDL results in the dissociation of PON1 from HDL. HDL (1300 μg protein/ml) was incubated for 24 hours at 37 °C with no addition (HDL), or with glucose (100 mg% or 200 mg%). The HDL were rum on 10% acrylamide gel containing 0.1% SDS. Western blot analysis was performed using mouse anti-human PON1 antibody. Densytometric analysis of the PON1 bands is provided, and a representative image of the bands is shown in the insert.. Results are the mean± SD of three different experiments. *p < 0.01 vs. HDL

we have demonstrated that increased PON1 lipo-lactonase activity is associated with the stimulatory effect of PON1 on HDL-mediated macrophage cholesterol efflux (Rosenblat et al.,[1] 2006).

Thus, we analyzed the effect of adding increasing glucose concentrations to isolated HDL particles, on PON1 lactonase activity, and in parallel on the ability of the HDL to induce cholesterol efflux from J774 A.1 macrophages (Fig. 5). Incubation of HDL with increasing glucose concentrations (0–400 mg%) for 24 hours at 37 °C, resulted in a glucose does-dependent decrement in PON1 lactonase activity towards thiobuthylbutyro lactone (TBBL), by up to 38% (Fig. 5A). In parallel, the HDLs ability to stimulate macrophage cholesterol efflux was also significantly decreased, in a glucose dose-dependent manner, by up to 44% (Fig. 5B).

2.4. Effect of Glucose on Purified PON1 Lactonase Activity, and on its Ability to Stimulate HDL-mediated Macrophage Cholesterol Efflux

We next determined the direct effect of glucose on free (not HDL-bound) recombinant PON1 (which was expressed in E.coli by a direct evolution process,

Glucose Dose-dependently Reduces HDL-Associated PON1 Lactonase Activity, and its Ability to Induce Cholesterol Efflux from Macrophages

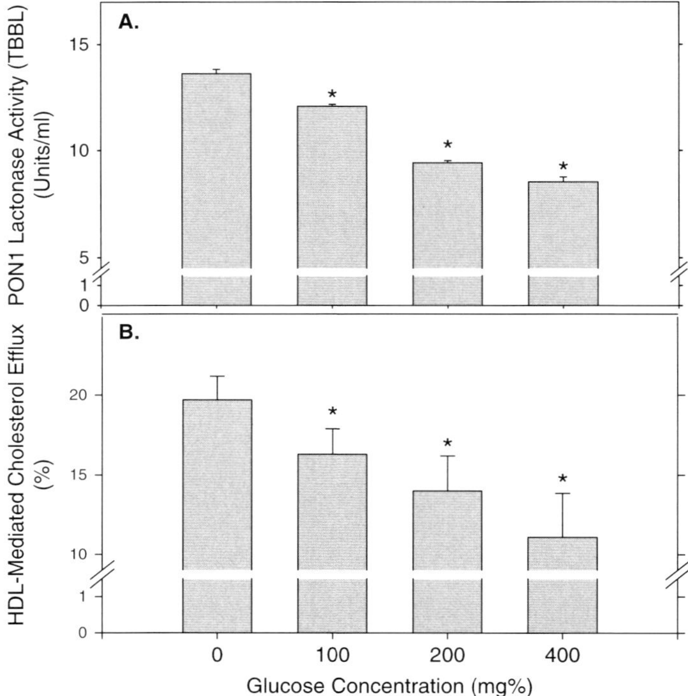

Figure 5. The effect of glucose on HDL- associated paraoxonase 1 (PON1) lactonase activity, and on its ability to induce cholesterol efflux from macrophages: concentration study. HDL was isolated from healthy subject by ultracentrifugation. The HDL was then incubated for 24 hours at 37 °C with increasing glucose concentrations (0–400 mg%). (A) PON1 lactonase activity towards thiobuthylbutyro lactone (TBBL) was determined in the HDL samples at the end of the incubation period. (B) The extent of cholesterol efflux from J774 A.1 macrophages prelabeled with ^{3}[H]-cholesterol by these HD L samples (100 μg of protein/ml) was determined after 3 hours incubation. Results are given as mean ±SD of three different experiments.*p < 0.01 vs. o concentration

Harel et al., 2004). Incubation of recombinant PON1 with increasing glucose concentrations (0–400 mg%) for 24 hours at 37 °C, resulted in a glucose dose-dependent decrease in PON1 lactonase activity towards TBBL, by up to 31% (Fig. 6A). The ability of these PON1 samples (20 μg/ml) to stimulated HDL-mediated macrophage cholesterol efflux was also studied (Fig. 6B). While the addition of untreated control PON1 to HDL, increased the macrophage cholesterol efflux rate by 45%, PON1 that was incubated with 100 mg% or 200 mg% of glucose, stimulated HDL-mediated macrophage cholesterol efflux by only 13% or 4%, respectively (Fig. 6A). PON1 that was incubated with the highest glucose concentration (400 mg%), was not able at all to stimulate HDL-mediated cholesterol efflux

**Glucose Reduces Lactonase Activity of Free PON1, and
its Ability to Stimulate HDL-Mediated Macrophage Cholesterol Efflux**

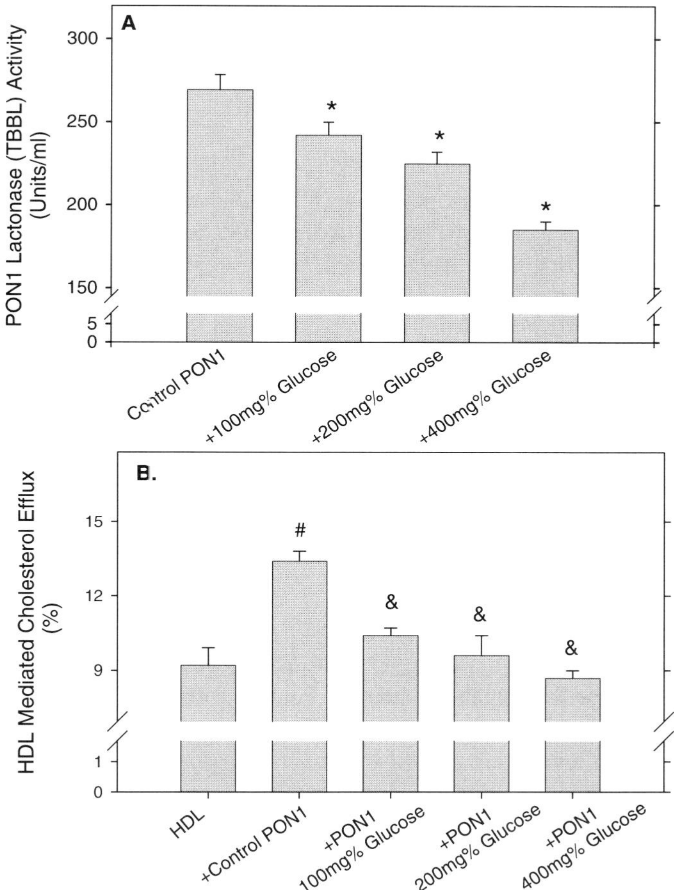

Figure 6. The effect of glucose on purified Paraoxonase 1 (PON1) lactonase activity, and on its ability to stimulate HDL-mediated macrophage cholesterol efflux. Recombinant PON1 (which was expressed in E.coli by a direct evolution process) was incubated for 24 hours at 37 °C with no addition (Control PON1) or with increasing glucose concentrations (100, 200 or 400 mg%). (A) PON1 lactonase activity towards thiobuthylbutyro lactone (TBBL) was determined at the end of the incubation period. (B) The above PON1 samples (20 μg/ml) were added to HDL. The extent of cholesterol efflux from J774 A.1 macrophages prelabeled with ^{3}[H]-cholesterol by these HDL samples, and by HDL that was not incubated with PON1 (HDL) was determined after 3 hours incubation. Results are given as mean ±SD of three different experiments. *p < 0.01 vs. Control PON1, #p < 0.01 vs. HDL, &p < 0.01 vs. HDL+ Control PON1

from macrophages (Fig. 6B). These results indicate that high glucose concentrations inactivate not only HDL-associated PON1 (Fig. 5), but also free PON1 (Fig. 6), thus affecting PON1 biological activity.

2.5.   PON1 Inactivation by Glucose in Diabetic Patients

2.5.1.   *Serum PON1 activities in diabetic patients vs. healthy subjects*

Serum PON1 activities (arylesterase and paraoxonase) in diabetic patients were found to be significantly reduced, in comparison to serum PON1 activities observed in control healthy subjects (Boemi et al., 2001; Letellier et al., 2002; Juretic et al., 2006). In serum from non insulin dependent diabetic mellitus (NIDDM) patients, arylesterase and paraoxonase activities were significantly lower, by 46% and by 51%, respectively, compared to the activities observed in serum from controls healthy subjects (Table 1). As it was recently became apparent that PON1 is in fact a lactonase with lipophylic lactones comprising its primary substrates (Teiber et al., 2003; Gaidukov and Tawfik, 2005; Khersonsky and Tawfik, 2005), we measured also PON1 lactonase activity towards TBBL. Indeed this activity was also markedly reduced, by up to 60%, in the diabetic patients vs. controls (Table 1).

2.5.2.   *PON1 stability in serum from diabetic patients vs. healthy subjects*

The low affinity calcium chelator nitrilotriacetic acid (NTA) was added to serum samples obtained from diabetic patients or healthy control subjects, and serum arylesterase activity was then determined along increasing time points interval (0–24 hours). PON1 kinetic curves were plotted and the inactivation rates calculated (Fig. 7). The free PON1 inactivation rate was significantly increased, by 2.4 or by 5.4 fold in serum from diabetic patients, compared to the free PON1 inactivation rate observed in a representative serum of control healthy subject (Fig. 7). The inactivation rate of the HDL-bound PON1 was also increased in the diabetic patients, but to a lesser extent than that of the free PON1, as noticed by only 1.9 or 2.9 fold increase, compared to HDL-bound PON1 inactivation rate that was noted in control

Table 1. Serum paraoxonase 1 (PON1) activities in Diabetic patients vs Control Healthy subjects

Serum PON1 Activity (Units/ml)	Healthy Subjects	Diabetic Patients
Paraoxonase	340 ± 37	$183 \pm 35^*$
Arylesterase	113 ± 16	$55 \pm 7^*$
Lactonase	75 ± 5	$30 \pm 3^*$

$^*\, p < 0.01$ *Patients vs. Control* $(n = 10)$

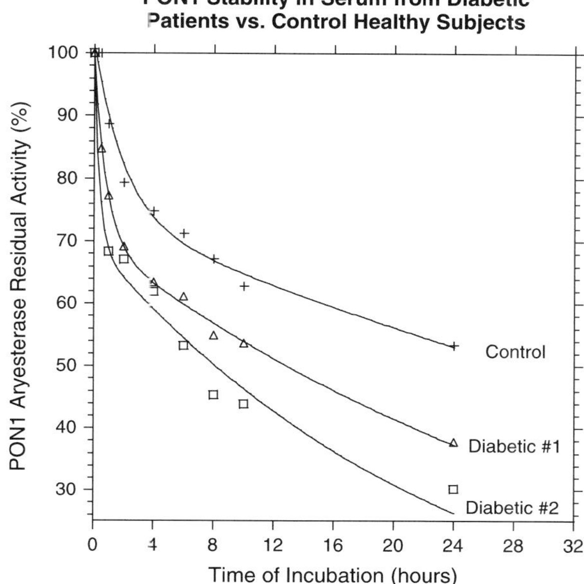

Figure 7. Paraoxonase 1 (PON1) stability in serum from diabetic patients vs. control healthy subject. The low affinity calcium chelator nitrilotriacetic acid (NTA) was added to serum samples obtained from two diabetic patients (glucose levels above 200 mg%), and from control healthy subject (glucose level of 86 mg%). Serum arylesterase activity was then determined along different time points interval (0–24 hours). PON1 kinetic curves are shown. The results of one representative experiment are shown

serum (Fig. 7). These results indicate that in diabetic PON1, both the free PON1 form and the HDL-bound PON1, are both more susceptible to inactivation, with the free PON1 being more susceptible than the HDL-bound PON1, as was also shown in vitro for the effect of high glucose on PON1 stability (Fig. 3).

2.5.3.   *HDL PON1 lactonase activity and the ability of HDL from diabetic patients vs. HDL from control healthy subjects, to stimulate macrophage cholesterol efflux*

Finally, we analyzed the relevance of the deleterious effects of glucose on HDL-associated PON1 activities, to the atherogenicity observed in diabetes. HDL was isolated from the sera samples of 5 healthy subjects and 5 diabetic patients by ultracentrifugation. Analyses of these HDLs revealed that HDL-associated lactonase activity towards TBBL, was significantly decreased, by 77%, in HDLs that were isolated from diabetic patients serum, compared to the controls' HDL (Fig. 8A). The ability of these HDLs to induce cholesterol efflux from J774A.1 macrophages was also studied. Cholesterol efflux rate by the diabetic patients HDL was significantly lower, by 53%, compared to the cholesterol efflux rate observed on using HDL from healthy subjects (Fig. 8B).

**HDL from Diabetic Patients has Lower PON1 Lactonase Activity,
and is Less Able to Induce Macrophage Cholesterol Efflux than
HDL from Healthy Subjects**

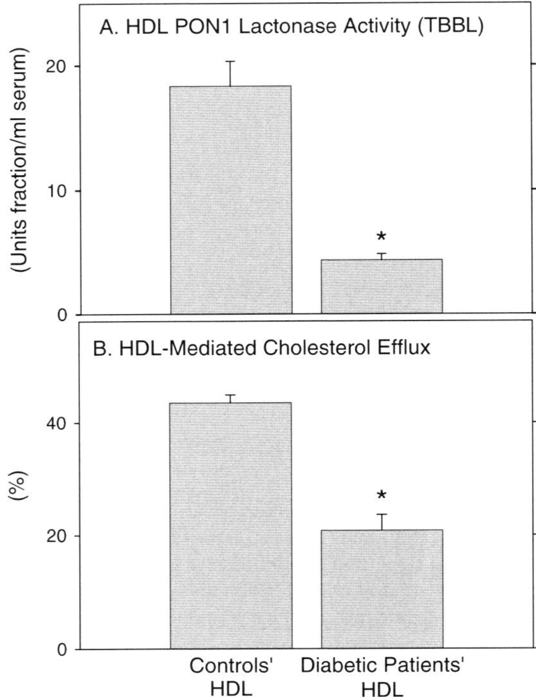

Figure 8. HDL from diabetic patients vs. HDL from healthy subjects: PON1 lactonase activity, and HDL-mediated macrophage cholesterol efflux. HDL was isolated by ultracentrifugation from the sera of 5 healthy subjects (Controls' HDL) and from 5 diabetic patients (glucose levels above 200 mg%). (A) PON1 lactonase activity towards thiobuthylbutyro lactone (TBBL) was measured in the HDL samples. (B) The HDL samples (100 μg protein/ml) were incubated for 3 hours at 37 °C with J774 A.1 macrophages prelabeled with ^{3}[H]-cholesterol, and the extent of HDL-mediated macrophage cholesterol efflux was determined. Results are given as mean ±SD (n = 5). *p < 0.01 vs. Controls' HDl

3. CONCLUSIONS

Our *in vitro* studies demonstrate the following observations:

1. Glucose inactivates serum PON1 (paraoxonase and lactonase activities), and this effect was glucose dose-dependent. The reduced serum PON1 activity by glucose results from a direct inactivation of both HDL-bound PON1, as well as free PON1.

2. The significant decrement in HDL-bound PON1 lactonase activity by glucose was associated with reduced ability of the HDL to remove cholesterol from macrophages (defected reverse cholesterol transport).

3. Glucose causes dissociation of PON1 from the HDL particle, thus making PON1 less stable.

Our *ex-vivo* studies with diabetic patients vs. controls healthy subjects demonstrate the following:

A. Serum PON1 lactonase activity was significantly reduced in the diabetic serum vs. control healthy subjects. The inactivation rates of both free PON1 and HDL-bound PON1 were significantly higher in the diabetic patients.

B. The decreased PON1 lactonase activity in HDL from diabetic patients vs. controls is associated with reduced HDL-mediated macrophage cholesterol efflux.

C. The results observed in diabetic patients may be related at least in part, to a direct effect of the high glucose concentration on PON1 activity.

4. The ability of glucose to cause a dissociation of PON1 from the HDL may explain our previous observations demonstrating that in diabetic patient's serum, HDL-associated PON1 decreased, whereas free PON1 in LPDS increased, in comparison to control healthy subjects (Rosenblat et al.,[2] 2006).

REFERENCES

Aviram M[1], Rosenblat M, 2004. Paraoxonases 1, 2, and 3, oxidative stress, and macrophage foam cell formation during atherosclerosis development. Free Radic Biol Med. 37:1304–16.

Aviram M[2], Rosenblat M, Bisgaier CL, Newton RS, Primo- Parmo SL, La Du BN, 1998. Paraoxonase inhibits high density lipoprotein (HDL) oxidation and preserves its functions: a possible peroxidative role for paraoxonase. J Clin Invest. 101:1581–1590.

Boemi M, Leviev I, Sirolla C, Pieri C, Marra M, James RW, 2001. Serum paraoxonase is reduced in type 1 diabetic patients compared to non-diabetic, first degree relatives; influence on the ability of HDL to protect LDL from oxidation. Atherosclerosis. 155:229–235.

Duell PB[3], 1991. High glucose levels do not directly impair cellular binding of HDL_3 or HDL-mediated efflux of cholesterol from human skin fibroblasts. Acta Diabetol. 28:174–178.

Duell PB[1], Oram JF, Bierman EL, 1990. Nonenzymatic glycosylation of HDL resulting in inhibition of high-affinity binding to cultured human fibroblasts. Diabetes. 39:1257–1263.

Duell PB[2], Oram JF, Bierman EL, 1991. Nonenzymatic glycosylation of HDL and impaired HDL-receptor- mediated cholesterol efflux. Diabetes. 40:377–384.

Ferretti G[1], Bacchetti T, Busni D, Rabini RA, Curatola G, 2004. Protective effect of paraoxonase activity in high-density lipoproteins against erythrocyte membranes peroxidation: a comparison between healthy subjects and type 1 diabetic patients. J Clin Endocrinol Metab. 89:2957–2962.

Ferretti G[2], Bacchetti T, Marchionni C, Caldarelli L, Curatola G, 2001. Effect of glycation of high density lipoproteins on their physiochemical properties and on paraoxonase activity. Acta Diabetol. 38:163–169.

Ferretti G[3], Bacchetti T, Negre- salvayre A, Salvayre R, Dousset N, Curatola G, 2006. Structural modification of HDL and functional consequences. Atherosclerosis. 184:1–7.

Fuhrman B, Volkova N, Aviram M, 2002. Oxidative stress increases the expression of the CD36 scavenger receptor and the cellular uptake of oxidized LDL in macrophages from atherosclerotic mice: protective role of antioxidants and of paraoxonase. Atherosclerosis. 161:307–16.

Gaidukov L, Tawfik DS, 2005. High affinity, stability, and lactonase activity of serum paraoxonase PON1 anchored on HDL with apoA-I. Biochemistry. 44:11843–11854.

Getz GS, Reardon CA, 2004. Paraoxonase, a cardioprotective enzyme: continuing issues. Curr Opin Lipidol. 15:261–67.

Gugliucci A, Hermo R, Tsuji M, Kimura S, 2006. Lower serum paraoxonase -1 activity in type 2 diabetic patients correlates with nitrated apolipoprotein A-I levels. Clin Chim Acta. 368–201–202.

Harel M, Aharoni A, Gaidukov L, Brumshtein B, Khersonsky O, Meged R, Dvir H, Ravelli RBG, McCarthy A, Toker L, Silman I, Sussman JL, Tawfik DS, 2004. Structure and evolution of the serum paraoxonase family of detoxifying and anti-atherosclerotic enzymes. Nat Struc Mol Biol. 11:412–419.

Hedrick CC, Thorpe SR, Fu MX, Harper CM, Yoo J, Kim SM, Wong H, Peters AL, 2000. Glycation impairs high-density lipoprotein function. Diabetologia. 43:312–320.

James RW, Deakin SP, 2004. The importance of high-density lipoprotein for paraoxonase-1 secretion, stability, and activity. Free Radic Biol Med. 37:1986–1994.

Juretic D, Motejlkova A, Kunovic B, Rekic B, Flegar-Mestric Z, Vujic L, Mesic R, Lukac-Bajalo J, Simeon-Rudolf V, 2006. Paraoxonase/arylesterase in serum of patients with type II diabetes mellitus. Acta Pharm. 56:59–68.

Karabina SA, Lehner AN, Frank E, Parthasarathy S, Santanam N, 2005. Oxidative inactivation of paraoxonase- implications in diabetes mellitus and atherosclerosis. Biochim Biophys Acta. 1725:213–221.

Khersonsky O, Tawfik DS, 2005. Structure- reactivity studies of serum paraoxonase 1 suggest that its native activity is lactonase. Biochemistry. 44:6371–6382.

Letellier C, Durou MR, Jouanolle AM, Le Gall JY, Poirier JY, Ruelland A, 2002. Serum paraoxonase activity and paraoxonase gene polymorphism in type 2 diabetic patients with or without vascular complications. Diabetes Metab. 28:297–304.

Li J, Wang X, Huo Y, Niu T, Chen C, Zhu G, Huang Y, Chen D, Xu X, 2005. PON1 polymorphism, diabetes mellitus, obesity, and risk of myocardial infarction: modifying effect of diabetes mellitus and obesity on the association between PON1 polymorphism and myocardial infarction. Genet Med. 7:58–63.

Mackness B, Hine D, McElduff P, Mackness M, 2006. High C-reactive protein and low paraoxonase 1 in diabetes as risk factors for coronary heart disease. Atherosclerosis. 186:396–401.

Mackness MI[2], Arrol S, Abbott CA, Durrington PN, 1993. Protection of low-density lipoprotein against oxidative modification by high-density lipoprotein associated paraoxonase. Atherosclerosis. 104:129–35.

Mackness MI[1], Durrington PN, Mackness B, 2004. Paraoxonase 1 activity, concentration and genotype in cardiovascular diseases. Curr Opin Lipidol. 15:399–404.

Nodecourt E, Jacqueminet S, Hansel B, Chantepie S, Grimaldi A, Chapman MJ, Kontush A, 2005. Defective antioxidative activity of small dense HDL3 particles in type 2 diabetes: relationship to elevated oxidative stress and hyperglycaemia. Diabetologia. 48:528–538.

Rashdani DL, Rifici VA, Schneider SH, Khachadurian AK, 1999. Glycation of high-density lipoprotein does not increase its susceptibility to oxidation or diminish its cholesterol efflux capacity. Metabolism. 48:139–143.

Rosenblat M[1], Gaidukov L, Khersonsky O, Vaya J, Oren R, Tawfik DS, Aviram M, 2006. The catalytic histidine dyad of high density lipoprotein associated paraoxonase 1 (PON1) is essential for PON1-mediated inhibition of low density lipoprotein oxidation and stimulation of macrophage cholesterol efflux. J Biol Chem. 281:7657–7665.

Rosenblat M[2], Karry R, Aviram M, 2006. Paraoxonase 1 (PON1) is a more potent antioxidant and stimulant of macrophage cholesterol efflux, when present in HDL than in lipoprotein- deficient serum: relevance to diabetes. Atherosclerosis. 187:74e1–74e10.

Rosenblat M[3], Vaya J, Shih DM, Aviram M, 2005. Paraoxonase 1 (PON1) enhances HDL-mediated macrophage cholesterol efflux via the ABCA1 transporter in association with increased HDL binding to the cells: a possible role for lysophosphatidylcholine. Atherosclerosis. 179:69–77.

Rozenberg O[1], Shih DM, Aviram M, 2003. Human serum paraoxonase (PON1) decreases macrophage cholesterol biosynthesis: a possible role for its phospholipase- A_2 activity and lysophosphatidylcholine formation. Arterioscler Thromb Vasc Biol. 23:461–67.

Rozenberg O[2], Shih SD, Aviram M, 2005. Paraoxonase 1 (PON1) attenuates macrophage oxidative status: studies in PON1 transfected cells and in PON1 transgenic mice. Atherosclerosis. 181:9–1818.

Sorenson RC, Bisgaier CL, Aviram M, Hsu C, Billecke S, La Du BN, 1999. Human serum paraoxonase/arylesterase's retained hydrophobic N-terminal leader sequence associates with HDLs by binding phospholipids: apolipoprotein A-I stabilizes activity. Arterioscler Thromb Vasc Biol. 19: 2214–25.

Teiber JF, Draganov DI, La Du BN, 2003 Lactonase and lactonizing activities of human serum paraoxonase 1 (PON1) and rabbit serum PON3. Biochem Pharmacol. 66:887–96.

Tsuzura S, Ikeda Y, Suchiro T, Ota K, Osaki F, Arii K, Kumon Y, Hashimoto K, 2004. Correlation of plasma oxidized low-density lipoprotein levels to vascular complications and human serum paraoxonase in patients with type 2 diabetes. Metabolism. 53:297–302.

Von Eckardstein A, Nofer JR, Assmann G, 2001. High density lipoproteins and arteriosclerosis. Role of cholesterol efflux and reverse cholesterol transport. Arterioscler Thromb Vasc Biol. 21:13–27.

CHAPTER 2

PARAOXONASE-1 AND CARDIOVASCULAR DISEASE

B. MACKNESS AND M.I. MACKNESS
*Division of Cardiovascular and Endocrine Sciences, University of Manchester, Department
of Medicine, Manchester Royal Infirmary, Oxford Road, Manchester M13 9WL
E-mail: mike.mackness@manchester.ac.uk*

Abstract: The oxidation of low-density lipoprotein (LDL) in the artery wall is responsible for
the initiation and progression of atherosclerosis. High-density lipoprotein (HDL) on
the other hand is protective against atherosclerosis development and can attenuate
the oxidation of LDL. The HDL-associated enzyme paraoxonase-1 (PON-1) is largely
responsible for the anti-oxidative action of HDL. This chapter reviews the *in vitro*
and *in vivo* evidence for this anti-oxidative activity of PON1 and provides a historical
perspective of this subject, that has placed this multi-faceted enzyme at the centre of
atherosclerosis development

Keywords: Paraoxonase-1, HDL, oxidised-LDL atherosclerosis

1.   INTRODUCTION

Clinical epidemiological evidence shows an inverse relationship between the
concentration of plasma HDL and the incidence of coronary heart disease (CHD)
indicating that HDL is protective against the development of atherosclerosis
(Gordon et al., 1989; Barter and Rye, 2001). Studies in animal models of atherosclerosis have invariably shown that increasing HDL concentration by a variety of
methods e.g. transgenic over expression of human apolipoprotein A1 attenuated
atherosclerosis development (Benoit et al., 1999). Perhaps, more importantly, intervention studies using fibrates in man have shown that for every 1% increase in
HDL caused by the therapy there is a 3% reduction in CHD incidence (Rubins
et al., 1999).

The central role of HDL in the process of reverse-cholesterol transport (RCT),
the process whereby excess cholesterol is removed from the peripheral tissues and
returned to the liver for excretion plays an important part in the protective effect
of HDL against atherosclerosis development. This has been the subject of several

*B. Mackness et al. (eds.), The Paraoxonases: Their Role in Disease Development
and Xenobiotic Metabolism, 51–60.*
© 2008 *Springer.*

recent reviews and, therefore, will not be dealt with here (Oram and Vaughan, 2006; Rader, 2006). However, HDL has a range of other anti-atherosclerotic functions which contribute to its protective effect. These include, anti-oxidative, anti-inflammatory and anti-fibrinolytic actions, as well as more selective actions such as the inhibition of matrix matalloproteinases in the plaque maintaining normal endothelial function and enhancing progenitor and mediated endothelial repair (Kontush and Chapman, 2006). All of these functions of HDL contribute to its antiatherosclerotic effect.

2.   PON1 INHIBITS LDL-OXIDATION *IN VITRO*

In the 1980's and early 1990's studies indicated that HDL was able to inhibit LDL-oxidation *in vitro* and *in vivo* (Parthasarathy et al., 1990; Navab et al., 1991; Klimov et al., 1993). However, a study from our laboratory was the first to show that HDL acted at a specific point in the oxidation cascade i.e. attenuating lipid-peroxide production on LDL (Mackness et al., 1993a). We further went on to show that HDL-associated PON1 was primarily responsible for the anti-oxidant function of HDL (Mackness et al., 1991, 1993b), (Fig. 1). Results of which have subsequently been confirmed and extended by several laboratories (Ahmed et al., 2001; Aviram et al., 1998; Watson et al., 1995). Although several other HDL-associated proteins such as apo AI, lecithin:cholesterol acyltransferase (LCAT) and platelet-activating factor acetyltransferase (PAFAH) also have antioxidant properties, PON1 seems to be the predominant antioxidant enzyme. Any interaction between these proteins requires further investigation (Mackness and Durrington, 1995)

HDL isolated from the blood of PON1 knock-out mice or from avian species which naturally lack PON1, has at best, no effect on LDL-oxidation and at worst promotes LDL-oxidation (Shih et al., 1998; Mackness et al., 1998a), (Fig. 2).

Conversely, HDL isolated from mice overexpressing human PON1 completely abolishes LDL-oxidation (Tward et al., 2002).

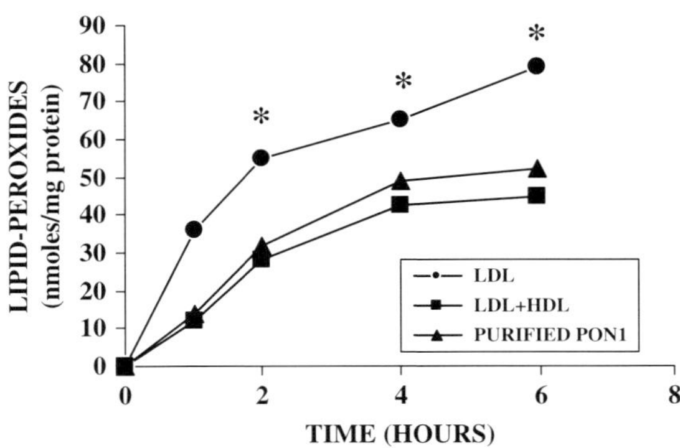

Figure 1. In vitro inhibition of LDL-oxidation by HDL and purified PON1
*Significantly different from ox-LDL+HDL or PON1 P < 0.01

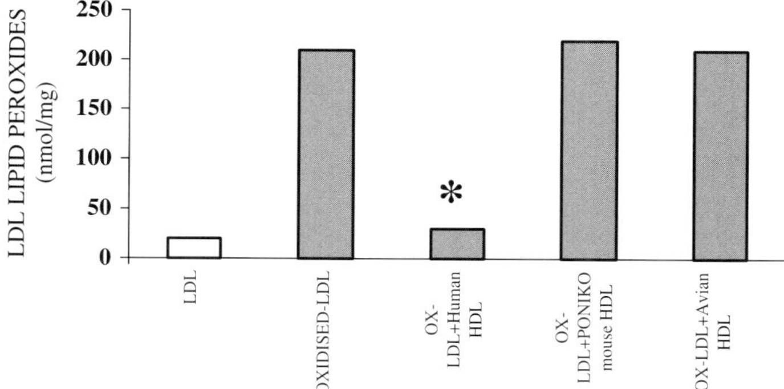

Figure 2. Effect of HDL lacking PON1 on LDL-oxidation (data derived from Shih et al., 1998; Mackness et al., 1998a)
*Significantly different from other treatments P < 0.001

The PON1 gene contains several functional polymorphisms in both the coding and promoter regions (Mackness and Mackness, 2004). In the coding region, the Q192R polymorphism determines isoforms of the enzyme which differ greatly in the rate of hydrolysis of certain substrates (Davies et al., 1996) e.g. paraoxon is hydrolysed at a far greater rate by the R192R isoform compared to the Q192Q isoenzyme whereas other substrates are hydrolysed oppositely. Whereas the L55M polymorphism, in conjunction with the promoter region C-108T polymorphism, effects PON1 levels (Leviev and James, 2000). We investigated the effect of the Q192R and L55M polymorphism on the ability of HDL to attenuate LDL-oxidation. HDL containing PON1 isoenzymes derived from the Q192Q/M55M genotypes were the most efficient at preventing LDL-oxidation and R192R/L55L homozygotes least efficient with heterozygotes having intermediate activity (Mackness et al., 1993b) (Fig. 3). These results were confirmed using purified PON1 isoenzymes by Aviram and colleagues (Aviram et al., 2000).

These findings have important implications for any potential relationship between PON1 and atherosclerosis development and has led to many genetic epidemiological studies on this relationship, which have proved inconclusive (Wheeler et al., 2004).

The problem with most of these studies was that the majority of them omitted any measure of PON1 quality i.e. activity and/or mass (Mackness and Mackness, 2004). As PON1 quality is determined by many factors other than genetics including environmental, nutritional, lifestyle, and hormonal factors as well as disease status (Mackness et al., 2004a), several laboratories have found that PON1 quality more accurately predicts atherosclerosis susceptibility than PON1 genotype (Jarvik et al., 2000, 2003; Mackness et al., 2001). The only prospective study to have reported so far found low PON1 activity to be an independent risk factor for future coronary events in men, and an even stronger risk factor in men who had a previous event (Mackness et al., 2003).

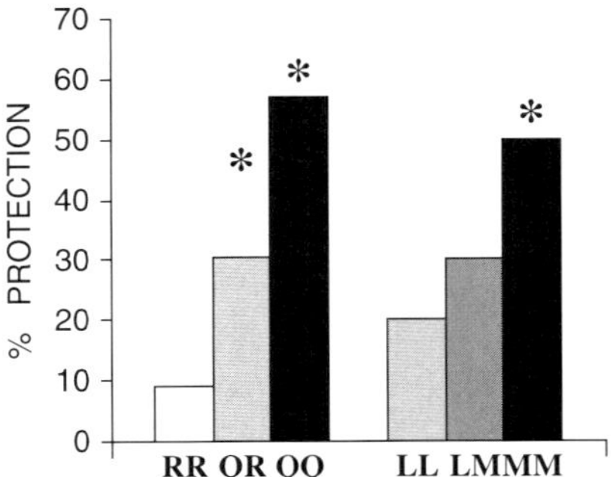

Figure 3. Effect of PON1-192 (left) and 55 (right) isoenzymes on the ability of HDL to inhibit LDL-oxidation
*Significantly different from other genotypes P < 0.01

Several diseases of an inflammatory nature and/or with an increased susceptibility to develop CHD have lower PON1 activity than healthy controls (Fig. 4).

In our laboratory we were particularly interested in the effect of low PON1 levels on the ability of HDL to retard LDL-oxidation. People with type 2 diabetes have a 3–4 fold increased risk of developing atherosclerosis compared to people without type 2 diabetes but with equivalent risk factors e.g. elevated total cholesterol, reduced HDL, smoking, etc. It is possible that HDL from people with type 2 diabetes is deficient in its antiatherosclerotic effect (Hedrick et al., 2000) and the possibility existed that HDL in type 2 diabetes would also be deficient in preventing lipid-peroxide formation.

In order to test this, we studied the ability of HDL from people with type 2 diabetes and no CHD, CHD without diabetes and control subjects to metabolise

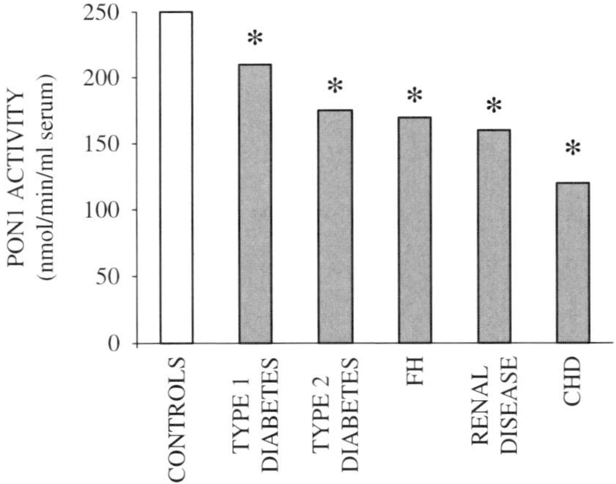

Figure 4. Effect of disease status on PON1 activity
*Significantly different from controls P < 0.01 minimum

oxidised palmitoyl, arachidonyl phosphatidylcholine (ox-PAPC), a common product of LDL-oxidation and a human PON1 substrate (Aviram et al., 1998). We found that HDL from people with type 2 diabetes and from those with CHD were equally defective in metabolising ox-PAPC compared to HDL from healthy controls leading to an increase in circulating oxidised LDL and the endothelial dysfunction typical of these diseases (Mastorikou et al., 2006).

In vitro, human PON1 attenuates oxidised-LDL induced monocyte/endothelial cell interactions in a co-culture model, therefore retarding one of the initiating steps in the inflammatory response leading to atherosclerosis (Mackness et al., 2004b). We showed that this was due to the suppression of oxidised-LDL stimulated monocyte-chemotatic protein 1 production by PON1. Ferretti et al., 2004 and Klimov et al., 2001 have also shown that PON1 can metabolise cell membrane associated phospholipid hydroperoxides.

3.   PON1 AND LDL-OXIDATION *IN VIVO*

In human arteries PON1 immunostaining increases with the progression of atherosclerosis (Fig. 5) probably as a protective response to increased oxidative stress (Mackness et al., 1997), although the source of the PON1, whether locally synthesised or imported from HDL is uncertain, there was a concomitant increase in apo AI and clusterin suggesting the latter.

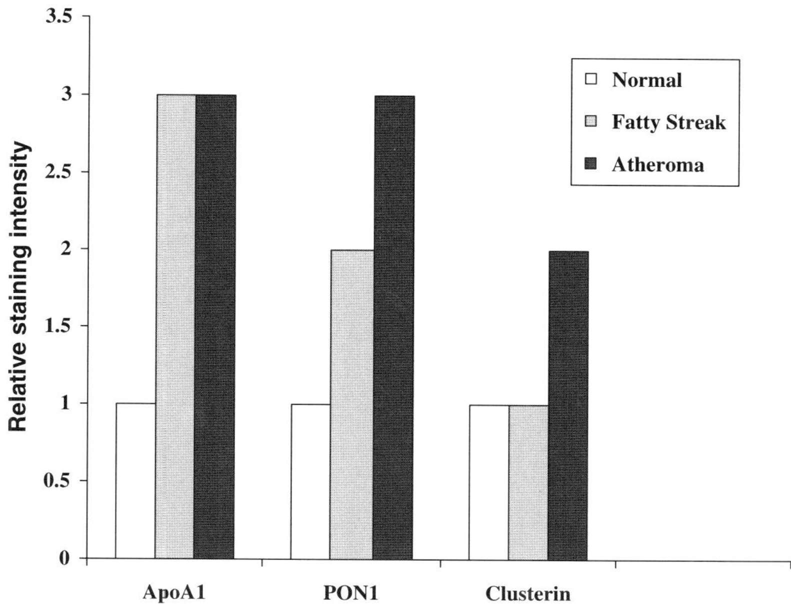

Figure 5. Immunostaining of human PON1 in human aorta with the progression of atherosclerosis

In mouse models of atherosclerosis such as apo-E$^-$/- and LDL-R$^-$/-, expression of human PON1, results in a decrease in ox-LDL but also a reduction in macrophage oxidative stress, reduced levels of peroxides, increased glutathione, reduced capacity to oxidise LDL and the promotion of RCT from the macrophages. These changes result in significantly decreased atherosclerosis compared to wild-type littermates (Rozenberg et al., 2005, Aviram and Rosenblat, 2004). Conversely, macrophages for PON1 knock-out mice show increased oxidative stress, increased lipid-peroxides, reduced glutathione and increased capacity to oxidise LDL resulting in macrophage foam cell formation and a marked increase in atherosclerosis development (Aviram and Rosenblat, 2004).

Several human studies have sown an inverse linear relationship between the concentration of oxidised-LDL in the circulation and PON1 activity, strongly implicating PON1 in the metabolism of oxidised-LDL *in vivo* (Sampson et al., 2005; Tsuzara et al., 2004).

We investigated the expression of human PON1 in a mouse model of the metabolic syndrome, a pre-diabetic condition (Mackness et al., 2006). LDL-R$^-$/-/leptin$^-$/- double knock-out mice were injected with either a human PON1 containing adenovirus (liver specific expression, adPON1) or a control adenovirus (AdRR5). PON1 activity was increased significantly ($P < 0.001$) by 4.4 fold in AdPON1 mice, whereas in AdRR5 mice activity did not change. Expressing PON1 decreased circulating oxidised-LDL levels progressively with time (Fig. 6).

Expressing human PON1 significantly reduced the total volume of plaque, the volume of macrophages in the plaques (i.e. macrophage size) and the volume of

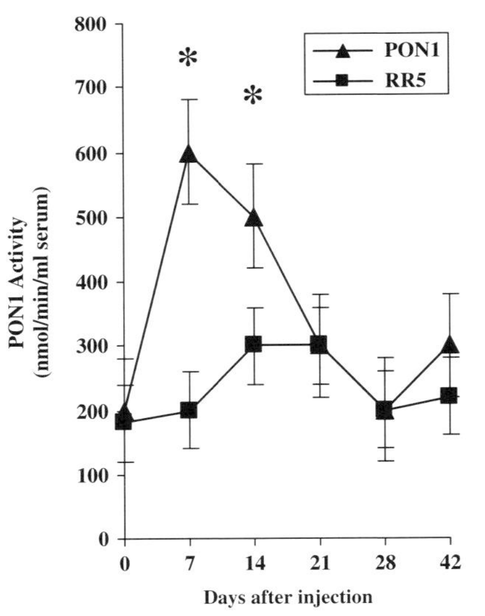

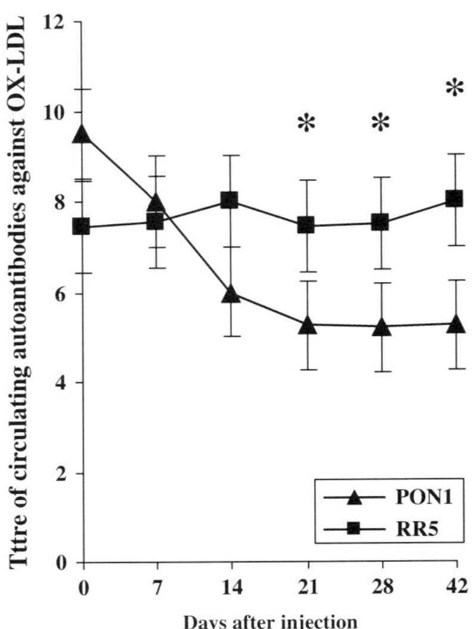

Figure 6. Effect of adenovirus mediated expression of human PON1 in a mouse model of metabolic syndrome on PON1 activity and circulating oxidised LDL
*Significantly different from RR5 injected mice, $P < 0.05$

plaque associated oxidised LDL. The percentage of smooth muscle cells in the plaques was increased. There was no effect of PON1 expression on plasma total cholesterol, triglyceride or HDL. Expressing human PON1 in this model, inhibited the development of atherosclerosis, probably by reducing the amount of oxidised LDL in plasma and in plaques, thereby preventing its proatherogenic effects and stabilising the plaque.

4. CONCLUSION

HDL-associated PON1 is responsible for attenuating the oxidation of LDL both *in vitro* and *in vivo*, mainly, by metabolising oxidised phospholipids therefore suppressing its proatherogenic effects. Low PON1 activity rather than genetic polymorphisms has been shown to be a strong independent risk factor of coronary events in the only prospective study to report so far. Further prospective studies in this area are required. PON1 is probably responsible for the majority of HDL's antioxidative/anti-inflammatory effects, although other HDL-associated enzymes such as LCAT and apo AI also contribute and require further investigation.

It has been suggested that PAFAH is responsible for the antioxidative properties of HDL and not PON1 (Marathe et al., 2003). However, several lines of evidence would suggest that this is not so, firstly the experimental conditions described in this publication (Marathe et al., 2003) would have resulted in a significant inhibition of PON1 as seen by the very low activity reported. Secondly, detailed studies carried out in our laboratory found no evidence of PAFAH on ultracentrifugally isolated HDL (Rodrigo et al., 2001). Thirdly, HDL from PON1 knock-out mice is pro-inflammatory despite the fact that PAFAH activity did not change. It is worth noting that the majority of studies have shown that LDL-associated PAFAH is pro-atherogenic and it is entirely possible that this is true of the HDL-associated enzyme (Packard et al., 2000).

Thus the experimental evidence continues to suggest that PON1 is a major player in the development of atherosclerosis and may result in novel strategies (both pharmacological and nutritional and possibly gene-therapy) to combat atherosclerosis.

ACKNOWLEDGEMENTS

The authors work is supported by the Medical Research Council (UK), British Heart Foundation, Diabetes UK and an International HDL Research Award (to BM). The authors thank Ms C. Price for expert typing of the manuscript.

REFERENCES

Ahmed, Z., Ravandi, A., Maguire, G.F., Emili, A., Draganov, D., La Du, B.N., Kuksis, A., Connelly, P.W. 2001. Apolipoprotein AI promotes the formation of phosphatidylcholine core aldehydes that are hydrolysed by paraoxonase (PON1) during high density lipoprotein oxidation with a peroxynitrite donor. *J. Biol. Chem.* 276: 24473–24481

Aviram, M., Billecke, S., Sorenson, R., Bisgaier, C., Newton, R., Rosenblat, M., Erogul, J., Hsu, C., Dunlop, C., La Du, B.N. 1998. Paraoxonase active site required for protection against LDL oxidation involves its free sulphydryl group and is different from that required for its arylesterase/paraoxonase activities: selective action of human paraoxonase alloenzymes Q and R. *Arterioscl. Thromb. Vasc. Biol.* 10: 1617–1624

Aviram, M., Hardak, E., Vaya, J., Mahmood, S., Milo, S., Hoffmann, A., Billicke, S., Draganov, D., Rosenblat, M. 2000. Human serum paraoxonase (PON1) Q and R selectively decrease lipid peroxides in human coronary and carotid atherosclerotic lesions: PON1 esterase and peroxidase-like activities. *Circulation* 101: 2510–2517

Aviram, M., Rosenblat, M. 2004. Paraoxonases 1, 2, and 3, oxidative stress, and macrophage foam cell formation during atherosclerosis development. *Free Radic Biol Med.* 37: 1304–16

Barter, P.J., Rye, K-A. 2001. Cholesteryl ester transfer protein, high density lipoprotein and arterial disease. *Curr. Opin. Lipidol* 12: 377–382

Benoit, P., Emmanuel, F., Caillund, J.M., Bassinet, L., Castro, G., Gallix, P, et al., 1999. Somatic gene transfer of human apo A1 inhibits atherosclerosis progression in mouse models. *Circulation* 99: 105–110

Davies, H.G., Richter, R.J., Keifer, M., Broomfield, C.A., Sowalla, J., Furlong, C.E. 1996. The effect of the human serum paraoxonase polymorphism is reversed with diazoxon, soman and sarin. *Nature Genetics* 14: 334–336

Ferretti, G., Bacchetti, T., Busni, D., Rabini, R.A., Curatola, G. 2004. Protective effect of paraoxonase activity in high-density lipoproteins against erythrocyte membranes peroxidation: a comparison between healthy subjects and type 1 diabetic patients. *J Clin Endocrinol Metab.* 89: 2957–2962.

Gordon, D.J., Probstfield, J.L., Garrison, R.J. 1989. High-density lipoprotein cholesterol and cardiovascular disease. Four prospective American studies. *Circulation* 79: 8–15

Hedrick, C.C., Thrope, S.R., Fu, M.X., Harper, C.M., Yoo, J., Kim, S.M., et al., 2000. Glycation impairs high-density lipoprotein function. *Diabetologia* 43: 312–320

Jarvik, G.P., Hatsukami, T.S., Carlson, C., Richter, R.J., Jampsa, R., Brophy, V.H. et al., 2003. Paraoxonase activity, but not haplotype utilizing the linkage disequilibrium structure, predicts vascular disease. *Artherioscler. Thromb. Vasc. Biol.* 231: 1465–1471

Jarvik, G.P., Rozek, L.S., Brophy, V.H., Hatsukami, T.S., Richter, R.J., Schellenberg, G.D. Furlong, C.E. 2000; Paraoxonase (PON1) phenotype is a better predictor of vascular disease than is PON1$_{192}$ or PON1$_{55}$ genotype. *Arterioscl. Thromb. Vasc. Biol.* 20: 2441–2447

Klimov, A.N., Gurevich, V.S., Nikiforova, A.A., Shatilina, L.V., Kuzmin, A.A., Plavinsky, S.L., Teryukova, N.P. 1993. Antioxidative activity of high-density lipoproteins in vivo. *Atherosclerosis* 100: 13–19

Klimov, A.N., Kozhevnikova, K.A., Kuzmin, A.A., Kuznetsov, A.S., Belova, E.V. 2001. On the ability of high density lipoproteins to remove phospholipid peroxidation products from erythrocyte membranes. *Biochemistry* (Moscow) 66: 300–304

Kontush, A., Chapman, M.J. 2006. Functionally defective high-density lipoprotein: a new therapeutic target at the crossroads of dyslipidemia, inflammation, and atherosclerosis. *Pharmacol Rev.* 58: 342–374

Leviev, I. and James, R.W. 2000. Promoter polymorphisms of human paraoxonase PON1 gene and serum paraoxonase activities and concentrations. *Arterioscler. Thromb. Vasc. Biol.* 20: 516–521

Mackness, B., Davies, G.K., Turkie, W., Lee, E., Roberts, D.H., Hill, E., Roberts, C., Durrington, P.N., Mackness, M.I. 2001. Paraoxonase status in coronary heart disease. Are activity and concentration more important than genotype? *Arterioscler. Thromb. Vasc. Biol.* 21: 1451–1457

Mackness, B., Durrington, P., McElduff, P., Yarnell, J., Azam, N., Watt, M., Mackness, M. 2003. Low paraoxonase activity predicts coronary events in the Caerphilly prospective Study. *Circulation* 107: 2775–2779

Mackness, B., Durrington, P.N., Mackness, M.I. 1998a. Lack of protection against oxidative modification of LDL by avian HDL. *Biochem. Biophys. Res. Comm.* 247: 443–446

Mackness, B., Hine, D., Liu, Y., Mastorikou, M., Mackness, M. 2004b. Paraoxonase 1 inhibits oxidised LDL-induced MCP-1 production by endothelial cells. *BBRC* 318: 680–683

Mackness, B., Hunt, R., Durrington, P.N., Mackness, M.I. 1997. Increased immunolocalisation of paraoxonase, clusterin and apolipoprotein AI in the human artery wall with progression of atherosclerosis. *Arterioscler. Thromb. Vasc. Biol.* 17: 1233–1238

Mackness, B., Mackness, M.I., Arrol, S., Turkie, W., Durrington, P.N. 1998b. Effect of the human serum paraoxonase 55 and 192 genetic polymorphisms on the protection by high density lipoprotein against low density lipoprotein oxidative modification. *FEBS Letts* 423: 57–60

Mackness, B., Quarck, R., Verreth, W., Mackness, M., Holvoet, P. 2006. Human paraoxonase-1 overexpression inhibits atherosclerosis in a mouse model of metabolic syndrome. *Arterioscler Thromb Vasc Biol.* 26:1545–1550

Mackness, M., Durrington, P., Mackness, B. 2004a. Role of Paraoxonase 1 activity in Cardiovascular disease: Potential for therapeutic intervention. *Am. J. Cardiovasc Drugs* 4: 211–217

Mackness, M., Mackness, B. 2004. Paraoxonase 1 and Atherosclerosis: Is the gene or the protein more important? *Free. Rad. Biol. Med.* 37: 1317–1323

Mackness, M.I., Abbott, C.A., Arrol, S., Durrington P.N. 1993a. The role of high density lipoprotein and lipid-soluble antioxidant vitamins in inhibiting low-density lipoprotein oxidation. *Biochem. J.* 294: 829–835

Mackness, M.I., Arrol, S., Abbott, C.A., Durrington, P.N. 1993b. Protection of low-density lipoprotein against oxidative modification by high-density lipoprotein associated paraoxonase. *Atherosclerosis* 104: 129–135

Mackness, M.I., Arrol, S., Durrington, P.N. 1991. Paraoxonase prevents accumulation of lipoperoxides in low-density lipoprotein. *FEBS Letts* 286: 152–154

Mackness, M.I., Durrington, P.N. 1995. High density lipoprotein, its enzymes and its potential to influence lipid peroxidation. *Atherosclerosis* 115:243–253

Marathe, G.K., Zimmerman, G.A., McIntyre, T.M. 2003. Platelet-activating factor acetylhydrolase, and not paraoxonase-1, is the oxidized phospholipid hydrolase of high density lipoprotein particles. *J. Biol. Chem.* 278: 3937–47

Mastorikou, M., Mackness, M., Mackness, B. 2006. Defective metabolism of oxidised-phospholipid by high-density lipoprotein from people with type 2 diabetes. *Diabetes* 55: 3099–3103

Navab, M., Imes, S.S., Hama, S.Y., Hough, G.P., Ross, L.A., Bork, R.W., Valente, A.J., Berliner, J.A., Drinkwater, D.C., Laks, H., Fogelman, A.M. 1991. Monocyte Transmigration induced by modification of low density lipoprotein in cocultures of human aortic wall cells is due to induction of monocyte chemotactic protein 1 synthesis and is abolished by high density lipoprotein. *J. Clin. Invest.* 88: 2039–2046

Oram, J.F., Vaughan, A.M. 2006. ATP-Binding cassette cholesterol transporters and cardiovascular disease. *Circ. Res.* 99: 1031–1043

Packard, C.J., O'Reilly, D.S.J., Caslake, M.J., McMahon, A.D., Ford, I., Cooney, J. et al., 2000. Lipoprotein-associated phospholipase A2 as an independent predictor of coronary heart disease. *N. Eng. J. Med.* 343: 1148–1155

Parthasarathy, S., Barnett, J., Fong, L.G. 1990. High-density lipoprotein inhibits the oxidative modification of low-density lipoprotein. *Biochim. Biophys Acta* 1044: 275–283

Rader, D.J. 2006. Molecular regulation of HDL metabolism and function: implications for novel therapies. *J. Clin. Invest.* 116: 3090–3100

Rozenberg, O., Shih, D.M., Aviram, M. 2005. Paraoxonase 1 (PON1) attenuates macrophage oxidative status: studies in PON1 transfected cells and in PON1 transgenic mice. *Atherosclerosis.* 181: 9–18.

Rubins, H.B., Robins, S.J., Collins, D., Fye, C.L., Anderson, J.W., Elam, M.B., Faas, F.H., Linares, E., Schaeffer, E.J., Schectman, G., Wilt, T.J., Wittes, J. 1999. for the Veterans Affairs High-Density Lipoprotein Cholesterol Intervention Trial Study Group. Gemfibrozil for the secondary prevention of coronary heart disease in men with low levels of high-density lipoprotein cholesterol. *N. Engl. J. Med.* 341: 410–418

Sampson, M.J., Braschi, S., Willis, G., Astley, S.B. 2005. Paraoxonase-1 (PON1) genotype and activity and in vivo oxidised, plasma low-density lipoprotein in type II diabetes. *Clin. Sci* 109: 189–197

Shih, D.M., Gu, L., Xia Y-R., Navab, M., Li, W-F., Hama, S., Castellani, L.W., Furlong, C.E., Costa, L.G., Fogelman, A.M., Lusis, A.J. 1998. Mice lacking serum paraoxonase are susceptible to organophosphate toxicity and atherosclerosis. *Nature*. 394: 284–287

Tsuzura, S., Ikeda, Y., Suehiro, T., Ota, K., Osaki, F., Arii, K., Kumon, Y., Hashimoto, K. 2004. Correlation of plasma oxidized low-density lipoprotein levels to vascular complications and human serum paraoxonase in patients with type 2 diabetes. *Metabolism* 53: 297–302

Tward, A., Xia, Y.R., Wang, X.P., Shi, Y.S., Park, C., Castellani, L.W., Lusis, A., Shih, D,H. 2002. Decreased atherosclerotic lesion formation in human serum paraoxonase transgenic mice. *Circulation* 106: 484–490

Watson, A.D., Berliner, J.A., Hama, S.Y., La Du, B.N., Fault, K.F., Fogelman, A.M., Navab, M. 1995. Protective effect of high density lipoprotein associated paraoxonase - Inhibition of the biological activity of minimally oxidised low-density lipoprotein. *J. Clin. Invest.* 96: 2882–2891

Wheeler, J.G., Keavney, B.D., Watkins, H., Collins, R., Danesh, J. 2004. Four paraoxonase gene polymorphisms in 11212 cases of coronary heart disease and 12786 controls: meta-analysis of 43 studies. *Lancet* 363: 689–95

CHAPTER 3

OXIDATIVE STRESS & ANTIOXIDANTS AND PON1 IN HEALTH AND DISEASE

ESER YILDIRIM SOZMEN[1], FERHAN GIRGIN SAGIN[1],
MERAL KAYIKCIOGLU[2] AND BULENT SOZMEN[3]

[1] *MD, PhD, Ege University Faculty of Medicine, Dept. of Biochemistry*
[2] *MD, specialist, Ege University Faculty of Medicine, Dept. of Cardiology*
[3] *MD, specialist, Atatürk Research and Trainig Hospital, Dept of Internal Medicine,*
Izmir/Republic of Turkiye

Abstract: Impairment in oxidative stress/antioxidant balance is an important trigger for a variety of diseases. As an antioxidant molecule on high-density lipoprotein (HDL), paraoxonase (PON1) contributes to the antioxidant mechanisms by removing oxidised lipids both on HDL and low-density lipoprotein (LDL). In this chapter, we will document and evaluate the results of our studies on healthy, atheroscleoric and diabetic cases which showed that (a) PON1, superoxide dismutase (SOD) and arylesterase probably work in a collaboration against oxidative stress, especially superoxide radical scavenging; (b) PON1 and SOD activities concomitantly decrease with the oxidative stress & severity of disease (higher HbA1c values in diabetics, more diseased vessels in atherosclerosis) while catalase (CAT) acts the opposite way; (c) Since PON1 activity and erythrocte thiobarbituric acid reactive substances (eTBARS) levels are affected by traditional risk factors (hypertension, aging and gender), determination of arylesterase activity might be a better indicator of antioxidant activity of PON1; (d) SOD activity has the greatest variability in regard to PON1 phenotype therefore it's important to define the PON1 polymorphism as well as PON1, arylesterase and other antioxidant enzyme activities

Keywords: paraoxonase, arylesterase, atherosclerosis, catalase, TBARS, superoxide dismutase, LDL oxidation

Enhancement of free radicals and impairment of antioxidant status are crucial processes underlying pathophysiologic mechanisms in a variety of diseases including atherosclerosis, diabetes mellitus and cancer (Mates et al., 1999; Parthasarathy et al., 1999; Aguirre et al., 1998). Enzymatic and nonenzymatic antioxidant systems (such as superoxide dismutase-SOD, catalase-CAT, glutathione peroxidase-GPx, paraoxonase-PON1 and vitamin E) are important in scavenging free radicals and their metabolic products as well as in maintaining normal cellular

61

B. Mackness et al. (eds.), The Paraoxonases: Their Role in Disease Development
and Xenobiotic Metabolism, 61–73.
© 2008 *Springer.*

physiology, promotion of immunity and prevention of various diseases (Mates et al., 1999). Experimental, clinical and epidemiological studies have shown the depletion of various antioxidants in a variety of diseases (Mates et al., 1999; Parthasarathy et al., 1999; Aguirre et al., 1998; Maxwell et al., 1997).

In this review, we focused on oxidative stress and antioxidant systems both in healthy humans and in patients with atherosclerosis and diabetes, emphasizing the changes in oxidant/antioxidant status in regard to PON1 phenotyping as well as PON1 genotyping in order to elucidate the antioxidant role of PON1.

1. RELATIONSHIP BETWEEN PON1 AND OTHER ANTIOXIDANT ENZYMES IN HEALTHY HUMANS AND IN DISEASES

According to the "oxidative modification hypothesis", atherogenesis is initiated by oxidation of the low-density lipoprotein (LDL) (Ross, 1999; Aviram, 1996; Steinberg, 1997; Chisolm and Steinberg, 2000) and increasing evidence suggests that this modification plays a central role in the further propagation of atherogenesis as well (Jialal and Devaraj, 1996; Chisolm and Steinberg, 2000; Kaplan and Aviram, 1999; Steinberg, 1997; Parthasarathy et al., 1999). The LDL oxidative state is elevated by increased ratio of poly/mono unsaturated fatty acids in LDL and it is reduced by enhanced LDL-associated antioxidant content such as vitamin E, beta-carotene, lycopene, polyphenolic flavonoids and other external antioxidants (Aviram and Rosenblat, 2005). Previously, it has been shown that PON1 prevents LDL from oxidation by removing oxidised phosholipids from LDL. This is supported by the finding in PON1-knock out mice in which PON1's preventive effect on LDL oxidation was not observed (Mackness et al., 1996; Laplaud et al., 1998; Durrington et al., 1999; Shih et al., 1996). Apart from PON1, other antioxidant systems are important in the prevention of various diseases by scavenging free radicals (Mates et al., 1999).

Previous research on PON1 and antioxidants (such as SOD, CAT, GPx, etc.) triggered our work on the role of antioxidant enzymes in the maintenance of PON1 activity during LDL oxidation in various groups namely, healthy, diabetic and atherosclerotic cases. Our first study indicated a negative correlation between PON1 activities and conjugated diene ($r = -0.297$, $p = 0.034$) & thiobarbituric acid reactive substance (TBARS) ($r = -0.265$, $p = 0.053$) levels of LDL at baseline (Sozmen et al., 2001b). Another important finding of our study was the positive correlation between SOD and PON1 activity in healthy cases ($n = 66$) (Fig. 1).

PON1 and SOD activities were decreased in type 2 DM ($n = 109$) patients while CAT activities were increased (Sozmen et al., 2001a). Another important finding was the positive correlation between CAT/SOD ratio & CAT/PON1 ratio and serum HbA1c. These data are in parallel to the previous findings: [a] It's known that enhanced oxidative stress such as in diabetes, and especially hydrogen peroxide induces CAT activity while reducing SOD activity (Freeman and Crapo, 1982). [b] Arai et al. (1987) indicated that glycosylation of SOD in poor glycemic control is another factor contributing to low SOD activity.

Our data also showed a significant increase in CAT/PON1 ratio in type 2 DM patients with complications compared to controls (Sozmen et al., 2001a). In the light of these results, we proposed that low PON1 activity together with enhanced oxidative stress may be a causative factor leading to complications in diabetes and CAT/SOD and CAT/PON1 ratio may be used as markers in management of glycaemic control.

Since the impairment in oxidative stress-antioxidant balance is regarded as the main factor in the pathophysiology of coronary heart disease, many clinical, epidemiological and experimental studies were conducted to investigate the oxidative stress and antioxidant enzymes in the atherosclerotic process (Jialal and Devaraj, 1996; Lankin et al., 1984; Steinberg, 1997; Parthasarathy et al., 1999; Azarsız et al., 2003). Under oxidative stress, proteins and lipids, especially LDL is prone to oxidation (Chisolm and Steinberg, 2000) and oxidatively modified LDL is recognized by the macrophage scavenger receptors (Henriksen et al., 1981). The propagation of LDL oxidation and thus the development of atherosclerotic processes is inhibited by some protective properties of high density lipoprotein (HDL). The proposed antioxidant effects of HDL are not only due to its reverse cholesterol transport activity but also may be due to the influence of several molecules on the lipoprotein such as apolipoprotein A-1 (apoA-1), platelet activating factor acetyl hydrolase and PON1 (Kaplan and Aviram, 1999; Mackness et al., 2002). However data on the relationship between PON1 and antioxidant enzymes are very limited. In order to evaluate the activities of PON1 and antioxidant enzymes through the stages of atherosclerotic process, we conducted a case-control study. Twenty-four healthy volunteers and 101 coronary artery disease (CAD) patients, among whom 68 had diagnosis confirmed by coronary angiography were included in the study.

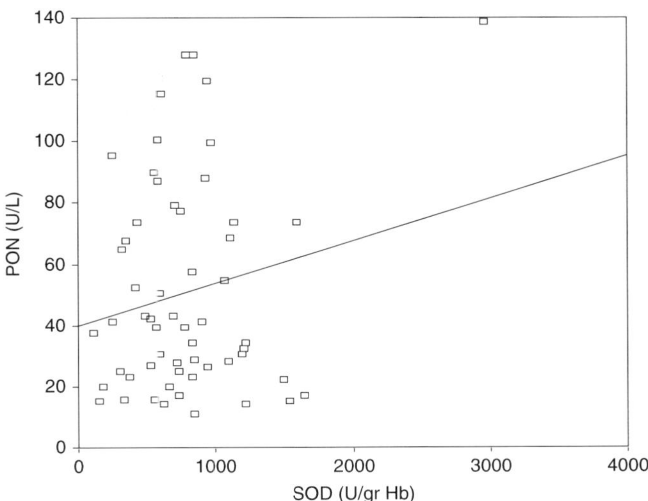

Figure 1. The correlation between SOD and PON1 activities in healthy subjects

Our data showed that PON1 activity in patients with $(41.6 \pm 26.8\,\text{U/L})$ and without $(50.1 \pm 37.2\,\text{U/L})$ angiographically assessed CAD was lower than the controls $(52.3 \pm 30.5\,\text{U/L})$. This finding indicated that a depletion of PON1 activity takes place during the propagation phase of the atherosclerotic process. PON1 activities of older CAD+ patients $(56.9 \pm 36.4\,\text{U/L}$ vs $38.8 \pm 24.7\,\text{U/L}$, $p = 0.087)$ were lower than controls and CAD-patients, however this difference was not statistically significant (Azarsız et al., 2003).

Although PON1 activities were observed to be closely related to HDL levels in all groups, it's known that changes in PON1 activity may occur independently of changes in HDL-cholesterol and apoA-1 (Mackness et al., 1996; Mackness et al., 1998). In accordance with this, Aviram et al. (1999) previously reported that protection against LDL oxidation is accompanied by PON1 inactivation and this finding was attributed to the interaction between PON's free sulfhydryl group and specific oxidized lipids in Ox-LDL. Based on these researches, we proposed that the depletion in PON1 activity in CAD+ patients in our work results from the increased production of free radicals (which we have assessed by the increase in TBARS levels) during the atherosclerotic process (Azarsız et al., 2003).

Since Watson et al. (1995) previously showed that treatment of mildly modified LDL with PON1 and HDL-associated esterase inhibited its ability to induce monocyte-endothelial interactions, it may be proposed that PON1 does not only have a role in inactivation of LDL-oxidation during the initiation phase of atherosclerosis, but also in the prevention of monocyte-endothelial interaction during the propagation phase of atherosclerosis. In accordance with our data, a number of reports later demonstrated a depletion in PON1 activity in CAD patients (Mackness et al., 2003; Graner et al., 2006; Jaouad et al., 2003) speculating on different mechanisms (such as attack by hydroxyl radicals, direct oxidation by peroxides and negatively charged lysophospholipids, alkylation by α, β-unsaturated lipid aldehydes, etc) to explain this depletion in PON1 activity. It is likely that hydroxyl radicals may be the active species primarily responsible for the oxidative inactivation of PON1 of *in vivo* systems (Nguyen and Sok, 2003). Nguyen and Sok (2003) suggested that ROS (hydrogen peroxide and superoxide anions) generated in the presence of copper or iron at submicromolar- or micromolar concentrations could cause the oxidative inactivation of HDL-PON1, which would result in the reduction of antioxidative function of HDL of in vivo systems. On the other hand, van Lenten et al. showed that oxidised phosholipids in ox-LDL decrease the expression of PON1 in liver and increase that of apolipoprotein J (van Lenten et al., 2001). It seems that further research may still unravel some other mechanisms in the depletion of PON1 in the atherosclerotic process besides the aforementioned ones.

Another finding of our work was the lower SOD activities in patients who had more severe disease and this is in line with other reports. We have previously observed a decrease in SOD activity in the collared arteries that might have resulted from the oxidative stress (Sozmen et al., 2000). Parallel to this work, it has been shown that antioxidant enzyme activities were greatly reduced in intima and media of the human aorta in different types of atherosclerotic lesions (Lankin et al., 1984).

To investigate the relationship between the oxidative stress markers and the severity of coronary disease, we categorized CAD+ patients based on the number of diseased coronaries: those with > 50% obstruction in one vessel (n = 22), two vessels (n = 26) and three vessels (n = 20). Basal and stimulated LDL-diene levels were higher in patients who had more diseased vessels than those who had less. Basal LDL-TBARS levels were higher in all patient groups compared to controls and stimulation of oxidation by copper led to a greater increase in LDL-TBARS and LDL-diene levels of patients compared to controls (Azarsız et al., 2003). Besides the oxidative stress parameters, we also investigated the antioxidant enzymes and PON1 activities in patient groups. There was a significant reduction in PON1 and arylesterase activities in regard to the severity of CAD, especially in patients who had 3 diseased vessels. While SOD activities were reduced as the involved vessel number increased, CAT activities showed an opposite trend (Fig. 2).

We determined a total stenosis degree in CAD patients calculated by adding the narrowing degrees of each vessel. There was a negative correlation between the total degree and arylesterase & SOD activities (Fig. 3).

There was a positive correlation between SOD and arylesterase activities in healthy subjects (Fig. 4) while PON1, arylesterase and SOD activities decreased in accordance with the progress in the atherosclerotic process in CAD patients. Therefore we suggested that these enzymes work in a collaboration against oxidative stress, especially superoxide radical scavenging. Our data supports the notion that the main substrates for PON1 are lipid hydroperoxides and PON1 activity is reduced through oxidative inactivation during the detoxification of lipid hydroperoxides by its esterolytic activity (Karabina et al., 2005). It has been suggested that other antioxidant enzymes might prevent this inhibition of PON1 activity. Our results support the view that antioxidant enzymes all have a concomitant role in all stages of atherosclerosis and elevation in oxidative stress might inhibit these enzymes.

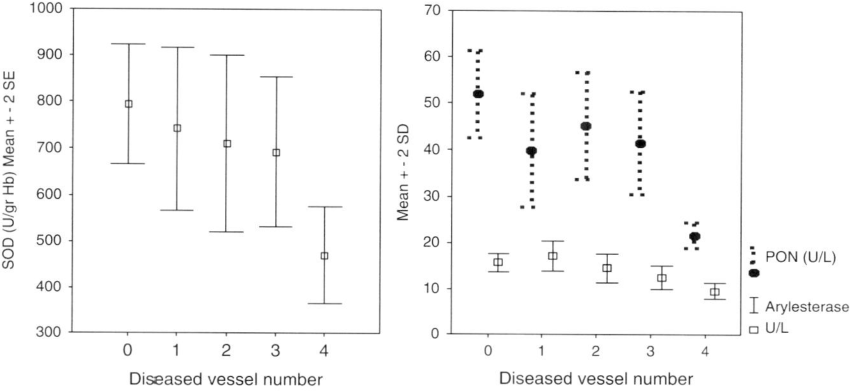

Figure 2. The activities of SOD, arylesterase and PON1 in regard to diseased vessel number

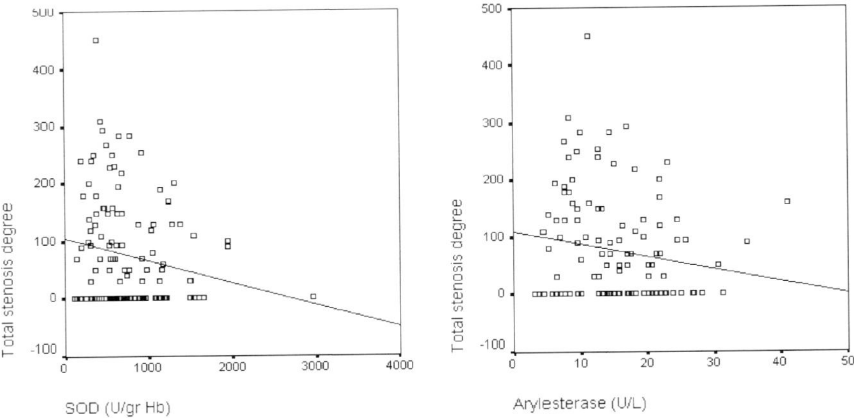

Figure 3. Correlation between the total stenosis degree and arylesterase & SOD activities in patients with CAD

Rozenberg et al. (2003) demonstrated that PON1 deficiency results in an increment in serum and in macrophage oxidative stress (increase in cellular super-oxide anion release) in PON1 knock-out mice and the addition of PON1 to macrophages reduces their oxidative stress. It has been suggested that PON1 is located in the external membrane of cells, thus it can hydrolyze lipid peroxides in macrophages resulting in a decrease in cellular oxidative stress.

PON1 activity and eTBARS levels were affected by risk factors (hypertension, aging and gender) but not arylesterase activity (Azarsız et al., 2003; Ferre et al., 2003) therefore the lactonase/arylesterase activities of PON1 are more important than the PON1 activity in the enzyme's physiological role (in oxidation protection and cholesterol efflux) (Rosenblat et al., 2006).

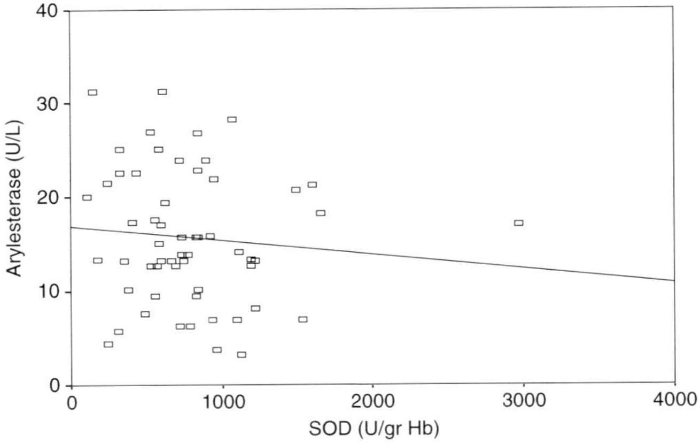

Figure 4. Correlation between the arylesterase and SOD activities in healthy human

2. OXIDANT/ANTIOXIDANT STATUS IN REGARD TO PON1 PHENOTYPING/GENOTYPING

Since PON1 activity has 40-fold interindividual variation due to environmental and nutritional factors (Mackness et al., 2002), genotyping/phenotyping studies attracted great interest for determination of the PON1 status.

The molecular basis of the PON1 activity lies on the polymorphism which has been shown to be an aminoacid substitution at position 192 (glutamine- Q→ arginine-R) and 55 (leucine-L → methionine-M) (Mackness et al., 1998). The R allele has several fold higher activity toward paraoxon hydrolysis than the Q allele (Adkins et al., 1993). PON1 exists in two genetically determined allozymic forms; A and B which possess both paraoxonase and arylesterase activities. The B-type esterase has relatively higher paraoxonase activity and is stimulated to a greater degree by 1 M NaCl than the A allozyme (LaDu et al., 1993). This characteristic of PON1 is used to determine the PON1 phenotyping. Adkins et al. (1993) also proposed to use the antimode of histogram of the ratio of PON1 stimulation by salt and the ratio of arylesterase to salt-stimulated PON1 activity to determine PON1 phenotyping.

Previous studies investigated the role of PON1 genotype in the susceptibility of LDL to oxidation and conflicting results have been reported (Sanghera et al., 1997; Serrato and Marian, 1995; Aviram, 1999; Aviram et al., 1999; Ng et al., 2005). A recent meta analysis of 43 studies involving 11 212 cases and 12 786 controls showed a weak association between PON1 R192 and CHD (Wheeler et al., 2004).

Recently, it has been proposed that PON1 phenotyping is a more predictive factor than PON1 genotyping for PON1 activity and CAD (Mackness et al., 2001; Jarvik et al., 2000). Therefore, we investigated PON1 and other antioxidant enzymes activities in regard to PON1 phenotypes in healthy human cases and in patients with CAD.

LDL samples obtained from subjects with AA allele were shown to be more prone to oxidation as observed by their higher stimulated conjugated diene (p = 0.041) and TBARS (p = 0.042) levels compared to samples from AB or BB alleles (Table 1).

Interestingly, the baseline TBARS levels were normal in these cases while there was a higher susceptibility of LDL to in vitro oxidation. The higher susceptibility of these samples to in vitro oxidation was shown by higher stimulated TBARS (p = 0.042) levels compared to samples from AB or BB alleles (Sozmen et al., 2001b). This striking finding may be explained by high cholesterol levels in LDL of these subjects.

In our study, subjects with BB allele had higher paraoxonase activities towards paraoxon hydrolysis as would be expected. The increased SOD (p = 0.021) and CAT (insignificant increase) activities determined in these subjects may be due to a compensatory induction of these antioxidants to increased oxidative stress (Sozmen et al., 2001b) (Fig. 5). In accordance with our data Agachan et al. (2005) showed elevated levels of glutathione in healthy subjects with BB allele. This co-activity of antioxidant enzymes with PON1 was previously demonstrated by Aviram et al. (1998) who showed that HDL-associated PON1 and purified PON1 is also able to hydrolyse hydrogen peroxide, the main substrate of CAT in vivo which is also a

Table 1. Serum and LDL parameters in regard to PON1 phenotypes

	AA Phenotype n = 40	AB Phenotype n = 15	BB Phenotype n = 11
Serum Trygliceride mg/dL	111 ± 45	109 ± 42	124 ± 40
Serum T.cholesterol mg/dL	190 ± 34	197 ± 50	166 ± 52
Serum HDL-cholesterol mg/dL	54 ± 7,8	56 ± 7,0	51 ± 8,2
Serum LDL-cholesterol mg/dL	114 ± 34	115 ± 44	102 ± 30
LDL-diene (basal) μmol/mg pr	207 ± 102*	179 ± 69	130 ± 13
LDL-dien (stimulated) μmol/mg pr	286 ± 93*	248 ± 71	229 ± 62
LDL-MDA (basal) nmol/mg pr	0.43 ± 0.33	0.34 ± 0.19	0.38 ± 0.28
LDL-MDA (stimulated) nmol/mg pr	13.3 ± 6.8*	12.4 ± 5.2	10.3 ± 5.9
LDL-phospholipid mmol/mg pr	0.52 ± 0.18	0.51 ± 0.11	0.59 ± 0.21
LDL-cholesterol μmol/mg pr	0.84 ± 0.5	0.87 ± 0.51	0.48 ± 0.43
LDL- vitamine E nmol/mg pr	14.4 ± 12.4	19.3 ± 19.8	10.4 ± 4.5

** $p < 0.05$ comparisons were made versus to BB phenotype (student's t test)*

potent inactivator of PON1. The induction in antioxidant enzymes found in our study might explain the decrease in conjugated diene levels as well as the preservation in PON1 activity in BB allele subjects.

The data obtained from patients with atherosclerotic disease were in agreement with the results from healthy human, that is the SOD activities were higher in patients with BB allele (Azarsız et al., 2003) (Fig. 6). Although there was an increase in oxidative stress and LDL oxidation in CAD patients, the ratio of

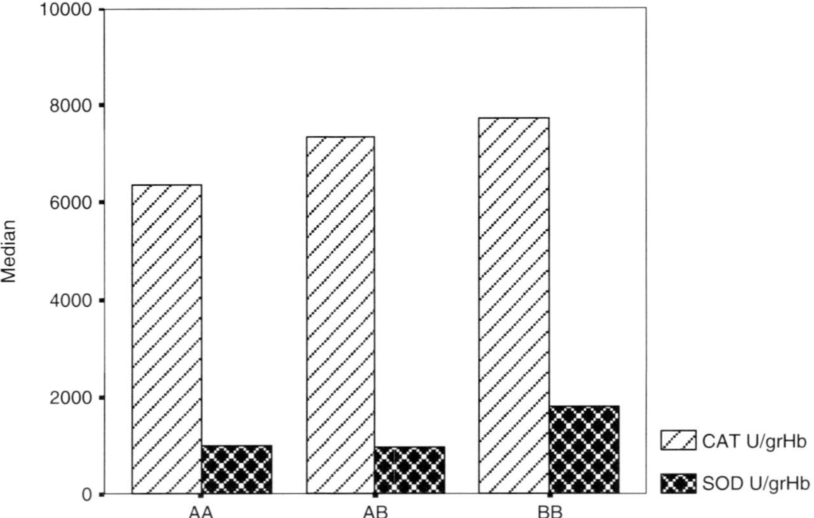

Figure 5. CAT and SOD activities of healthy subjects in regard to PON1 phenotypes. SOD activities showed significant differences between groups (p = 0.0032, one way ANOVA test)

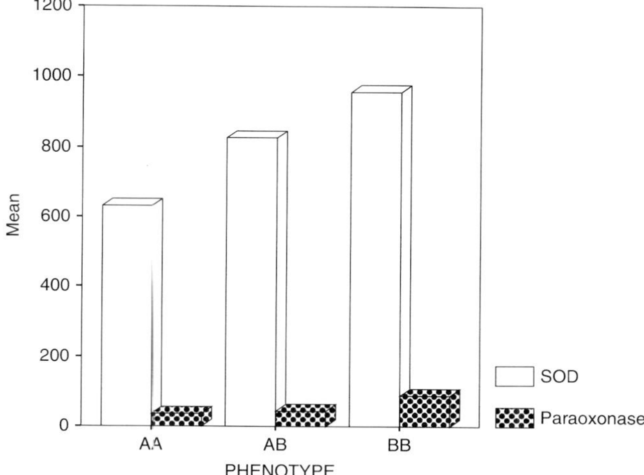

Figure 6. SOD and PON1 activities in regard to PON1 phenotypes in patients with CAD

antioxidant enzymes preserved, this data support the hypothesis that antioxidant enzymes (especially SOD and PON1) might affect the same substrate and the changes in their activities are closely related.

3. ANTIOXIDANT ROLE OF PARAOXONASE DURING LDL OXIDATION

As it has been summarised in Fig. 7, PON1 acts as antioxidant molecule under physiologic conditions and at several steps of LDL oxidation.

During LDL oxidation in the presence of PON1, the enzyme is partially inactivated. This effect can be possibly related to displacement of calcium ions by copper ions. Removal of calcium ion from PON1 abolishes its arylesterase/paraoxonase activities, but not its ability to protect LDL from oxidation. Oxidised LDL phospholipids and cholesteryl esters are physiological substrates for serum PON1 and PON1's reaction with peroxides results in PON1 inactivation (Aviram, 1999).

PON1 conformational changes when present in lipoprotein-deficient serum (LPDS) versus HDL result in the loss of its arylesterase and lactonase activities, but stimulate its paraoxonase activity. It might also be that the lactonase/arylesterase activities of PON1 are more important than the paraoxonase activity in its physiological roles (in oxidation protection and cholesterol efflux), which are probably related to PON1-association with HDL (Rosenblat et al., 2006). PON1 prevents the production of reactive aldehydes resulting from lipid peroxidation through hydrolyzing oxidised lipids, thus it prevents the interaction between the reactive aldehydes and oxidised LDL receptors (Sangvanich et al., 2003; Mackness et al., 2003).

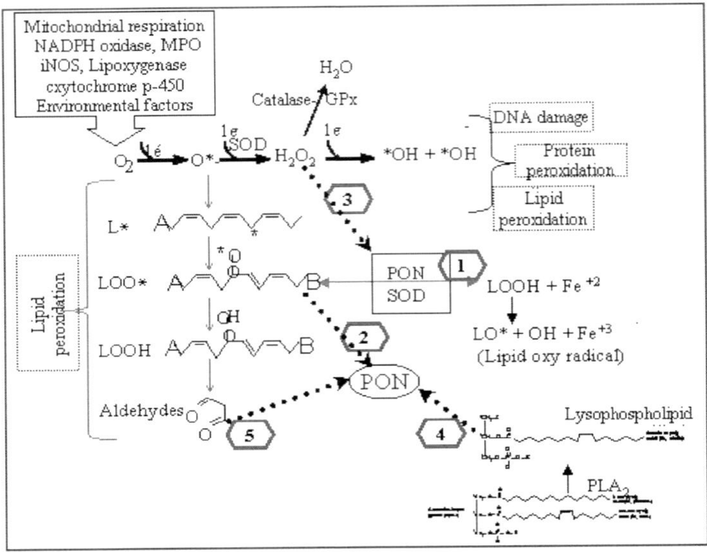

Figure 7. During lipid peroxidation, PON1 shows oxidative-peroxidative activity together with other antioxidant enzymes (Mackness et al., 2002; Aviram et al., 1998, 1999; Watson et al., 1995; Nguyen and Sok, 2003, 2006). PON1 and SOD affect lipid peroxy radicals while lipid hydroxides result from lipid peroxy radicals, activities of PON1 and SOD decrease stoichiometrically during this reaction (1). Lipid peroxy radicals may directly inhibit PON1 activity (2). An oxidative molecule, H_2O_2 which is scavenged by CAT, also directly inhibits PON1 (3). Lysophospholipids, which are separated from phospholipids through phospholipase A2 action, may inhibit PON1 activity (4). Aldehydes (malonyldialdehyde, 4-hydroxy nonenal), also contribute to the inactivation of PON1 (5)

PON1 induces the LPC (lysophosphatidylcholine) formation which might act in both direction as prooxidant and antioxidant. While the oxidant effect of LPC is based on the upregulated effects on adhesive molecules, it shows an antioxidant effect by increasing the expression of extracellular SOD in monocyte-macrophages (Rosenblat et al., 2006). This finding provides another evidence for the relation between PON1 and SOD activities.

CONCLUSION

In the light of the above data, we could suggest that; SOD, PON1, arylesterase and CAT all show an antioxidant co-activity in healthy human. PON1, arylesterase and SOD activities decrease while atherosclerosis progresses, therefore these enzymes might work in a collaboration against oxidative stress, especially superoxide radical scavenging. A possible explanation to this inactivation is the inhibition through superoxide radicals. The role of CAT and hydrogen peroxide during this process and the relation between CAT and PON1 are not clear except from the data that hydrogen peroxide inhibits PON1 directly. PON1 activity and eTBARS levels are affected by traditional risk factors (hypertension, aging and gender) but not arylesterase activity. Determination of arylesterase activity might be a better indicator of antioxidant activity of PON1. Since SOD activity has the greatest variability in regard to PON1 phenotype, it's important to define the PON1 polymorphism as well as PON1, arylesterase and other antioxidant enzyme activities.

REFERENCES

Adkins S., Gan K.N., Mody M., LaDu B.N., 1993, Molecular basis for the polymorphic forms of human serum paraoxonase/arylesterase; glutamine or arginine at position 191, for respective A or B allozymes. *Am J Hum Genet* 52: 598–608.

Agachan B., Yılmaz H., Ergen H.A., Karaali Z.E., Isbir T., 2005, Paraoxonase (PON1) 55 and 192 polymorphism and its effects to oxidant-antioxidant system in Turkish patients with type 2 diabetes mellitus. *Physiol Res* 54: 287–93.

Aguirre F., Martin I., Grinpson D., Ruiz M., Hager A., De Paoli T., Ihlo J., Farach H.A., Poole C.P., 1998, Oxidative damage, plasma antioxidant capacity and glycemic control in elderly NIDDM patients. *Free Rad Biol Med* 24: 580–5.

Arai K., Iizuka S., Tada Y., Oikawa K., Taniguchi N., 1987, Increase in the glycosylated form of erythrocyte Cu-Zn superoxide dismutase in diabetes and close associaton of the nonenzymatic glucosylation with the enzyme activity. *Biochim Biophys Acta* 924: 292–6.

Aviram M., 1996, Interaction of oxidized low density lipoprotein with macrophages in atherosclerosis and the atherogenicity of antioxidants. *Eur J Clin Biochem* 34: 599–608.

Aviram M., 1999, Does paraoxonase play a role in susceptibility to cardiovascular disease. *Mol Med Today* 5: 381–6.

Aviram M., Rosenblat M., 2005, Paraoxonases and cardiovascular diseases: pharmacological and nutritional influences. *Curr Opin Lipidol* 16: 393–9.

Aviram M., Rosenblat M., Bisgaier C.L., Newton R.S., Primo-Parma S.L., LaDu B.N., 1998, Paraoxonase inhibits high-density lipoprotein oxidation and preserves its functions. A possible peroxidative role for paraoxonase. *J Clin Invest* 101(8): 1581–90.

Aviram M., Rosenblat M., Scott B., Drogul J., Sorenson R., Bisgaier C.L., Newton R.S., La Du B., 1999, Human serum paraoxonase (PON1) is inactivated by oxidised low density lipoprotein and preserved by antioxidants. *Free Rad Biol Med* 26(7/8): 892–904.

Azarsiz E., Kayikcioglu M., Payzin S., Sozmen E.Y., 2003, PON1 Activities and Oxidative Markers of LDL in Patients With Angiographically Proven Coronary Artery Disease. *Int J Cardiol* 91: 43–51.

Chisolm G.M., Steinberg D., 2000, The oxidative modification hypothesis of atherogenesis: an overview. *Free Rad Biol Med* 28(12): 1815–26.

Durrington P.N., Mackness B., Mackness M.I., 1999, Role of HDL in preventing atherogenic modification of LDL. *Atherosclerosis* 146 (suppl): 813.

Ferre N., Camps J., Fernandez-Ballart J., Arija V., Murphy M.M., Ceruleo S., Biarnes E., Vilella E., Tous M., Joven J., 2003, Regulation of serum paraoxonase activity by genetic, nutritional and lifestyle factors in the general population. *Clin Chem* 49(9): 1491–7.

Freeman B.A., Crapo J.D., 1982, Biology of disease: free radicals and tissue injury. *Lab Invest* 47(5): 412–25.

Graner M., James R.W., Kahri J., Nieminen M.S., Syvanne M., Taskinen M.R., 2006, Association of paraoxonase-1 activity and concentration with angiographic severity and extent of coronary artery disease. *J Am Coll Cardiol* 47(2): 2429–35.

Henriksen T., Mahoney E.M., Steinberg D., 1981, Enhanced macrophage degradation of low density lipoprotein previously incubated with cultured endothelial cells; recognition by receptors for acetylated low density lipoproteins. *Proc Natl Acad Sci USA* 78: 6499–503.

Jaouad L., Milochevitch C., Khalil A., 2003, PON1 activity is reducued during HDL oxidation and is an indicator of HDL antioxidant capacity. *Free Rad Res* 37(1): 77–83.

Jarvik G.P., Rozek L.S., Brophy V.H., Hatsukami T.S., Richter R.J., Schellenberg G.D., Furlong C.E., 2000, Paraoxonase phenotype is a beter predictor of vascular disease than is PON192 or PON 55 genotype. *Arterioscler Thromb Vasc Biol* 20: 2441–7.

Jialal I., Devaraj S., 1996, Low density lipoprotein oxidation, antioxidants and atherosclerosis. A clinical biochemistry perspective. *Clin Chem* 42(4): 498–506.

Kaplan M., Aviram M., 1999, Oxidized low density lipoprotein: atherogenic and proinflamatory characteristics during macrophage foam cell formation. An inhibitory role for nutritional antioxidants and serum paraoxonase. *Clin Chem Lab Med* 37(8): 777–87.

Karabina S.A.P., Lehner A.N., Frank E., Parthasarathy S., Santanam N., 2005, Oxidative inactivation of paraoxonase- implications in diabetes mellitus and atherosclerosis. *Biochim Biophys Acta* 1725: 213–21.

La Du B.N., Adkins S., Kuo C.L., Lipsig D., 1993, Studies on human serum paraoxonase/arylesterase. *Chem Biol Interact* 87: 25–34.

Lankin V.Z., Vikhert A.M., Kosykh V.A., Tikhaze A.K., Galakhov L.E., Orekhov A.N., 1984, Enzymatic detoxication of superoxide anion radicals and lipoperoxides in intima and media of atherosclerotic aorta. *Biomed Biochim Acta* 43: 797–802.

Laplaud R.M., Dantoine T., Chapman M.J., 1998, Paraoxonase as a risk marker for cardiovascular disease: facts and hypotheses. *Clin Chem Lab Med* 36(7): 431–41.

Mackness B., Davies G.K., Turkie W., Lee E., Roberts D.H., Roberts C., Durrington P.N., Mackness M.I., 2001, Paraoxonase status in coronary heart disease: are activity and concentration more important than genotype. *Arterioscler Thromb Vasc Biol* 21: 1451–7.

Mackness B., Durrington P., McElduff P., Yarnell J., Azam N., Watt M., Mackness M., 2003, Low paraoxonase activity predicts coronary events in the Caerphilly prospective study. *Circulation* 107: 2775–9.

Mackness B., Durrington P.N., Mackness M.I., 1998, Human serum paraoxonase. *Gen Pharm* 31(3): 329–36.

Mackness M.I., Mackness B., Durrington P.N., Connelly P.W., Hegele R.A., 1996, Paraoxonase: biochemistry, genetics and relationship to plasma lipoproteins. *Curr Opin Lipidol* 7: 69–76.

Mackness M.I., Mackness B., Durrington P.N., 2002, Paraoxonase and coronary heart disease. *Atherosclerosis* suppl 3: 49–55.

Mates J.M., Perez-Gomez C., De Castro I.N., 1999, Antioxidant enzymes and human diseases. *Clin Biochem* 32 (8): 595–603.

Maxwell S.R.J., Thomason H., Sandler D., Leguen C., Baxter M.A., Thorpe G.H.G., Jones A.F., Barnett A.H., 1997, Poor glycaemic control is associated with reduced serum free radical scavenging (antioxidant) activity in non-insulin-dependent diabetes mellitus. *Ann Clin Biochem* 34: 638–44.

Ng C.J., Shih D.M., Hama S.Y., Villa N., Navab M., Reddy S.T., 2005, The paraoxonase gene family and atherosclerosis. *Free Rad Biol Med* 38: 153–63.

Nguyen S.D., Sok D.E., 2003, Oxidative inactivation of paraoxonase-1, an antioxidant protein and its effect on antioxidant action. *Free Rad Res* 37(12): 1319–30.

Nguyen S.D., Sok D.E., 2006, Preferable stimulation of PON1 arylesterase activity by phosphatidylcholines with unsaturated acyl chains or oxidized acyl chains at sn-2 position. *Biochim Biophys Acta* 1758: 499–508.

Parthasarathy S., Santanam N., Ramachandran S., Meilhac O., 1999, Oxidants and antioxidants in atherogenesis; an appraisal. *J Lipid Res* 40: 2143–57.

Rosenblat M., Oren R., Aviram M., 2006, Lysophosphatidylcholine (LPC) attenuates macrophage mediated oxidation of LDL. *Biochem Biophys Res Comm* 344: 1271–7.

Ross R., 1999, Atherosclerosis: an inflammatory disease. *NEJM* 340(2): 115–26.

Rozenberg O., Rosenblat M., Coleman R., Shih D.M., Aviram M., 2003, Paraoxonase deficiency is associated with increased macrophage oxidative stress: studies in PON1-knockout mice. *Free Rad Biol Med* 34(6): 774–84.

Sanghera D.K., Saha N., Aston C.E., Kamboh M.I., 1997, Genetic polymorphism of paraoxonase and the risk of coronary heart disease. *Arterioscler Thromb Vasc Biol* 17: 1067–73.

Sangvanich P., Mackness B., Gaskill S., Durrington P.N., Mackness M.I., 2003, The effect of HDL on the formation of lipid/protein conjugates during in vitro oxidation of LDL. *Biochem Biophys Res Comm* 300: 501–6.

Serrato M., Marian A.J., 1995, A variant of human paraoxonase/arylesterase (HUMPONA) gene is a risk factor for coronary artery disease. *J Clin Invest* 96: 3005–8.

Shih D.M., Gu L., Hama S., Xia Y., Navab M., Fogelman A.M., Lusis A.J., 1996, Genetic-dietary regulation of serum paraoxonase expression and its role in atherogenesis in a mouse model. *J Clin Invest* 97(7): 1630–9.

Sozmen E.Y., Kerry Z., Uysal F., Yetik G., Yasa M., Ustunes L., Onat T., 2000, Antioxidant enzyme activities and total nitrite/nitrate levels in the collar model: effect of nicardipine. *Clin Chem Lab Med* 38(1): 21–5.

Sozmen E.Y., Sozmen B., Delen Y., Onat T., 2001a, Catalase/superoxide dismutase and catalase/paraoxonase ratios may implicate poor glycemic control. *Arch Med Res* 32: 283–7.

Sozmen E.Y., Sozmen B., Girgin F.K., Delen Y., Azarsiz E., Erdener D., Ersoz B., 2001b, Antioxidant enzymes and paraoxonase show co-activity in preserving LDL from oxidation. *Clin Exp Med* 1: 195–9.

Steinberg D., 1997, Oxidative modification of LDL and atherogenesis. *Circulation* 95: 1062–71.

Van Lenten B.J., Wagner A.C., Navab M., Fogelman A.M., 2001, Oxidized phospholipids induce changes in hepatic paraoxonase and ApoJ but not monocyte chemoattractant protein-1 via interleukin-6. *J Biol Chem* 19; 276 (3): 1923–9.

Watson A.D., Berliner J.A., Hama S.Y., La Du B.N., Fauli F.K., Fogelman A.M., Navab M., 1995, Protective effect of high density lipoprotein associated paraoxonase-inhibition of the biological activity of minimally oxidized low density lipoprotein. *J Clin Invest* 96: 2882–91.

Wheeler J.G., Keavney B.D., Watkins H., Collins R., Danesh J., 2004, Four paraoxonase gene polymorphisms in 11212 cases of coronary heart disease and 12786 controls: meta analysis of 43 studies. *Lancet* 363: 689–95.

CHAPTER 4

D-4F INCREASES PARAOXONASE 1 ACTIVITY IN HDL

GREG HOUGH

From the David Geffen School of Medicine at UCLA Los Angeles, California

Abstract: HDL, apoA-I, and apoA-I mimetic peptides have been shown to prevent LDL oxidation in cell-free systems and in the artery wall coculture studies. Moreover, HDL, apoA-I, and apoA-I mimetics have been shown to decrease lesions and improve vascular reactivity in animal models of atherosclerosis and in humans. The primary mechanism by which HDL and apoA-I and apoA-I mimetic peptides exert their beneficial effect has been presumed to be the enhancement of reverse cholesterol transport. However, apoA-I has also been shown to be capable of removing "seeding molecules" from LDL, thus preventing the oxidation of LDL-derived phospholipids to those that are thought to be responsible for the inflammatory response characteristic of atherosclerosis. D-4F is an apoA-I mimetic peptide that demonstrates many of the properties of apoA-I itself. Oral D-4F increases paraoxonase activity in monkeys and causes the formation of pre-ß HDL. When added in nanomolar amounts to normal human plasma, D-4F reduces lipoprotein lipid hydroperoxide content, increases paraoxonase activity, and converts proinflammatory HDL to anti-inflammatory. Adding the combination of D-4F and pravastatin to mouse chow synergized to result in a significant increase in apoA-I levels, HDL-cholesterol, and paraoxonase activity in old apoE-null mice. We have shown that incubation of HDL from human patients or apoE deficient mouse with D-4F results in reduction of lipid hydroperoxides and an increase in paraoxonase activity in HDL

Keywords: HDL, apoA-I, peptide mimetics, D-4F, lipid hydroperoxides, paraoxonase

1. INTRODUCTION

One of the first indications that HDL might be a marker for inflammation came from our early studies (Van Lenten et al., 1995). We reported that the acute phase response in humans converted HDL from anti-inflammatory to proinflammatory.

Correspondence to: Greg Hough, Room BH-307 CHS, Division of Cardiology, Department of Medicine, David Geffen School of Medicine at UCLA, 10833 Le Conte, Avenue, Los Angeles, CA 90095-1679, USA. E-mail: ghough@mednet.ucla.edu

B. Mackness et al. (eds.), The Paraoxonases: Their Role in Disease Development and Xenobiotic Metabolism, 75–86.
© 2008 *Springer*.

These studies compared HDL taken from humans before and after elective surgery. Before surgery, HDL was anti-inflammatory in a human artery wall cell coculture model (i.e., HDL inhibited LDL oxidation and LDL-induced monocyte chemotactic activity). However, at the peak of the acute phase response, 3 days after surgery, HDL from the same patient was proinflammatory (i.e., it promoted LDL oxidation and monocyte chemotactic activity in the human artery wall coculture). One week after surgery, the HDL returned to an anti-inflammatory state. These changes in HDL are consistent with a classic acute phase response. Gabay and Kushner emphasized that the acute phase response can become chronic. We (Navab et al., 1997) demonstrated that such a chronic acute phase response was present in apoE null mice on a chow diet and in LDL receptor null mice on a high-fat diet and was accompanied by proinflammatory HDL.

Sevanian and colleagues noted that a subpopulation of freshly isolated LDL contains lipid hydroperoxides (Sevanian et al., 1997). We found that freshly isolated LDL from normal individuals always contained small amounts of lipoxygenase pathway products (e.g., HPODE and HPETE) (Navab et al., 2000). These were present even when blood was collected into tubes containing potent antioxidants. The levels of HPODE and HPETE did not increase during in vitro incubations in the presence of these antioxidants. This indicated that these seeding molecules were present in LDL in vivo. However, when the freshly isolated LDL was incubated with apoA-I in the presence of antioxidants and the LDL and the apoA-I were then rapidly separated, the LDL treated with apoA-I contained only approximately 30% to 50% as much HPODE and HPETE as was initially present. Before the incubation, the apoA-I contained no detectable HPODE or HPETE, but after the incubation with LDL, one-half to two-thirds of the HPODE and HPETE that had been present in the LDL was transferred to the apoA-I, along with some cholesterol and phospholipid. The LDL treated with apoA-I was unable to generate lipid hydroperoxides, nor was it able to induce monocyte adherence or monocyte chemotactic activity when added to human artery wall cocultures. If the apoA-I that was incubated with the LDL was subjected to lipid extraction and the extracted lipids were added back to the LDL that had been treated with apoA-I, the reconstituted LDL was able to induce lipid hydroperoxide formation and induce monocyte adherence and monocyte chemotactic activity. Consistent with these properties of apoA-I, others have reported that HDL is a major carrier of lipid hydroperoxides in humans (Bowry et al., 1992). HDL appears to be the major carrier of lipid hydroperoxides in mice, and the concentration of lipid hydroperoxides in HDL taken from the atherosclerosis-susceptible C57BL/6J mice either on a low-fat chow diet or on an atherogenic diet was significantly greater than the lipid hydroperoxide levels found in the HDL of the atherosclerosis-resistant C3H/HeJ mice.

In addition to apolipoproteins such as apoA-I, HDL also contains enzymes that can prevent the formation of or destroy the oxidized phospholipids that mediate the inflammatory response induced by MM-LDL. These enzymes include paraoxonase, platelet-activating factor acetylhydrolase, lecithin:cholesterol acyltransferase

(LCAT), and possibly glutathione peroxidase (Navab et al., 2004). *In vivo*, the absence of the HDL-associated enzyme paraoxonase was demonstrated to result in increased LDL oxidation and increased atherosclerosis (Watson et al., 1995). We have demonstrated that the acute phase response in rabbits and humans resulted in decreased activities of HDL-associated paraoxonase and platelet-activating factor acetylhydrolase.

We (Navab et al., 2001) have reported that the HDL inflammatory index measures the net action of a large number of factors in HDL. These factors include oxidized phospholipids, lipid hydroperoxides, paraoxonase activity, platelet-activating factor acetylhydrolase activity, lecithin:cholesterol acyl transferase activity, possibly GSH peroxidase activity, apoA-I, apoJ, serum amyloid A, ceruloplasmin, antioxidant vitamins, and probably products such as nitrotyrosine, which can be generated by myeloperoxidase. Thus, the HDL inflammatory index likely represents the net effect of all of these factors. The lipid hydroperoxide content and the other factors in HDL are likely interdependent.

The correlation of lipid hydroperoxide content with HDL that is less able to prevent the formation of the LDL-derived proinflammatory oxidized phospholipids suggests that oxidation impairs HDL function.

2. APOA-I MIMETIC PEPTIDES

D-4F is an apoA-I mimetic peptide that demonstrates many of the properties of apoA-I itself. Despite identical amino acid composition, differences in class A amphipathic helical peptides caused by differences in the order of amino acids on the hydrophobic face (Anantharamaiah et al., 1985) results in substantial differences in antiinflammatory properties (Navab et al., 2003). One of these peptides the apolipoprotein A-I (apoA-I) mimetic, D-4F, when given orally to mice and monkeys, caused the formation of pre-ß HDL, improved HDL-mediated cholesterol efflux, reduced lipoprotein lipid hydroperoxides, increased paraoxonase activity, and converted HDL from pro-inflammatory to antiinflammatory. In apolipoprotein E (apoE)-null mice, D-4F increased reverse cholesterol transport from macrophages. Oral D-4F reduced atherosclerosis in apoE-null and LDL receptor-null mice. *In vitro,* when added to human plasma at nanomolar concentrations, D-4F caused the formation of pre-ß HDL, reduced lipoprotein lipid hydroperoxides, increased paraoxonase activity, and converted HDL from pro-inflammatory to antiinflammatory. Physical-chemical properties and the ability of various class A amphipathic helical peptides to activate LCAT *in vitro* did not predict biologic activity *in vivo*. In contrast, the use of cultured human artery wall cells in evaluating these peptides was more predictive of their efficacy *in vivo*. Therefore the antiinflammatory properties of different class A amphipathic helical peptides depends on subtle differences in the configuration of the hydrophobic face of the peptides, which determines the ability of the peptides to sequester inflammatory lipids. These differences appear to be too subtle to predict efficacy based on physical-chemical properties alone. However, understanding these physical-chemical properties provides an explanation for the mechanism of action of the active peptides (Navab et al., 2001).

The data reviewed here suggests that the mechanism of action of apoA-I mimetic peptide D-4F involves the formation of small cholesterol-containing particles with pre-ß mobility, that are enriched in apoA-I and paraoxonase activity, removes the oxidized lipids in dysfunctional HDL, that is, oxidized lipids inhibiting PON activity, and HDL paraoxonase regains activity, resulting in the conversion of HDL from proinflammatory to antiinflammatory.

3.   RESULTS

We have demonstrated that addition of D-4F to apoE-null mouse plasma in vitro rapidly caused the movement of apoA-I from α-migrating HDL to ß-migrating HDL HDL (Navab et al., 2004). Twenty minutes after 500 µg of D-4F was given orally as a bolus by stomach tube to apoE-null mice, the plasma contained 138 to 322 ng of D-4F/mL and 85% was associated with. Twenty minutes after 500 µg of D-4F was given orally as a bolus by stomach tube to apoE-null mice, small cholesterol-containing particles 7 to 8 nm in size with pre-ß mobility and enriched in apoA-I and paraoxonase activity were found in plasma. Before the administration of D-4F, mature HDL and fast protein liquid chromatography fractions containing the cholesterol-containing particles were pro-inflammatory when assayed in the human artery wall coculture system for monocyte chemotactic activity. Twenty minutes after oral D-4F, both HDL and the cholesterol-containing particles became antiinflammatory and HDL-mediated cholesterol efflux from macrophages in vitro increased by 2.2-fold ($p < 0.05$). Oral D-4F also promoted reverse cholesterol transport by 1.3-fold ($p < 0.05$) from intraperitoneally injected cholesterol-loaded macrophages in vivo. After oral D-4F, lipoprotein lipid hydroperoxides decreased 67% ($p < 0.05$) in very-low-density lipoprotein/intermediate density lipoprotein, decreased 73% ($p < 0.05$) in LDL, decreased in mature HDL, but increased by 1.9-fold ($p < 0.05$) in pre-ß HDL. Both lipid hydroperoxide content and paraoxonase activity increased in pre-ß HDL after oral D-4F. Before D-4F, the pre-ß HDL fractions from apoE-null mice were very pro-inflammatory, however, after oral D-4F, despite the increase in lipid hydroperoxide content, the pre-ß HDL fractions were antiinflammatory when assayed in the human artery wall coculture system for monocyte chemotactic activity. Thus, whereas oral D-4F caused the movement of lipid hydroperoxides to pre-ß HDL, the increase in antioxidant enzyme activities such as paraoxonase must have more than compensated to render the pre-ß HDL antiinflammatory.

Oral D-4F also caused the formation of pre-ß HDL in wild-type C57BL/6J mice on a chow diet and decreased lipid hydroperoxides in HDL while increasing the content of lipid hydroperoxides in pre-ß HDL, indicating that the absence of apoE was not required for these actions of D-4F.

4. ORAL D-4F INCREASES PARAOXONASE ACTIVITY IN MONKEYS

Oral administration of D-4F resulted in an increase in paraoxonase activity in monkeys (Fig. 1A) and also caused the formation of pre-ß HDL (Fig. 1B). As previously reported, oral D-4F reduced lipid hydroperoxide levels, converted HDL from pro-inflammatory to antiinflammatory, and enhanced HDL-mediated cholesterol efflux in monkeys (Navab et al., 2005). Thus, oral D-4F appears to act similarly in mice and monkeys.

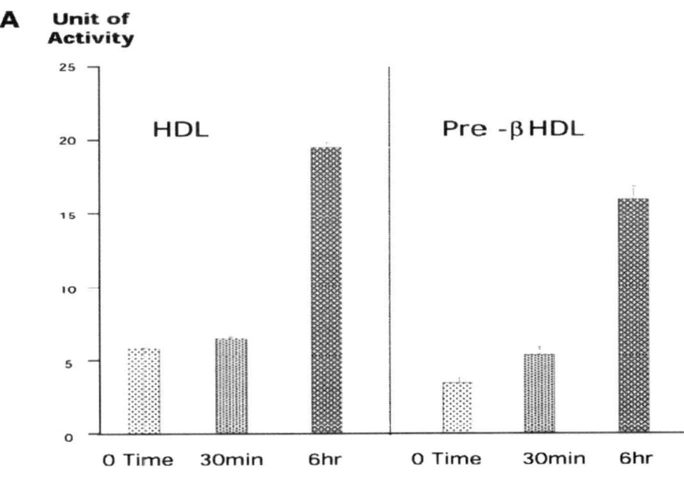

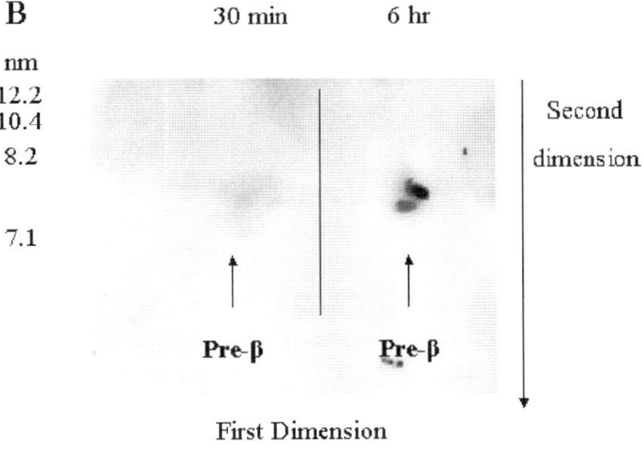

Figure 1. (a) Oral D-4F increases paraoxonase activity in monkeys. After approval by the UCLA Animal Research Committee blood was obtained from 2 male ($\approx$ 4 kg) and 2 female ($\approx$ 4 kg) Cynomolgus monkeys with conscious sedation (ketamine 10 mg/kg, intramuscularly) before (time zero) and after the administration of a banana shake (by gavage) containing 40 mg of D-4F. (b) FPLC fractions immediately after the main peak of HDL analyzed by 2-D gels and probed for apoA-I by Western blot

5. ADDING NANOMOLAR AMOUNTS OF D-4F TO NORMAL HUMAN PLASMA REDUCES LIPOPROTEIN LIPID HYDROPEROXIDE CONTENT, INCREASES PARAOXONASE ACTIVITY, AND CONVERTS PROINFLAMMATORY HDL TO ANTIINFLAMMATORY

The plasma concentration of D-4F after oral administration of a single dose of 500 µg to apoE-null mice was in the 100 to 300 ng/mL range (Navab et al., 2004). Adding 250 ng/mL of D-4F to normal human plasma caused the movement of apoA-I to smaller particles, which had pre-ß mobility. Adding 250 ng/mL of D-4F to human plasma reduced HDL lipid hydroperoxide content and increased paraoxonase activity (Fig. 2). Adding 250 ng/mL of D-4F converted pro-inflammatory HDL to antiinflammatory. Thus, adding D-4F to human plasma in vitro produced results similar to that seen in mice and monkeys in vivo. These results raise 2 questions. First, if apoA-I concentrations in normal human plasma are on the order of 1 mg/mL (1, 000 µg/mL or 1,000,000 ng/mL), how could adding 25 µg/mL or 2.5 µg/mL or 250 ng/mL of D-4F cause the movement of apoA-I from larger to smaller particles? The answer probably relates to the ability of D-4F to interact with lipids. As noted, D-4F, 3F-1, and 3F-2 have the ability to separate cholesterol from phospholipids in membranes and to penetrate into those membranes (Datta et al., 2004; Epand et al., 2004). Mature α-migrating HDL particles are constantly changing because some portion of apoA-I is always leaving the large HDL particle, generating smaller lipid-poor apoA-I particles. The data reviewed here suggest that apoA-I mimetic peptides such as D-4F can dramatically accelerate this process, probably as a result of their ability to bind to HDL and separate cholesterol from phospholipids, which may facilitate the movement of apoA-I to smaller particles (Fig. 3).

Second, how could D-4F decrease HDL lipid hydroperoxide content and increase paraoxonase activity *in vitro* (Navab et al., 2004)? Again the physical-chemical properties of D-4F, 3F-1, and 3F-2 may provide the explanation. Probably because of the structural characteristics of these peptides when they form lipid-peptide complexes, they allow some water to penetrate the complexes, which may facilitate the ability of the complexes to effectively sequester lipid hydroperoxides (Datta et al., 2004; Epand et al., 2004). Forte and colleagues have shown that a number of enzymes including paraoxonase are reversibly inhibited by lipid hydroperoxides. The ability of peptides such as D-4F to effectively sequester lipid hydroperoxides may lead to the activation of enzymes such as paraoxonase. We have shown that activated paraoxonase can destroy such lipid hydroperoxides (Watson et al., 1995). Thus, the effective sequestration of a very small quantity of lipid hydroperoxides by peptides such as D-4F may activate enzymes such as paraoxonase, leading to further lipid hydroperoxide destruction and providing a positive feedback loop. The physical-chemical characteristics of the peptides that determine their structure after binding lipids, and hence their interaction with the lipid acyl chains of membranes, could lead to a series of events that would appear to be catalytic, whereas in fact the peptides themselves are not catalysts by the usual definition.

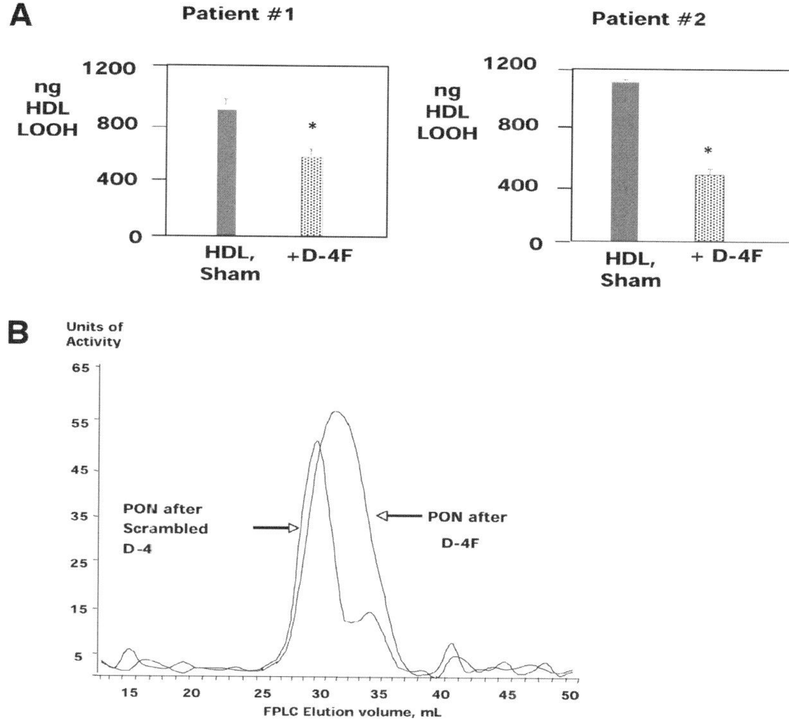

Figure 2. D-4F decreases HDL lipid hydroperoxides and increases paraoxonase activity. A, Plasma from 2 patients from the study described were incubated with 250 ng of D-4F/mL (+D-4F) or the same volume of vehicle in which the D-4F was added (HDL Sham) at 37 °C for 30 minutes in a Millipore Amicon Ultra Centrifugal Device with a 100-kDa molecular weight cutoff filter (catalog number UFC 810096), followed by centrifugation at 2100 g. After centrifugation, the supernatant was separated by FPLC and analyzed for HDL lipid hydroperoxides (ng HDL LOOH). Patient 1's HDL became antiinflammatory after treatment with simvastatin.[22] Patient 2's HDL remained pro-inflammatory despite simvastatin treatment. *P < 0.003. The experiment shown is representative of 3 of 3 experiments. B, D-4F or scrambled D-4F (250 ng/mL) were added to normal human plasma and incubated at 37 °C for 30 minutes as described in (A). After centrifugation, the supernatant was fractionated by FPLC and paraoxonase activity was determined in the fractions. The area under the curve after D-4F treatment was 190% of the area under the curve after treatment with scrambled D-4F, which was not different from plasma treated without additions (data not shown). The experiment shown is representative of 6 of 6 experiments

In additional studies, we observed that when D-4F was given orally at a dose of 50 µg D-4F per mouse per day in combination with pravastatin, doses of pravastatin and D-4F which, by themselves were ineffective, there was a remarkable synergy resulting in increased intestinal synthesis of apoA-I with increased plasma levels of apoA-I including apoA-I particles with pre-ß mobility. The combination treatment also increased HDL–cholesterol levels, increased PON activity, rendered HDL antiinflammatory, lesion formation was prevented in young mice, and there was

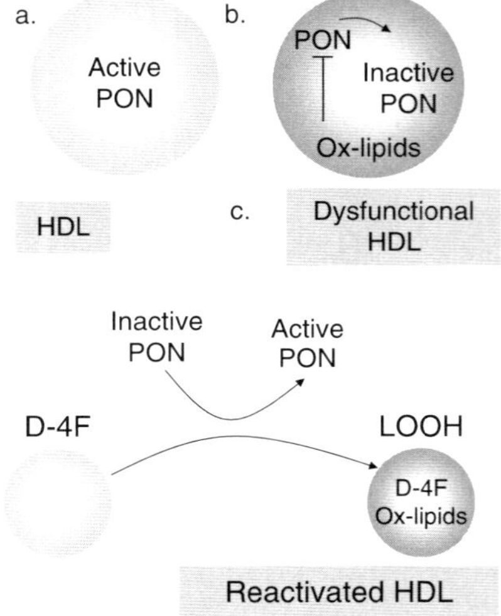

Figure 3. a- Normal HDL with paraoxonase activity, b- Dysfunctional HDL with oxidized lipids inhibiting PON activity c, D-4F removes the oxidized lipids in HDL, c. HDL paracxonase regains in activity

significant regression of established lesions in old apoE null mice suggesting that the combination of a statin and an HDL-based therapy may be a particularly potent treatment strategy.

As shown in Table 1 adding the combination of pravastatin and D-4F to mouse chow resulted in a significant increase in HDL–cholesterol (22%), paraoxonase activity (33%) and plasma apoA-I levels (19%). Administration of D-4F or pravastatin alone at these doses did not increase HDL–cholesterol, paraoxonase activity, or plasma apoA-I levels (Table 1).

Table 1. Plasma Lipids, Paraoxonase Activity, and ApoA-I There was no significant difference in the consumption of water or chow, body weight, liver weight, or heart weight between groups

	Total Cholesterol, mg/dL (Mean $\pm$ SD)	HDL Cholesterol, mg/dL (Mean $\pm$ SD)	Triglycerides, mg/dL (Mean $\pm$ SD)	Paraoxonase Activity, U/mL	Plasma ApoA-I, mg/dL (Mean $\pm$ SD)
Chow	508 ± 47	18 ± 1.7	111 ± 8	24 ± 4.2	99 ± 13
D-4F	476 ± 32	19 ± 1.3	113 ± 11	23 ± 2.1	102 ± 11
Pravastatin	469 ± 27	17 ± 1.9	123 ± 8	26 ± 4.7	104 ± 8
D-4F together with Pravastatin	457 ± 21	$22 \pm 1.8^*$	120 ± 15	32 ± 3.4	$118 \pm 8^*$

* $P < 0.03$ compared to Chow
$P < 0.01$ compared to Chow

Table 2. Synergy between pravastatin and D-4F increasing Apo A-I, HDL and Paraoxonase activity in old apoE deficient mouse

	Total Cholesterol, mg/dL (Mean ± SD)	HDL Cholesterol, mg/dL (Mean ± SD)	Triglycerides, mg/dL (Mean ± SD)	Paraoxonase Activity, U/mL	Plasma ApoA-I, mg/dL (Mean ± SD)
Time Zero	453 ± 21	20 ± 2.2	123 ± 11	26 ± 3.8	105 ± 5
Plus 6 months Chow	521 ± 30	20 ± 3.7	119 ± 10	27 ± 4.2	104 ± 4
Plus 6 months D-4F	493 ± 22	21 ± 2.3	111 ± 11	29 ± 2.5	106 ± 4
Plus 6 months Pravastatin	481 ± 25	22 ± 2.3	129 ± 7	26 ± 4.7	107 ± 2
Plus 6 months Pravastatin together with D-4F	461 ± 22*	26 ± 1.9*	129 ± 13	34 ± 3.8*	119 ± 8

* $P < 0.02$ vs + 6 months Chow
$P < 0.01$ vs + 6 months Chow

6. WHEN ADDED TO MOUSE CHOW, D-4F AND PRAVASTATIN SYNERGIZE TO CAUSE REGRESSION OF EXISTING LESIONS IN OLD APOE NULL MICE

Regression of existing lesions has not been previously demonstrated with D-4F. Although the combination of 50 μg pravastatin per mouse per day together with 12.5 μg D-4F per mouse per day prevented lesions, we reasoned that a higher dose would likely be required to induce regression of existing lesions.

The addition of D-4F and pravastatin to mouse chow at doses of 50 μg D-4F per mouse per day and 50 μg pravastatin per mouse per day resulted in synergy in old apoE null mice. The combination treatment rendered apoE null HDL antiinflammatory (Table 2).

7. OTHER STUDIES

Rozenberg and Aviram (2006) have reported that PON1 activity is regulated by S-glutathionylation. Incubation of PON1 or HDL with GSSG resulted in a dose-dependent inactivation of PON1 activities, including its physiological activity to increase HDL-mediated macrophage cholesterol efflux. This PON1 inactivation was associated with the formation of a mixed disulfide bond between GSSG and PON1's cysteine residue(s), as detected by immunoblotting with anti-glutathione IgG. PON1 activity was recovered following the addition of a reducing agent, DL-Dithiothreitol (DTT), to the PON1-SSG complex. They have concluded that HDL-associated serum PON1 can undergo S-glutathionylation under oxidative stress with a consequent reversible inactivation. They propose that S-Glutathionylation regulates HDL-associated PON1 activity. It would be interesting to see if treatment with D-4F has any effect on the glutathione system and thus reactivation of PON1.

Kassai and colleagues (2007) showed that in thirty-three patients with types II.a and II.b primary hyperlipoproteinemia, when treated with atorvastatin, significant increases in serum PON-specific activity and PON/HDL ratio were observed. It would be interesting to determine if there are similarities between the mechanism involved here and that related to increase in PON activity by D-4F. Ekmekci and colleagues (2006) have reported that PON 1 activity is low in asthmatic patients. Condouri and colleagues have reported that D-4F reduces the symptoms in a mouse model of asthma. It would be interesting to determine if the PON activity is normalized after D-4F administration and if that has a role in improvement observed.

8.   CONCLUSION

There is continuing evidence of a role for lipid oxidation in atherogenesis. The oxidation hypothesis of atherogenesis has evolved to focus on specific proinflammatory oxidized phospholipids that result from the oxidation of LDL phospholipids containing arachidonic acid and that are recognized by the innate immune system in animals and humans (Navab et al., 2004; Navab et al., 2005). These oxidized phospholipids are generated via the lipoxygenase and myeloperoxidase pathways. Among many possible reasons, the failure of antioxidant vitamins to influence clinical outcomes may result from the failure of vitamin E to prevent the formation of these proinflammatory, oxidized phospholipids. Preliminary data suggest that the oxidation hypothesis of atherogenesis and the reverse cholesterol transport hypothesis of atherogenesis may be linked. Measurements of the HDL inflammatory index, of the levels of oxidized lipids in lipoproteins, paraoxonase activity, and products of the myeloperoxidase pathway may predict susceptibility to atherogenesis (Ansell et al., 2003; Barter et al., 2004). ApoA-I and apoA-I mimetic peptides may reduce oxidized lipids and also improve reverse cholesterol transport and, therefore, may have therapeutic potential (Navab et al., 2006; Ou et al., 2003). As for paraoxonase activity, when reduced, it may affect the ability of HDL to defend itself against oxidation. Agents that prevent the reduction in PON 1 activity or reverse the loss of activity can result in improved HDL function and thus normal HDL physiology and metabolism (Fogelman, 2004; Navab et al., 2006). The apoA-I mimetic D-4F that is currently under investigation may prove to be of benefit in relation to activating PON and thus HDL. The well being of the endothelial cells is also a highly important requirement and it is likely that PON and HDL play major roles in maintaining the balance needed.

ACKNOWLEDGMENTS

This work was supported in part by USPHS grants HL 30568, HL 34343, the Laubisch, Castera, and M.K. Grey Funds at UCLA.

REFERENCES

Anantharamaiah GM, Jones JL, Brouillette CG, Schmidt CF, Chung BH, Hughes TA, Bhown AS, Segrest JP. *Studies of synthetic peptide analogs of amphipathic helix I: Structure of peptide/DMPC complexes.* J Biol Chem. 1985; 260: 10248–10255.

Ansell BJ, Navab M, Hama S, Kamranpour N, Fonarow G, Hough G, Rahmani S, Mottahedeh R, Dave R, Reddy ST, Fogelman AM. *Inflammatory/antiinflammatory properties of high-density lipoproteins distinguish patients from control subjects better than high-density lipoprotein cholesterol levels and are favorably impacted by simvastatin treatment.* Circulation. 2003; 108: 2751–2756.

Barter PJ, Nicholls S, Rye K-A, Anantharamaiah GM, Navab M, Fogelman AM. *Antiinflammatory properties of HDL.* Circ Res. 2004; 95: 764–772.

Datta G, Epand RF, Epand RM, Chaddha M, Kirksey MA, Garber DW, Lund-Katz S, Phillips MC, Hama S, Navab M, Fogelman AM, Palgunachari MN, Segrest JP, Anantharamaiah GM. *Aromatic Residue Position on the Nonpolar Face of Class A Amphipathic Helical Peptides Determines Biological Activity.* J Biol Chem. 2004; 279: 26509–26517.

Bowry VW, Stanley KK, Stocker R. *High density lipoprotein is the major carrier of lipid hydroperoxides in human blood plasma from fasting donors.* Proc Natl Acad Sci U S A. 1992 Nov 1; 89(21): 10316–20.

Epand RM, Epand RF, Sayer BG, Melacini G, Palgunachari MN, Segrest JP, Anantharamaiah GM. *An apolipoprotein AI mimetic peptide: membrane interactions and the role of cholesterol.* Biochemistry. 2004; 43: 5073–5083.

Epand RM, Epand RF, Sayer BG, Datta G, Chaddha M, Anantharamaiah GM. *Two homologous apolipoprotein AI mimetic peptides. Relationship between membrane interactions and biological activity.* J Biol Chem. 2004; 279: 51404–51414.

19-Ekmekci OB, Donma O, Ekmekci H, Yildirim N, Uysal O, Sardongan E, Demrel H, Demir, T. *Plasma paraoxonase activities, lipoprotein oxidation, and trace element interaction in asthmatic patients.* Biol Trace Elem Res. 2006; 111: 41–52.

Fogelman AM. *When good cholesterol goes bad.* Nature Medicine 2004; 10: 902–3.

Gabay C, Kushner I. *Acute-phase proteins and other systemic responses to inflammation.* N. Engl. J. Med. 1999; 340: 448–454.

Kassai A, Ilyes L, Mirdamadi HZ, Seres I, Kalmar T, Audikovsky M, Paragh G. *The effect of atorvastatin therapy on lecithin:cholesterol acyltransferase, cholesteryl ester transfer protein and the antioxidant paraoxonase.* Clin Biochem. 2007; 40: 1–5.

Navab M, Hama-Levy S, Van Lenten BJ, Fonarow GC, Cardinez, CJ, Castellani LW, Brennan M-L, Lusis AJ, Fogelman AM. *Mildly oxidized LDL induces an increased apolipoprotein J/paraoxonase ratio.* J. Clin. Invest. 1997; 99: 2005–2019.

Navab M, Hama SY, Anantharamaiah GM, Hassan K, Hough GP, Watson AD, Reddy ST, Sevanian A, Fonarow GC, Fogelman AM. *Normal high density lipoprotein inhibits three steps in the formation of mildly oxidized low density lipoprotein: steps 2 and 3.* J Lipid Res. 2000 Sep; 41(9): 1495–508.

Navab M, Anantharamaiah GM, Reddy ST, VanLenten BJ, Ansell BJ, Fonarow GC, Vahabzadeh K, Hama S, Hough G, Kamranpour N, Berliner JA, Lusis AJ, Fogelman AM. *The oxidation hypothesis of atherogenesis: the role of oxidized phospholipids and HDL.* J Lipid Res. 2004; 45: 993–1007.

Navab M, Berliner JA, Subbanagounder G, Hama S, Lusis AJ, Castellani LW, Reddy S, Shih D, Shi W, Watson, AD, Van Lenten BJ, Vora D, Fogelman AM. *HDL and the inflammatory response induced by LDL-derived oxidized phospholipids.* Arterioscler. Thromb. Vasc. Biol. 2001; 21: 481–488.

Navab M, Anantharamaiah GM, Reddy ST, Van Lenten BJ, Hough G, Wagner A, Nakamura K, Garber DW, Datta G, Segrest JP, Hama S, Fogelman AM. *Human apolipoprotein AI mimetic peptides for the treatment of atherosclerosis.* Curr Opin Investig Drugs. 2003; 4: 1100–4. Review.

Navab M, Van Lenten BJ, Reddy ST, Fogelman AM. *High-density lipoprotein and the dynamics of atherosclerotic lesions.* Circulation. 2001; 104: 2386–7.

Navab M, Anantharamaiah GM, Reddy ST, Hama S, Hough G, Grijalva VR, Wagner AC, Frank JS, Datta G, Garber D, Fogelman AM. *Oral D-4F causes formation of pre-beta high-density lipoprotein and improves high-density lipoprotein-mediated cholesterol efflux and reverse cholesterol transport from macrophages in apolipoprotein E-null mice.* Circulation. 2004 Jun 29; 109(25): 3215–20.

Navab M, Hama S, Reddy ST, Fogelman AM. *Paraoxonase (PON1) in Health and Disease, Basic and Clinical Aspects.* LG Costa and CE Furlong, editors, Kluwer Academic Publishers, Norwell, MA. 2002. PP. 125–136.

Navab M, Anantharamaiah GM, Reddy ST, Hama S, Hough G, Grijalva VR, Yu N, Ansell BJ, Datta G, Garber DW, Fogelman AM. *Apolipoprotein A-I mimetic peptides.* Arterioscler Thromb Vasc Biol. 2005; 25: 1325–31.

Navab M, Anantharamaiah GM, Hama S, Hough G, Reddy ST, Frank JS, Garber DW, Handattu S, Fogelman AM. *D-4F and statins synergize to render HDL antiinflammatory in mice and monkeys and cause lesion regression in old apolipoprotein E-null mice.* Arterioscler Thromb Vasc Biol. 2005; 25: 1426–32.

Navab M, Anantharamaiah GM, Reddy ST, Van Lenten BJ, Ansell BJ, Fonarow GC, Vahabzadeh K, Hama S, Hough G, Kamranpour N, Berliner JA, Lusis AJ, Fogelman AM. *The oxidation hypothesis of atherogenesis: the role of oxidized phospholipids and HDL.* J Lipid Res. 2004; 45: 993–1007.

Navab M, Anantharamaiah G, Reddy TS, Van Lenten BJ, Ansell BJ, Fogelman AM. *Pro-atherogenic HDL: an evolving field.* Nature Reviews Clinical Practice, Endocrinology and Metabolism. Sept 2006; Vol.2 No. 9 pp. 504–511.

Navab M, Anantharamaiah G, Reddy ST, Fogelman AM. *Apolipoprotein A-I mimetic peptides and their role in atherosclerosis prevention.* Nat Clin Pract Cardiovasc Med 2006; 3: 540–7.

Ou J, Ou Z, Jones DW, Holzhaure S, Hatoum OA, Ackerman AW, Weihrauch DW, Gutterman DD, Guice K, Oldham KT, Hillery CA, Pritchard KA, Jr. *L-D-4F, an apolipoprotein A-1 mimetic, dramatically improves vasodilation in hypercholesterolemia and sickle cell disease.* Circulation. 2003; 107: 2337–2341.

Rozenberg O, Aviram M. *S-Glutathionylation regulates HDL-associated paraoxonase 1 (PON1) activity.* Biochem Biophys Res Commun. 2006; 15; 351: 492–8.

Sevanian A, Bittolo-Bon G, Cazzolato G, Hodis H, Hwang J, Zamburlini A, Maiorino M, Ursini F. *LDL- is a lipid hydroperoxide-enriched circulating lipoprotein.* J Lipid Res. 1997 Mar; 38(3): 419–28.

Van Lenten BJ, Hama SY, deBeer FC, Stafforini DM, McIntyre TM, Prescott SM, La Du BN, Fogelman AM, Navab M. *Anti-inflammatory HDL becomes pro-inflammatory during the acute phase response.* J. Clin. Invest. 1995; **96**: 2758–2767.

Watson AD, Berliner JA, Hama SY, La Du BN, Faull KF, Fogelman AM, Navab M. *Protective effect of high density lipoprotein associated paraoxonase. Inhibition of the biological activity of minimally oxidized low density lipoprotein.* J Clin Invest. 1995 Dec; 96(6): 2882–91.

CHAPTER 5

PARAOXONASE 1 (PON1), A JUNCTION BETWEEN THE METABOLISMS OF HOMOCYSTEINE AND LIPIDS

H. JAKUBOWSKI

Department of Microbiology & Molecular Genetics, UMDNJ-New Jersey Medical School, International Center for Public Health, 225 Warren Street, Newark, NJ 07101, USA and Institute of Bioorganic Chemistry, Polish Academy of Sciences, 61-704 Poznań, Poland
E-mail: jakubows@umdnj.edu

Abstract: Homocysteine and cholesterol are linked to the development of atherothrombotic disease. Distinct homocysteine-specific and lipid-specific mechanisms contribute to the pathogenicity resulting from excesses of these metabolites. However, it is becoming increasingly clear that there are important interactions between the homocysteine and lipoprotein metabolic pathways. This article reviews evidence documenting interactions between metabolisms of homocysteine and high-density lipoproteins

Keywords: Homocysteine thiolactone; protein N-homocysteinylation; high-density lipoprotein; low-density lipoprotein; cholesterol; paraoxonase; thiolactonase; gene expression; metabolism; atherosclerosis

1.   HOMOCYSTEINE METABOLISM – AN OVERVIEW

Homocysteine (Hcy) is a sulfur-containing amino acid that is found as an intermediary metabolite in all three domains of life. In mammals Hcy is formed from dietary methionine (Met) as a result of cellular methylation reactions (Mudd et al., 2001). In this pathway dietary Met is taken up by cells and then activated by ATP to yield S-adenosylmethionine (AdoMet), a universal methyl donor (Fig. 1). As a result of the transfer of its methyl group to an acceptor, AdoMet is converted to S-adenosylhomocysteine (AdoHcy). *The reversible enzymatic hydrolysis of AdoHcy is the only known source of Hcy in the human body.* Levels of Hcy are regulated by remethylation to Met, catalyzed by the enzyme Met synthase (MS), and transsulfuration to cysteine, the first step of which is catalyzed by the enzyme cystathionine β-synthase (CBS). The remethylation requires vitamin B_{12} and 5,10-methyl-tetrahydrofolate (CH_3-THF), generated by 5,10-methylene-THF reductase (MTHFR). The transsulfuration requires vitamin B_6.

87

B. Mackness et al. (eds.), The Paraoxonases: Their Role in Disease Development and Xenobiotic Metabolism, 87–102.

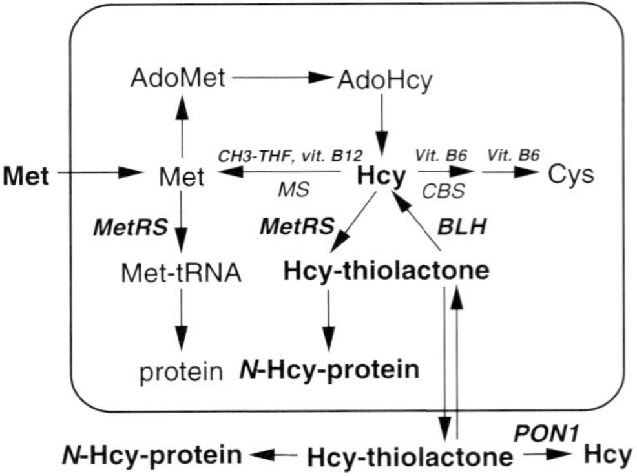

Figure 1. Scheme of Hcy metabolism in humans

A small fraction of Hcy is also metabolized by methionyl-tRNA synthetase (MetRS) to a thioester, Hcy-thiolactone, (Fig. 1) in an error-editing reaction in protein biosynthesis when Hcy is mistakenly selected in place of Met (Jakubowski, 2004, 2005a, b, 2006a). The flow through the Hcy-thiolactone pathway is increased by a high-Met diet (Chwatko et al., 2007), inadequate supply of CH_3-THF (Jakubowski, 1997; Jakubowski et al., 2000), or impairment of re-methylation or trans-sulfuration reactions by genetic alterations of enzymes, such as CBS (Chwatko et al., 2007; Jakubowski, 1991, 1997, 2002a), MS (Jakubowski, 1991, 2002a), and MTHFR (Chwatko et al., 2007). Because of its exceptionally low pK_a value (Table 1),

Table 1. Physical-chemical properties of Hcy-thiolactone and Hcy (Jakubowski, 2004)

Property	Hcy-thiolactone	Hcy
Chemical character	Aminoacyl-thioester	Mercaptoamino acid
UV/Vis light absorption	Yes, a maximum at $\lambda = 240$ nm	No significant absorption at $\lambda > 220$ nm
Half-life at pH 7.4, 37 °C		
phosphate buffered saline	~ 30 h	2 h
human serum	~ 1 h	2 h
pK_a of amino group	6.67*	9.04. 9.71** 9.02, 9.69 (thiol group)**
Chemical reactivity	- Resistant to oxidation - Reacts with protein lysine ε-amino groups - Prone to base-catalyzed hydrolysis to Hcy - Reacts with aldehydes, forming tetrahydrothiazines*	- Oxidized to disulfides - Reacts with nitric oxide to form *S-nitroso*-Hcy - Converts to Hcy-thiolactone with an acid - Reacts with aldehydes, forming tetrahydrothiazines*

* (Jakubowski, 2006b)
** There are four ionic species of Hcy (Reuben and Bruice, 1976)

Hcy-thiolactone is neutral at physiological pH and thus can diffuse out of the cell (Fig. 1) and accumulate in the extracellular fluids (Chwatko and Jakubowski, 2005a). Hcy-thiolactone is hydrolyzed to Hcy by intracellular (Zimny et al., 2006) and extracellular Hcy-thiolactonases (Jakubowski, 2000a), previously known as bleomycin hydrolase (BLH) and paraoxonase 1 (PON1), respectively (Fig. 1).

2. HCY AND CARDIOVASCULAR DISEASE

Hcy has become a focus of intense studies in human pathophysiology. Severe hyperhomocysteinemia due to mutations in the *CBS, MTHFR*, or *MS* genes causes pathological changes in multiple organs and leads to premature death due to vascular complications (Kluijtmans et al., 1999; Mudd et al., 2001; Yap et al., 2001). Advanced arterial lesions occur in children with inborn errors in Hcy metabolism, which led to a hypothesis that Hcy causes vascular disease (McCully, 1969). Although severe hyperhomocysteinemia is rare, mild hyperhomocysteinemia is quite prevalent in the general population and is associated with an increased risk of vascular events and predicts mortality in heart disease patients (Anderson et al., 2000; Wald et al., 2002). The strongest evidence that Hcy plays a causal role in atherothrombosis comes from studies of severe genetic hyperhomocysteinemia in humans (Kluijtmans et al., 1999; Mudd et al., 2001; Yap et al., 2001), as well as genetic and nutritional hyperhomocysteinemia in animals (Lawrence de Koning et al., 2003; Lentz, 2005). Lowering plasma Hcy by vitamin-B supplementation leads to a 21–24% reduction of vascular outcomes in high risk stroke (Lonn et al., 2006; Spence et al., 2005), but not myocardial infarction (MI) patients (Bonaa et al., 2006; Lonn et al., 2006). The results of ongoing Hcy-lowering trials are required before making recommendations on the use of vitamins for prevention of vascular disease (Clarke et al., 2007).

3. THE HCY-THIOLACTONE HYPOTHESIS

Mechanisms underlying Hcy involvement in cardiovascular disease are subjects of intense investigation (Jakubowski, 2004, 2005c, 2006a; Lawrence de Koning et al., 2003; Lentz, 2005). One hypothesis states that a pathway initiated by metabolic conversion of Hcy to Hcy-thiolactone contributes to pathohysiologies associated with Hcy excess (Fig. 2) (Jakubowski, 1997; Jakubowski, 2005c, 2006a; Jakubowski et al., 2000). Consistent with this hypothesis, plasma Hcy-thiolactone is elevated under conditions predisposing to atherosclerosis, such as hyperhomocysteinemia caused by mutations in the *CBS* or *MTHFR* genes in humans or a high-Met diet in mice (Chwatko et al., 2007). Also consistent with the Hcy-thiolactone hypothesis (Fig. 2) are observations that chronic treatments of animals with Hcy-thiolactone cause pathophysiological changes similar to those observed in human genetic hyperhomocysteinemia. For example, Hcy-thiolactone infusions or Hcy-thioactone-supplemented diet produce atherosclerosis in baboons (Harker et al., 1974) and rats (Endo et al., 2006), whereas treatment with Hcy-thiolactone

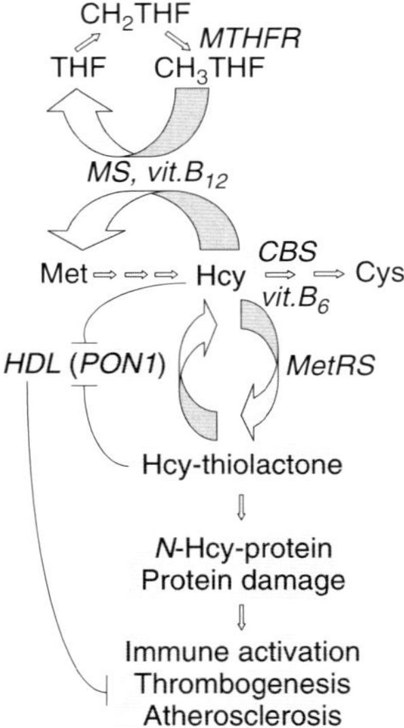

Figure 2. The pathophysiologic hypothesis of Hcy-thiolactone-mediated vascular disease

causes developmental abnormalities in chick embryos (Rosenquist et al., 1996), including optic lens dislocation (Maestro de las Casas et al., 2003) characteristic of the CBS-deficient humans (Kluijtmans et al., 1999; Mudd et al., 2001; Yap et al., 2001).

Hcy-thiolactone is a chemically reactive thioester (Table 1) that causes protein N-homocysteinylation through a facile formation of amide bonds with ε-amino groups of protein lysine residues (Jakubowski, 1997, 1999, 2004, 2006a) (Fig. 3). As predicted by the Hcy-thiolactone hypothesis (Fig. 2), N-Hcy-protein is elevated in hyperhomocysteinemic subjects (Glowacki & Jakubowski, 2004; Jakubowski, 2000b, 2001b, 2002b). N-Linked protein Hcy is present in hemoglobin, albumin, γ-globulin, transferrin, high-density lipoproteiein (HDL), low-density lipoprotein (LDL), α1-antitrypsin, and fibrinogen. *N-Hcy-Hemoglobin, present at 12.7 μM (or 0.9 mg/ml), constitutes the largest known Hcy pool in human blood* (Jakubowski, 2002b). The inability to detect N-linked Hcy in transthyretin (Sass et al., 2003) is most likely due to inadequate assay sensitivity.

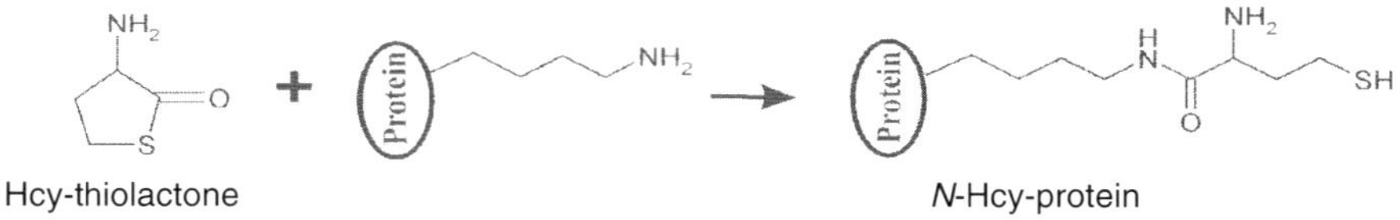

Hcy-thiolactone   *N*-Hcy-protein

Figure 3. N-Homocysteinylation of a protein lysine residue

Hcy incorporation creates altered proteins with newly acquired interactions, which can be detrimental to human body. For example, *N*-Hcy-proteins tend to form aggregates *in vitro* (Jakubowski, 1999, 2000b), which can be toxic *in vivo*. This has been demonstrated for *N*-Hcy-LDL, which, in contrast to native LDL, induces cell death in cultured human endothelial cells (Ferretti et al., 2004), a finding consistent with the inherent toxicity of protein aggregates (Stefani, 2004).

Fifbrinogen is *N*-homocysteinylated by Hcy-thiolactone *in vitro* (Jakubowski, 1999, 2000b) and *in vivo* in humans (Jakubowski, 2002b). Clots formed from Hcy-thiolactone-treated normal human plasma or fibrinogen lyse slower than clots from untreated controls (Sauls et al., 2006). Some of the ten Lys residues susceptible to *N*-homocysteinylation, located in the D and αC domains of fibrinogen, are close to tPA and plasminogen binding or plasmin cleavage sites, which can explain abnormal characteristics of clots formed from *N*-Hcy-fibrinogen (Sauls et al., 2006). Detrimental effects of elevated plasma tHcy on clot permeability and resistance to lysis in humans are consistent with a mechanism involving fibrinogen modification by Hcy-thiolactone (Undas et al., 2006b). These results support the Hcy-thiolactone hypothesis and suggest that fibrinogen *N*-homocysteinylation leads to abnormal resistance of fibrin clots to lysis and contributes to increased risk of thrombogenesis (Sauls et al., 2006; Undas et al., 2006b) (Fig. 2).

N-Hcy-Proteins elicit an auto-immune response in humans, which is enhanced in stroke and cardiovascular disease patients, and is a general feature of atherosclerosis (Jakubowski, 2005c, 2006a). Anti-*N*ε-Hcy-Lys-protein auto-antibody (Perla et al., 2004; Undas et al., 2004; Undas et al., 2005; Undas et al., 2006a) and *N*-Hcy-protein levels (Glowacki & Jakubowski, 2004; Jakubowski, 2000b, 2001b, 2002b) vary considerably among individuals and are strongly correlated with plasma Hcy, but not with Cys or Met. Such correlations can be explained by direct mechanistic links between Hcy-related species, outlined by the Hcy-thiolactone hypothesis (Fig. 2): elevation in Hcy leads to inadvertent elevation in Hcy-thiolactone, observed in human endothelial cells (Jakubowski, 2002a; Jakubowski et al., 2000) and in human and mouse plasma (Chwatko & Jakubowski, 2005a, b; Chwatko et al., 2007; Jakubowski, 2002a). Hcy-Thiolactone mediates Hcy incorporation into proteins and the formation of neo-self antigens, *N*ε-Hcy-Lys-protein (Fig. 3). Raising levels of neo-self *N*ε-Hcy-Lys-protein triggers an auto-immune response (Fig. 2). Auto-antibodies recognizing the *N*ε-Hcy-Lys epitope react with any *N*ε-Hcy-Lys-protein (Perla et al., 2004; Undas et al., 2004) in many tissues, contributing to deleterious effects of hyperhomocysteinemia on many organs (Kluijtmans et al., 1999; Mudd et al., 2001; Yap et al., 2001). If the neo-self *N*ε-Hcy-Lys epitopes were present on endothelial cell membrane proteins, anti-*N*ε-Hcy-Lys-protein auto-antibodies would form antigen-antibody complexes on the surface of the vascular vessel. Endothelial cells coated with anti-*N*ε-Hcy-Lys-protein auto-antibodies would be taken up by the macrophage *via* the Fc receptor, resulting in injury to vascular surface. Under chronic exposures to excess Hcy the neo-self epitopes *N*ε-Hcy-Lys, which initiate the injury, are formed continuously, and the repeating attempts to repair the damaged vascular wall would lead to an atherosclerotic lesion (Fig. 2) (Chwatko et al., 2007; Jakubowski, 2005c, 2006a).

4.   ACUTE TOXICITY OF HCY-THIOLACTONE

Excess Hcy-thiolactone is also acutely toxic in experimental animals and cell cultures. For example, acute infusions into mice or rats cause seizures and death within minutes (Folbergrova, 1997; Langmeier et al., 2003; Spence et al., 1995). Exposure of mouse (Greene et al., 2003) or rat (Van Aerts et al., 1993) embryos to Hcy-thiolactone causes lethality, growth retardation, and develop-mental abnormalities. In one study Hcy-thiolactone was reported to be non-teratogenic in mouse embryos, but the maximum dose used in that study (Hansen et al., 2001) was lower than those used in the other studies, so that an embryotoxic dose had not been reached. Hcy-thiolactone induces apoptotic death in cultured human vascular endothelial (Kerkeni et al., 2006; Mercie et al., 2000) and promyeloid cells (Huang et al., 2001), placental trophoblasts (Kamudhamas et al., 2004), and inhibits insulin signaling in rat hepatoma cells (Najib & Sanchez-Margalet, 2005). Hcy-thiolactone was also shown to induce endoplasmic reticulum stress and unfolded protein response in retinal epithelial cells (Roybal et al., 2004).

5.   PON1 IS AN HCY-THIOLACTONASE

As reviewed briefly above, Hcy-thiolactone is a toxic metabolite linked to immune activation and thrombogenesis in human cardiovascular disease. Thus, it may not be surprising that mammalian organisms evolved the ability to eliminate Hcy-thiolactone. Indeed, studies of Hcy-thiolactone elimination mechanisms have led to discoveries of two major enzymes that can clear Hcy-thiolactone in humans (Fig. 1): extracellular calcium-dependent Hcy-thiolactonase identical to serum paraoxonase 1 (PON1) synthesized exclusively in the liver and carried on high-density lipoproteins (HDL) in the circulation (Domagała et al., 2006; Jakubowski, 2000a; Jakubowski et al., 2001; Lacinski et al., 2004) and intracellular thiol-dependent Hcy-thiolactonase identical to bleomycin hydrolase (BLH) present ubiquitously in human tissues (Zimny et al., 2006). The *in vitro* catalytic efficiency of the enzymatic hydrolysis of Hcy-thiolactone by human intracellular BLH is greater than the efficiency of human extracellular PON1. Thus, BLH is expected to hydrolyze most of Hcy-thiolactone produced in the human body. Hcy thiolactone that escapes intracellular hydrolysis by BLH, passes into the circulation. Once in the circulation, Hcy-thiolactone is cleared by the kidney (Chwatko & Jakubowski, 2005a) and mopped up by PON1.

PON1, named for its ability to hydrolyze the organophosphate paraoxon, has been studied in the field of toxicology since 1960s. More recent studies have implicated PON1 in the pathogenesis of cardiovascular disease. For example, transgenicPON1-/- mice are more sensitive to a high-fat diet-induced atherosclerosis and to organophosphate poisoning than wild type littermates (Shih et al., 1998; Shih et al., 2000). *In vitro* studies indicate that HDL from PON1-/- animals is unable to prevent LDL oxidation. Human PON1 expressed in transgenic mice protects against a high-fat diet induced atherosclerosis (Tward et al., 2002). Human *PON1* has several genetic polymorphisms, e.g. *M55L*, *R192Q*, which affect paraoxonase activity. Human clinical studies suggest that PON1 phenotype, i.e., paraoxonase activity is a much stronger predictor of cardiovascular disease status than *PON1* genetic polymorphisms (Jarvik et al., 2000; Jarvik et al., 2003; Mackness et al., 2001; Mackness et al., 2003; Mackness et al., 2004a). The cardio protective role of the PON1 protein is believed to be due in part to its anti-oxidative properties. However, recent data show that the PON1 protein has no intrinsic ability to protect against lipid oxidation or inactivate harmful oxidation products (Connelly et al., 2005; Marathe et al., 2003; Teiber et al., 2004). Studies of the physiological role of PON1 were also hampered by the uncertainty regarding the identity of its pathophysiologically relevant lipid-related substrate. Although numerous non-naturally occurring substrates of PON1 have been reported (James, 2006; Jarvik et al., 2000; Mackness et al., 2004a) and highly speculative mechanisms proposed (Rosenblat et al., 2006), the only naturally occurring substrate known is Hcy-thiolactone (Domagała et al., 2006; Jakubowski, 2000a; Jakubowski et al., 2001; Lacinski et al., 2004).

PON1 has been found to hydrolyze Hcy-thiolactone (Jakubowski, 2000b, 2001a) and to protect proteins against *N*-homocysteinylation in cultured human cells (Jakubowski et al., 2000) and serum *in vitro* (Jakubowski et al., 2001). Serum Hcy-thiolactonase activity is carried on HDL, and is the major activity that metabolizes Hcy-thiolactone in human and other mammalian sera. However, Hcy-thiolactonase activity is not present in chicken serum (Jakubowski, 2000a, 2001a), which makes chicken embryos particularly sensitive to Hcy-thiolactone toxicity (Maestro de las Casas et al., 2003; Rosenquist et al., 1996). Nonionic detergents were required for purification of the Hcy-thiolactonase activity to homogeneity from the HDL fraction of serum lipoproteins. Homogenous Hcy-thiolactonase requires calcium for activity and stability, and has *N*-terminal sequence identical to that of paraoxonase (PON1) (Jakubowski, 2000a, 2001a). Additional proof that PON1 is an Hcy-thiolactonase came from the finding that *PON1* knockout mice, which lack serum paraoxonase activity (Shih et al., 1998), are also devoid of serum Hcy-thiolactonase activity (Jakubowski, 2000a, 2001a). Other investigators confirmed the ability of PON1 to hydrolyze Hcy-thiolactone (Khersonsky and Tawfik, 2005; Teiber et al., 2003).

6. *PON1* GENOTYPE AFFECTS HCY-THIOLACTONASE ACTIVITY

The two major polymorphic sites in the *PON1* coding region, *PON1-192* and *PON1-55*, known to affect paraoxonase activity (Jarvik et al., 2000; Mackness et al., 2001), also affect Hcy-thiolactonase activity. Hcy-Thiolactonase activity is strongly associated with *PON1* genotype in diverse human populations: young (16–18 years-old) white and black North American populations (Jakubowski et al., 2001), adult (54–59-years old) population from the western part of Poland (Lacinski et al., 2004), and adult (50–68-years old) population from the Manchester and Blackpool regions of the UK (Domagała et al., 2006). High Hcy-thiolactonase activity is associated with L55 and R192 alleles, more frequent in blacks than in whites; low

Hcy-thiolactonase activity is associated with M55 and Q192 alleles, more frequent in whites than in blacks. The hydrolytic activities of PON1 towards Hcy-thiolactone and the organophosphate paraoxon are strongly correlated in the three studied populations. The strong correlation between paraoxonase and Hcy-thiolactonase activities suggests that the paraoxonase activity can be a useful surrogate for the physiological Hcy-thiolactonase activity.

7.   *PON1* PROTECTS AGAINST HCY-THIOLACTONE ACCUMULATION AND PROTEIN N-HOMOCYSTEINYLATION *IN VITRO*

Supplementation of cultured human endothelial cells with 1 mg/ml HDL lowers Hcy-thiolactone accumulation to 21% of the level observed in the absence of HDL. In HDL-supplemented cell cultures, protein N-homocysteinylation is diminished to 34% of that observed in the absence of HDL (Jakubowski et al., 2000). In *in vitro* experiments with human serum, half-life of Hcy-thiolactone was found to depend on the *PON1* genotype of the donor. In serum from *PON1 LL55/RR192* and *PON1 MM55/QQ192* donors Hcy-thiolactone half-life was 30 min and 60 min, respectively (Jakubowski et al., 2001). When PON1 is inactivated by removal of calcium with EDTA, Hcy-thiolactone half-life is 90 min. The extent of protein N-homocysteinylation by Hcy-thiolactone in sera from *PON1 LL55/RR192* donors is lower compared to that in sera from *PON1 MM55/QQ192* donors (Jakubowski et al., 2001). Taken together, these results show that high Hcy-thiolactonase activity affords a better protection than the low Hcy-thiolactonase activity against protein *N*-homocysteinylation. Because rabbits have several fold higher levels of serum Hcy-thiolactonase activity than most humans, rabbit serum protein is much less susceptible than human serum protein to *N*-homocysteinylation by Hcy-thiolactone. This might explain the failure to induce atherosclerosis by infusions of Hcy-thiolactone into rabbits (Donahue et al., 1974; Makheja et al., 1978).

8.   HCY-THIOLACTONASE ACTIVITY PREDICTS CARDIOVASCULAR DISEASE

The first study to address the role of Hcy-thiolactonase activity of PON1 in cardiovascular disease was carried out with a group from the Western Poland region that included 184 subjects 32.6% of whom were healthy, 27.7% had angiographically proven coronary artery disease (CAD) but did not have myocardial infarction (MI), and 39.7% had MI (Lacinski et al., 2004). The frequency of the high Hcy-thiolactonase activity *PON1-192-RR* genotype tended to be lower in CAD subjects than in controls (2% *vs.* 10.0%, p = 0.057) and higher in MI subjects than in CAD subjects (10.9% vs. 2.0%, p = 0.001). Hcy-thiolactonase activity was lower in CAD patients, and higher in MI patients, than in controls, but the differences were not statistically significant, most likely due to low numbers of subjects. The high Hcy-thiolactonase activity R-allele was marginally associated with CAD

(26.7% in controls *vs.* 17.6% in CAD, p = 0.146) and significantly associated with MI (17.6% in CAD *vs.* 31.5% in MI, p = 0.018). A tendency for lower frequency of the *PON1-192-RR* genotype in CAD subjects, compared to healthy subjects, suggests that this genotype may be protective against CAD. Because MI patients included only survivors of a myocardial infarction, higher frequency of the *PON1-192-RR* genotype in MI subjects, compared with CAD subjects, is probably due to higher MI mortality in subjects with the *PON1-192-QQ* genotype relative to the *PON1-192-R* allele carriers, and suggests that the *PON1-192-RR* genotype protects against mortality associated with MI. According to International Cardiovascular Disease Statistics, mortality rates after MI can be as high as 75% (http://www.americanheart.org/). Such mortality rate among *PON1-192-QQ* carriers with CAD would account for the observed increase in *PON1-192-RR* carriers among MI survivors (Lacinski et al., 2004).

Using a much larger group of Manchester-Blackpool, UK subjects (272 cases and 202 controls) we found that the CAD status is a determinant of Hcy-thiolactonase activity: serum Hcy-thiolactonase activity was significantly higher in the CAD patient group than in the control group (Domagała et al., 2006). Although the CAD group had a greater proportion of males, was significantly older, had a larger average body mass index, and included a large proportion of statin users, these variables did not affect Hcy-thiolactonase activity (Domagała et al., 2006). Higher Hcy-thiolactonase activity in CAD subjects is surprising and suggests that Hcy-thiolactonase may not have a protective role as hypothesized, but may be associated with CAD by an unknown mechanism. However, the higher Hcy-thiolactonase activity in CAD cases than in controls may be accounted for by the survivor bias. Because the studied group included survivors of an MI, higher Hcy-thiolactonase activity in CAD subjects, compared with control subjects, is probably caused by higher MI mortality in subjects with low Hcy-thiolactonase activity. Because mortality rates after MI can be as high as 75% (*International Cardiovascular Disease Statistics*, http://www.americanheart.org/), we hypothesize that higher mortality among low Hcy-thiolactonase activity carriers with CAD than in high Hcy-thiolactonase activity carriers would account for the increase in high Hcy-thiolactonase activity carriers among MI survivors observed in our study population.

The finding that Hcy-thiolactonase activity was higher in the CAD group while paraoxonase activity was lower is surprising (Domagała et al., 2006). The different behaviour of the two activities could be caused by a post-translational modification (*N*-homocysteinylation, glycation, or modification by products of lipid peroxidation) of PON1 in CAD patients. Because Hcy-thiolactone and paraoxon appear to be hydrolyzed at different sites of PON1 (Jakubowski, 2000a; Khersonsky and Tawfik, 2005), such modification may have different effects on the two activities. Lower PON1 protein concentration in the CHD group compared with the control group suggests that the turnover of the PON1 protein may be accelerated in CHD patients, possibly because of increased modifications of the PON1 protein.

Despite the large impact of *PON1* genetic variation on the natural Hcy-thiolactonase activity (Domagała et al., 2006; Jakubowski et al., 2001; Lacinski et al., 2004) and other PON1 activities (paraoxonase, arylesterase, diazooxonase) (Jarvik et al., 2000; Jarvik et al., 2003; Mackness et al., 2001; Mackness et al., 2003; Mackness et al., 2004a; Mackness et al., 2004b), the *PON1* genetic variation alone is not associated with CAD status. Instead, the PON1 phenotype is a better predictor of CAD than are the *PON1-M55L* or *PON1-R192Q* genotypes. Why Hcy-thiolactonase and other PON1 activities predict vascular disease and its important genetic variability does not, remains unexplained. One possibility is that PON1 activities are modified by epigenetic factors. Another possibility is that the relationship between CAD and Hcy-thiolactonase activity may be influenced by the relationship of Hcy-thiolactonase activity to lipid and lipoprotein levels. The physical association of Hcy-thiolactonase activity with HDL (Jakubowski, 2000a, 2001a) and the presence of an important cholesterol transcriptional regulator site, SREB-2 (Deakin et al., 2003; Gouedard et al., 2003), suggest important Hcy-thiolactonase activity-lipoprotein relationships. However, the nature of this relationship is unclear.

9. CHOLESTEROL AND HCY-THIOLACTONASE ACTIVITY

Hcy-thiolactonase activity is associated with total cholesterol in populations from Western Poland (Lacinski et al., 2004) and Manchester-Blackpool, UK (Domagała et al., 2006), which suggests that such association is general, and not limited to a specific population. Positive correlation between Hcy-thiolactonase activity and total cholesterol suggests that cholesterol might regulate PON1/Hcy-thiolactonase at transcriptional or posttranscriptional level. Transcriptional regulation of PON1 protein expression is observed in several model systems, although the direction of this regulation varies depending on the system. For instance, in a human hepatoma HuH7 cultured cell model, statins decrease both PON1 mRNA levels and aryl esterase activity of the PON1 protein (with phenyl acetate as substrate), which is consistent with the presence of the liver oxysterol receptor (LXR) binding element in the human *PON1* gene promoter region (Gouedard et al., 2003). Statins also decrease the *PON1* gene promoter activity in HuH-7 cells (Gouedard et al., 2003), although the opposite effect was reported in HepG2 cells (Deakin et al., 2003), suggesting that the effect may be cell line-dependent. In a rat model, statins decrease liver tissue paraoxonase activity (Beltowski et al., 2004). However, a modest but significant increase in serum PON1 protein concentration, paraoxonase and aryl esterase activities is observed in simvastatin-treated human patients (Deakin et al., 2003). Simvastatin therapy also results in increased paraoxonase activity in hyper-cholesterolemic subjects (Tomas et al., 2000). The finding that statin therapy does not affect Hcy-thiolactonase activity in CAD patients suggests that statins may not affect each of the different activities of the PON1 protein in the same manner.

Findings that the Hcy-thiolactonase activity is positively correlated with LDL, HDL, and their components in CAD cases, but not in controls (Domagała et al., 2006), are surprising, but reminiscent of the results of another study that found significant correlations between arylesterase, diazooxonase, or paraoxonase activity and LDL-related measures in carotid artery disease patients but not in controls (Rozek et al., 2005). These findings suggest that regulation of Hcy-thiolactonase activity is different in subjects, compared to controls.

10. HCY AFFECTS PON1 AND APOA-1 GENE EXPRESSION

During the past decade, the development of animal models has contributed to rapid progress in defining the pathophysiological consequences of Hcy excess (Lentz, 2005). Dietary (e.g., a high-Met diet) as well as genetic approaches (knockout mice with a targeted disruption of the *CBS, MTHFR,* and *MTR* genes) are used to induce hyperhomocysteinemia in animals. Many pathophysiological features (endothelial dysfunction, hepatic lipid accumulation, and decreased plasma HDL cholesterol) observed in these animal models are common with human hyperhomnocysteinemia. Hyperhomocysteinemic mice exhibit accelerated atherosclerosis when crossed with apoE-deficient mice.

The epidemiological data indicate that plasma Hcy is negatively associated with HDL cholesterol (Anderson et al., 2000; Qujeq et al., 2001; Stampfer et al., 1992). Hcy-thiolactonase activity of PON1 is also negatively correlated with plasma Hcy in humans (Lacinski et al., 2004), which can be explained by the inhibition of hepatic PON1 synthesis by Hcy, demonstrated in hyperhomocysteinemic mouse models (Robert et al., 2003). Genetic hyperhomocsteinemia caused by the CBS deficiency leads to 10- and 3-fold down-regulation of PON1 mRNA and activity, respectively, in liver of CBS−/− mice, compared to wild type mice. That PON1 down-regulation is caused by excess Hcy is confirmed with mice fed a hyperhomocysteinemic high-Met diet: a significant decrease of hepatic PON1 activity is observed in mice fed a high-Met diet compared to mice fed a standard diet (Robert et al., 2003). However, because a high-Met diet increases Hcy and Hcy-thiolactone levels in mice (Chwatko et al., 2007), it is not clear which Hcy species is responsible for the decrease in PON1 expression.

Plasma total Hcy and apoA-1 levels are also negatively correlated in human subjects (Liao et al., 2006; Mikael et al., 2006). Such correlations are explained by findings that excess Hcy inhibits hepatic apoA-1 synthesis in mice and in cultured mouse hepatocytes. Mikael et al found that mildly hyperhomocysteinemic MTHFR+/− and CBS+/− mice have lower apoA-1 mRNA expression. The decreased expression of apoA-1 was confirmed at the protein level in the liver and plasma of MTHFR+/− and CBS+/− mice. High concentrations of exogenous Hcy also inhibited apoA-1 expression in *ex vivo* cultures of mouse hepatocytes. A similar decrease in apoA-1 levels was described by Liao et al using CBS−/−apoE−/− mouse model, although in this model Hcy appears to affect apoA-1 expression at translational, but not transcriptional level. Because genetic deficiencies in the MTHFR and CBS genes increase Hcy *and* Hcy-thiolactone (Chwatko et al., 2007), and Hcy supplementation increases Hcy-thiolactone levels in cultured cells (Jakubowski, 2002a; Jakubowski et al., 2000), it is not clear whether the decrease in apoA-1 expression is caused by Hcy or Hcy-thiolactone.

11.   CONCLUSIONS

Elevated plasma Hcy predicts cardiovascular disease. Treatments that decrease plasma Hcy also decrease negative outcomes in stroke but not in MI patients. Some proatherogenic effects of Hcy, such as immune activation and thrombogenesis, are mediated by Hcy-thiolactone. Low HDL is highly predictive of cardiovascular disease and treatments that increase plasma HDL cholesterol in animals and humans have been shown to reduce the progression or even promote regression of atherosclerosis. Thus, the influence on HDL metabolism could contribute to proatherogenic effects of Hcy or Hcy-thiolactone (Fig. 2). Antiatherogenic properties of HDL are due to its ability to facilitate a reverse cholesterol transport from cells, including macrophages in the artery wall. HDL may also protect by virtue of antioxidant, antithrombotic, and anti-inflammatory properties, in part due to PON1's abilities to protect against oxidation and protein modification (Fig. 2). The findings that PON1 is an Hcy-thiolactonase and that Hcy decreases PON1 and apoA-1 expression suggest that the decrease in HDL activities might contribute to the pathogenicity of Hcy excess and explain why Hcy is a risk factor for atherosclerosis as well as for thrombothic and fibrotic vascular disease, as originally noted (McCully, 1969).

REFERENCES

Anderson, J.L., Muhlestein, J.B., Horne, B.D., Carlquist, J.F., Bair, T.L., Madsen, T.E., and Pearson, R.R., 2000, Plasma homocysteine predicts mortality independently of traditional risk factors and C-reactive protein in patients with angiographically defined coronary artery disease. *Circulation*, **102**, 1227–32.

Beltowski, J., Wojcicka, G., and Jamroz, A., 2004, Effect of 3-hydroxy-3-methylglutarylcoenzyme A reductase inhibitors (statins) on tissue paraoxonase 1 and plasma platelet activating factor acetylhydrolase activities. *J Cardiovasc Pharmacol*, **43**, 121–7.

Bonaa, K.H., Njolstad, I., Ueland, P.M., Schirmer, H., Tverdal, A., Steigen, T., Wang, H., Nordrehaug, J.E., Arnesen, E., and Rasmussen, K., 2006, Homocysteine lowering and cardiovascular events after acute myocardial infarction. *N Engl J Med*, **354**, 1578–88.

Chwatko, G., and Jakubowski, H., 2005a, Urinary excretion of homocysteine-thiolactone in humans. *Clin Chem*, **51**, 408–15.

Chwatko, G., and Jakubowski, H., 2005b, The determination of homocysteine-thiolactone in human plasma. *Anal Biochem*, **337**, 271–7.

Chwatko, G., Boers, G.H., Strauss, K.A., Shih, D.M., and Jakubowski, H., 2007, Mutations in methylenetetrahydrofolate reductase or cystathionine β-syntase gene and high methionine diet increase homocysteine-thiolactone levels in humans and mice. *FASEB J*, **21**, in press.

Clarke, R., Lewington, S., Sherliker, P., and Armitage, J., 2007, Effects of B-vitamins on plasma homocysteine concentrations and on risk of cardiovascular disease and dementia. *Curr Opin Clin Nutr Metab Care*, **10**, 32–9.

Connelly, P.W., Draganov, D., and Maguire, G.F., 2005, Paraoxonase-1 does not reduce or modify oxidation of phospholipids by peroxynitrite. *Free Radic Biol Med*, **38**, 164–74.

Deakin, S., Leviev, I., Guernier, S., and James, R.W., 2003, Simvastatin modulates expression of the PON1 gene and increases serum paraoxonase: a role for sterol regulatory element-binding protein-2. *Arterioscler Thromb Vasc Biol*, **23**, 2083–9.

Domagała, T.B., Łacinski, M., Trzeciak, W.H., Mackness, B., Mackness, M.I., and Jakubowski, H., 2006, The correlation of homocysteine-thiolactonase activity of the paraoxonase (PON1) protein with coronary heart disease status. *Cell Mol Biol (Noisy-le-grand)*, **52**, 3–9.

Donahue, S., Struman, J.A., and Gaull, G., 1974, Arteriosclerosis due to homocyst (e) inemia. Failure to reproduce the model in weanling rabbits. *Am J Pathol*, **77**, 167–3.

Endo, N., Nishiyama, K., Otsuka, A., Kanouchi, H., Taga, M., and Oka, T., 2006, Antioxidant activity of vitamin B6 delays homocysteine-induced atherosclerosis in rats. *Br J Nutr*, **95**, 1088–93.

Ferretti, G., Bacchetti, T., Moroni, C., Vignini, A., Nanetti, L., and Curatola, G., 2004, Effect of homocysteinylation of low density lipoproteins on lipid peroxidation of human endothelial cells. *J Cell Biochem*, **92**, 351–60.

Folbergrova, J., 1997, Anticonvulsant action of both NMDA and non-NMDA receptor antagonists against seizures induced by homocysteine in immature rats. *Exp Neurol*, **145**, 442–50.

Glowacki, R., and Jakubowski, H., 2004, Cross-talk between Cys34 and lysine residues in human serum albumin revealed by N-homocysteinylation. *J Biol Chem*, **279**, 10864–71.

Gouedard, C., Koum-Besson, N., Barouki, R., and Morel, Y., 2003, Opposite regulation of the human paraoxonase-1 gene PON-1 by fenofibrate and statins. *Mol Pharmacol*, **63**, 945–56.

Greene, N.D., Dunlevy, L.E., and Copp, A.J., 2003, Homocysteine is embryotoxic but does not cause neural tube defects in mouse embryos. *Anat Embryol (Berl)*, **206**, 185–91.

Hansen, D.K., Grafton, T.F., Melnyk, S., and James, S.J., 2001, Lack of embryotoxicity of homocysteine thiolactone in mouse embryos in vitro. *Reprod Toxicol*, **15**, 239–44.

Harker, L.A., Slichter, S.J., Scott, C.R., and Ross, R., 1974, Homocystinemia. Vascular injury and arterial thrombosis. *N Engl J Med*, **291**, 537–43.

Huang, R.F., Huang, S.M., Lin, B.S., Wei, J.S., and Liu, T.Z., 2001, Homocysteine thiolactone induces apoptotic DNA damage mediated by increased intracellular hydrogen peroxide and caspase 3 activation in HL-60 cells. *Life Sci*, **68**, 2799–811.

Jakubowski, H., 1991, Proofreading in vivo: editing of homocysteine by methionyl-tRNA synthetase in the yeast Saccharomyces cerevisiae. *Embo J*, **10**, 593–8.

Jakubowski, H., 1997, Metabolism of homocysteine thiolactone in human cell cultures. Possible mechanism for pathological consequences of elevated homocysteine levels. *J Biol Chem*, **272**, 1935–42.

Jakubowski, H., 1999, Protein homocysteinylation: possible mechanism underlying pathological consequences of elevated homocysteine levels. *Faseb J*, **13**, 2277–83.

Jakubowski, H., 2000a, Calcium-dependent human serum homocysteine thiolactone hydrolase. A protective mechanism against protein N-homocysteinylation. *J Biol Chem*, **275**, 3957–62.

Jakubowski, H., Zhang, L., Bardeguez, A., and Aviv, A., 2000, Homocysteine thiolactone and protein homocysteinylation in human endothelial cells: implications for atherosclerosis. *Circ Res*, **87**, 45–51.

Jakubowski, H., 2000b, Homocysteine thiolactone: metabolic origin and protein homocysteinylation in humans. *J Nutr.* **130**, 377S–81S.

Jakubowski, H., 2001a, Biosynthesis and reactions of homocysteine thiolactone. In *Homocysteine in Health and Disease* (D. Jacobson, and R. Carmel, eds.), Cambridge University Press, Cambridge, UK, pp. 21–31.

Jakubowski, H., 2001b, Translational accuracy of aminoacyl-tRNA synthetases: implications for atherosclerosis. *J Nutr*, **131**, 2983S–7S.

Jakubowski, H., Ambrosius, W.T., and Pratt, J.H., 2001, Genetic determinants of homocysteine thiolactonase activity in humans: implications for atherosclerosis. *FEBS Lett*, **491**, 35–9.

Jakubowski, H., 2002a, The determination of homocysteine-thiolactone in biological samples. *Anal Biochem*, **308**, 112–9.

Jakubowski, H., 2002b, Homocysteine is a protein amino acid in humans. Implications for homocysteine-linked disease. *J Biol Chem*, **277**, 30425–8.

Jakubowski, H., 2004, Molecular basis of homocysteine toxicity in humans. *Cell Mol Life Sci*, **61**, 470–87.

Jakubowski, H., 2005a, tRNA synthetase editing of amino acids. In *Encyclopedia of Life Sciences*, John Wiley & Sons, Ltd, Chichester, UK, p. http://www.els.net/doi:10.1038/npg.els.0003933.

Jakubowski, H., 2005b, Accuracy of Aminoacyl-tRNA Synthetases: Proofreading of Amino Acids. In *The Aminoacyl-tRNA Synthetases* (M. Ibba, C. Francklyn, and S. Cusack, eds.), Landes Bioscience/Eurekah.com Georgetown, TX, pp. 384–96.

Jakubowski, H., 2005c, Anti-N-homocysteinylated protein autoantibodies and cardiovascular disease. *Clin Chem Lab Med*, **43**, 1011–4.

Jakubowski, H., 2006a, Pathophysiological consequences of homocysteine excess. *J Nutr*, **136**, 1741S–9S.

Jakubowski, H., 2006b, Mechanism of the condensation of homocysteine thiolactone with aldehydes. *Chemistry*, **12**, 8039–43.

James, R.W., 2006, A long and winding road: defining the biological role and clinical importance of paraoxonases. *Clin Chem Lab Med*, **44**, 1052–9.

Jarvik, G.P., Rozek, L.S., Brophy, V.H., Hatsukami, T.S., Richter, R.J., Schellenberg, G.D., and Furlong, C.E., 2000, Paraoxonase (PON1) phenotype is a better predictor of vascular disease than is PON1(192) or PON1(55) genotype. *Arterioscler Thromb Vasc Biol*, **20**, 2441–7.

Jarvik, G.P., Hatsukami, T.S., Carlson, C., Richter, R.J., Jampsa, R., Brophy, V.H., Margolin, S., Rieder, M., Nickerson, D., Schellenberg, G.D., Heagerty, P.J., and Furlong, C.E., 2003, Paraoxonase activity, but not haplotype utilizing the linkage disequilibrium structure, predicts vascular disease. *Arterioscler Thromb Vasc Biol*, **23**, 1465–71.

Kamudhamas, A., Pang, L., Smith, S.D., Sadovsky, Y., and Nelson, D.M., 2004, Homocysteine thiolactone induces apoptosis in cultured human trophoblasts: a mechanism for homocysteine-mediated placental dysfunction? *Am J Obstet Gynecol*, **191**, 563–71.

Kerkeni, M., Tnani, M., Chuniaud, L., Miled, A., Maaroufi, K., and Trivin, F., 2006, Comparative study on in vitro effects of homocysteine thiolactone and homocysteine on HUVEC cells: evidence for a stronger proapoptotic and proinflammative homocysteine thiolactone. *Mol Cell Biochem*, **291**, 119–26.

Khersonsky, O., and Tawfik, D.S., 2005, Structure-reactivity studies of serum paraoxonase PON1 suggest that its native activity is lactonase. *Biochemistry*, **44**, 6371–82.

Kluijtmans, L.A., Boers, G.H., Kraus, J.P., van den Heuvel, L.P., Cruysberg, J.R., Trijbels, F.J., and Blom, H.J., 1999, The molecular basis of cystathionine beta-synthase deficiency in Dutch patients with homocystinuria: effect of CBS genotype on biochemical and clinical phenotype and on response to treatment. *Am J Hum Genet*, **65**, 59–67.

Lacinski, M., Skorupski, W., Cieslinski, A., Sokolowska, J., Trzeciak, W.H., and Jakubowski, H., 2004, Determinants of homocysteine-thiolactonase activity of the paraoxonase-1 (PON1) protein in humans. *Cell Mol Biol (Noisy-le-grand)*, **50**, 885–93.

Langmeier, M., Folbergrova, J., Haugvicova, R., Pokorny, J., and Mares, P., 2003, Neuronal cell death in hippocampus induced by homocysteic acid in immature rats. *Epilepsia*, **44**, 299–304.

Lawrence de Koning, A.B., Werstuck, G.H., Zhou, J., and Austin, R.C., 2003, Hyperhomocysteinemia and its role in the development of atherosclerosis. *Clin Biochem*, **36**, 431–41.

Lentz, S.R., 2005, Mechanisms of homocysteine-induced atherothrombosis. *J Thromb Haemost*, **3**, 1646–54.

Liao, D., Tan, H., Hui, R., Li, Z., Jiang, X., Gaubatz, J., Yang, F., Durante, W., Chan, L., Schafer, A.I., Pownall, H.J., Yang, X., and Wang, H., 2006, Hyperhomocysteinemia decreases circulating high-density lipoprotein by inhibiting apolipoprotein A-I Protein synthesis and enhancing HDL cholesterol clearance. *Circ Res*, **99**, 598–606.

Lonn, E., Yusuf, S., Arnold, M.J., Sheridan, P., Pogue, J., Micks, M., McQueen, M.J., Probstfield, J., Fodor, G., Held, C., and Genest, J., Jr., 2006, Homocysteine lowering with folic acid and B vitamins in vascular disease. *N Engl J Med*, **354**, 1567–77.

Mackness, B., Davies, G.K., Turkie, W., Lee, E., Roberts, D.H., Hill, E., Roberts, C., Durrington, P.N., and Mackness, M.I., 2001, Paraoxonase status in coronary heart disease: are activity and concentration more important than genotype? *Arterioscler Thromb Vasc Biol*, **21**, 1451–7.

Mackness, B., Durrington, P., McElduff, P., Yarnell, J., Azam, N., Watt, M., and Mackness, M., 2003, Low paraoxonase activity predicts coronary events in the Caerphilly Prospective Study. *Circulation*, **107**, 2775–9.

Mackness, M., Durrington, P., and Mackness, B., 2004a, Paraoxonase 1 activity, concentration and genotype in cardiovascular disease. *Curr Opin Lipidol*, **15**, 399–404.

Mackness, M.I., Durrington, P.N., and Mackness, B., 2004b, The role of paraoxonase 1 activity in cardiovascular disease: potential for therapeutic intervention. *Am J Cardiovasc Drugs*, **4**, 211–7.

Maestro de las Casas, C., Epeldegui, M., Tudela, C., Varela-Moreiras, G., and Perez-Miguelsanz, J., 2003, High exogenous homocysteine modifies eye development in early chick embryos. *Birth Defects Res A Clin Mol Teratol*, **67**, 35–40.

Makheja, A.N., Bombard, A.T., Randazzo, R.L., and Bailey, J.M., 1978, Anti-inflammatory drugs in experimental atherosclerosis. Part 3. Evaluation of the atherogenicity of homocystine in rabbits. *Atherosclerosis*, **29**, 105–12.

Marathe, G.K., Zimmerman, G.A., and McIntyre, T.M., 2003, Platelet-activating factor acetylhydrolase, and not paraoxonase-1, is the oxidized phospholipid hydrolase of high density lipoprotein particles. *J Biol Chem*, **278**, 3937–47.

McCully, K.S., 1969, Vascular pathology of homocysteinemia: implications for the pathogenesis of arteriosclerosis. *Am J Pathol*, **56**, 111–28.

Mercie, P., Garnier, O., Lascoste, L., Renard, M., Closse, C., Durrieu, F., Marit, G., Boisseau, R.M., and Belloc, F., 2000, Homocysteine-thiolactone induces caspase-independent vascular endothelial cell death with apoptotic features. *Apoptosis*, **5**, 403–11.

Mikael, L.G., Genest, J., Jr., and Rozen, R., 2006, Elevated homocysteine reduces apolipoprotein A-I expression in hyperhomocysteinemic mice and in males with coronary artery disease. *Circ Res*, **98**, 564–71.

Mudd, S.H., Levy H.L., and J.P., K., 2001, Disorders of transsulfuration. In *The metabolic and Molecular Bases of Inherited Disease* (C. Scriver, Beaudet al., Sly WS et al., eds.) Mc Graw-Hill, New York, Vol. 2, pp. 2007–56.

Najib, S., and Sanchez-Margalet, V., 2005, Homocysteine thiolactone inhibits insulin-stimulated DNA and protein synthesis: possible role of mitogen-activated protein kinase (MAPK), glycogen synthase kinase-3 (GSK-3) and p70 S6K phosphorylation. *J Mol Endocrinol*, **34**, 119–26.

Perla, J., Undas, A., Twardowski, T., and Jakubowski, H., 2004, Purification of antibodies against N-homocysteinylated proteins by affinity chromatography on Nomega-homocysteinyl-aminohexyl-Agarose. *J Chromatogr B Analyt Technol Biomed Life Sci*, **807**, 257–61.

Qujeq, D., Omran, T.S., and Hosini, L., 2001, Correlation between total homocysteine, low-density lipoprotein cholesterol and high-density lipoprotein cholesterol in the serum of patients with myocardial infarction. *Clin Biochem*, **34**, 97–101.

Reuben, D.M., and Bruice, T.C., 1976, Reaction of thiol anions with benzene oxide and malachite green. *J Am Chem Soc*, **98**, 114–21.

Robert, K., Chasse, J.F., Santiard-Baron, D., Vayssettes, C., Chabli, A., Aupetit, J., Maeda, N., Kamoun, P., London, J., and Janel, N., 2003, Altered gene expression in liver from a murine model of hyperhomocysteinemia. *J Biol Chem*, **278**, 31504–11.

Rosenblat, M., Gaidukov, L., Khersonsky, O., Vaya, J., Oren, R., Tawfik, D.S., and Aviram, M., 2006, The catalytic histidine dyad of high density lipoprotein-associated serum paraoxonase-1 (PON1) is essential for PON1-mediated inhibition of low density lipoprotein oxidation and stimulation of macrophage cholesterol efflux. *J Biol Chem*, **281**, 7657–65.

Rosenquist, T.H., Ratashak, S.A., and Selhub, J., 1996, Homocysteine induces congenital defects of the heart and neural tube: effect of folic acid. *Proc Natl Acad Sci U S A*, **93**, 15227–32.

Roybal, C.N., Yang, S., Sun, C.W., Hurtado, D., Vander Jagt, D.L., Townes, T.M., and Abcouwer, S.F., 2004, Homocysteine increases the expression of vascular endothelial growth factor by a mechanism involving endoplasmic reticulum stress and transcription factor ATF4. *J Biol Chem*, **279**, 14844–52.

Rozek, L.S., Hatsukami, T.S., Richter, R.J., Ranchalis, J., Nakayama, K., McKinstry, L.A., Gortner, D.A., Boyko, E., Schellenberg, G.D., Furlong, C.E., and Jarvik, G.P., 2005, The correlation of paraoxonase (PON1) activity with lipid and lipoprotein levels differs with vascular disease status. *J Lipid Res*, **46**, 1888–95.

Sass, J.O., Nakanishi, T., Sato, T., Sperl, W., and Shimizu, A., 2003, S-homocysteinylation of transthyretin is detected in plasma and serum of humans with different types of hyperhomocysteinemia. *Biochem Biophys Res Commun*, **310**, 242–6.

Sauls, D.L., Lockhart, E., Warren, M.E., Lenkowski, A., Wilhelm, S.E., and Hoffman, M., 2006, Modification of fibrinogen by homocysteine thiolactone increases resistance to fibrinolysis: a potential mechanism of the thrombotic tendency in hyperhomocysteinemia. *Biochemistry*, **45**, 2480–7.

Shih, D.M., Gu, L., Xia, Y.R., Navab, M., Li, W.F., Hama, S., Castellani, L.W., Furlong, C.E., Costa, L.G., Fogelman, A.M., and Lusis, A.J., 1998, Mice lacking serum paraoxonase are susceptible to organophosphate toxicity and atherosclerosis. *Nature*, **394**, 284–7.

Shih, D.M., Xia, Y.R., Wang, X.P., Miller, E., Castellani, L.W., Subbanagounder, G., Cheroutre, H., Faull, K.F., Berliner, J.A., Witztum, J.L., and Lusis, A.J., 2000, Combined serum paraoxonase knockout/apolipoprotein E knockout mice exhibit increased lipoprotein oxidation and atherosclerosis. *J Biol Chem*, **275**, 17527–35.

Spence, A.M., Rasey, J.S., Dwyer-Hansen, L., Grunbaum, Z., Livesey, J., Chin, L., Nelson, N., Stein, D., Krohn, K.A., and Ali-Osman, F., 1995, Toxicity, biodistribution and radioprotective capacity of L-homocysteine thiolactone in CNS tissues and tumors in rodents: comparison with prior results with phosphorothioates. *Radiother Oncol*, **35**, 216–26.

Spence, J.D., Bang, H., Chambless, L.E., and Stampfer, M.J., 2005, Vitamin Intervention For Stroke Prevention trial: an efficacy analysis. *Stroke*, **36**, 2404–9.

Stampfer, M.J., Malinow, M.R., Willett, W.C., Newcomer, L.M., Upson, B., Ullmann, D., Tishler, P.V., and Hennekens, C.H., 1992, A prospective study of plasma homocyst(e)ine and risk of myocardial infarction in US physicians. *Jama*, **268**, 877–81.

Stefani, M., 2004, Protein misfolding and aggregation: new examples in medicine and biology of the dark side of the protein world. *Biochim Biophys Acta*, **1739**, 5–25.

Teiber, J.F., Draganov, D.I., and La Du, B.N., 2003, Lactonase and lactonizing activities of human serum paraoxonase (PON1) and rabbit serum PON3. *Biochem Pharmacol*, **66**, 887–96.

Teiber, J.F., Draganov, D.I., and La Du, B.N., 2004, Purified human serum PON1 does not protect LDL against oxidation in the in vitro assays initiated with copper or AAPH. *J Lipid Res*, **45**, 2260–8.

Tomas, M., Senti, M., Garcia-Faria, F., Vila, J., Torrents, A., Covas, M., and Marrugat, J., 2000, Effect of simvastatin therapy on paraoxonase activity and related lipoproteins in familial hypercholesterolemic patients. *Arterioscler Thromb Vasc Biol*, **20**, 2113–9.

Tward, A., Xia, Y.R., Wang, X.P., Shi, Y.S., Park, C., Castellani, L.W., Lusis, A.J., and Shih, D.M., 2002, Decreased atherosclerotic lesion formation in human serum paraoxonase transgenic mice. *Circulation*, **106**, 484–90.

Undas, A., Perla, J., Lacinski, M., Trzeciak, W., Kazmierski, R., and Jakubowski, H., 2004, Autoantibodies against N-homocysteinylated proteins in humans: implications for atherosclerosis. *Stroke*, **35**, 1299–304.

Undas, A., Jankowski, M., Twardowska, M., Padjas, A., Jakubowski, H., and Szczeklik, A., 2005, Antibodies to N-homocysteinylated albumin as a marker for early-onset coronary artery disease in men. *Thromb Haemost*, **93**, 346–50.

Undas, A., Stepien, E., Glowacki, R., Tisonczyk, J., Tracz, W., and Jakubowski, H., 2006a, Folic acid administration and antibodies against homocysteinylated proteins in subjects with hyperhomocysteinemia. *Thromb Haemost*, **96**, 342–7.

Undas, A., Brozek, J., Jankowski, M., Siudak, Z., Szczeklik, A., and Jakubowski, H., 2006b, Plasma homocysteine affects fibrin clot permeability and resistance to lysis in human subjects. *Arterioscler Thromb Vasc Biol*, **26**, 1397–404.

Van Aerts, L.A., Klaasboer, H.H., Postma, N.S., Pertijs, J.C., Peereboom, J.H., Eskes, T.K., and Noordhoek, J., 1993, Stereospecific in vitro embryotoxicity of L-homocysteine in pre- and post-implantation rodent embryos. *Toxic in Vitro*, **7**, 743–9.

Wald, D.S., Law, M., and Morris, J.K., 2002, Homocysteine and cardiovascular disease: evidence on causality from a meta-analysis. *Bmj*, **325**, 1202.

Yap, S., Boers, G.H., Wilcken, B., Wilcken, D.E., Brenton, D.P., Lee, P.J., Walter, J.H., Howard, P.M., and Naughten, E.R., 2001, Vascular outcome in patients with homocystinuria due to cystathionine beta-synthase deficiency treated chronically: a multicenter observational study. *Arterioscler Thromb Vasc Biol*, **21**, 2080–5.

Zimny, J., Sikora, M., Guranowski, A., and Jakubowski, H., 2006, Protective mechanisms against homocysteine toxicity: the role of bleomycin hydrolase. *J Biol Chem*, **281**, 22485–92.

CHAPTER 6

THE ROLE OF PON2 AND PON3 IN ATHEROSCLEROSIS AND RELATED TRAITS

N. BOURQUARD[1], D.M. SHIH[1], C.J. NG[1], N. VILLA-GARCIA[1],
K. NAKAMURA[3], D.A. STOLTZ[2], E. OZER[2], V. GRIJALVA[1], N.
ROZENGURT[4], S.Y. HAMA, J. ZABNER[2], M. NAVAB[1],
A.M. FOGELMAN[1] AND S.T. REDDY[1]

[1] *Atherosclerosis Research Unit, Department of Medicine, David Geffen School of Medicine at UCLA, Los Angeles, CA, USA*
[2] *Department of Internal Medicine, Roy J. and Lucille A. Carver College of Medicine, University of Iowa, Iowa City, IA, USA*
[3] *School of Medicine, UCSF, San Francisco, CA, USA*
[4] *Department of Pathology and Laboratory Medicine, UCLA, Los Angeles, CA, USA*

Abstract: The paraoxonase (PON) gene family consists of three members, PON1, PON2 and PON3. All PON proteins have been implicated in the pathogenesis of several inflammatory diseases and all share a capacity to protect cells from oxidative stress. However, their mechanism of action is currently unknown and has been the focus of a great deal of research in recent years. The aim of this review is to summarize some of the recent findings on the antioxidant properties, localization, and regulation of the PON proteins, with an emphasis on the protective roles of PON2 and PON3 in mouse models of atherosclerosis. In addition, we will describe a potential novel role for PON proteins in host defense against gram-negative microbial infection.

Epidemiological and biochemical studies have long associated PON2 and PON3 with inflammatory diseases, in particular atherosclerosis. Several transgenic mouse models have recently been developed that not only reinforce the antioxidant properties demonstrated in *in vitro* studies, but offer new clues as to their physiological function(s). Diet-induced atherosclerosis in C57Bl6/J mice results in a dramatic increase in lesion development in PON2 deficient mice whereas adenoviral-mediated human PON2 expression in apoE null mice effectively protects against atherosclerosis. Not only was there increased macrophage trafficking in the artery wall of PON2-def mice, but the macrophages exhibited exacerbated oxidative stress and inflammatory response. In addition, the increase in atherosclerosis in PON2-def mice was associated with an uncharacteristic decrease in secretion of apoB containing VLDL/LDL.

Address correspondence to: Srinivasa T. Reddy, PhD. Atherosclerosis Research Unit, David Geffen School of Medicine at UCLA, Los Angeles, CA, 90095. Fax: 310-206-3605; Tel: 310-206-3915; E-mail: sreddy@mednet.ucla.edu.

B. Mackness et al (eds.), The Paraoxonases: Their Role in Disease Development and Xenobiotic Metabolism, 103–128.
© 2008 *Springer.*

Although PON2 and PON3 are highly conserved and share similar anti-oxidant capacities, differences in enzymatic activities, localization, and regulation suggest that their physiological function is far from redundant. Further, while both human and mouse PON3 are highly expressed in the liver, human PON3 associates with HDL in the circulation while mouse PON3 appears to remain cell-associated with an expression pattern that overlaps that of PON2. Recently, an *in vivo* athero-protective role for PON3 has been demonstrated in several mouse models, including adenoviral-mediated delivery and a transgenic mouse model. Interestingly, the significant decrease in atherosclerosis observed in the latter model was only associated with male cohorts and was additionally associated with decreased obesity.

Finally, we review recent evidence of the potential novel role of PON proteins in host defense based on the recent discovery that PON proteins hydrolyze a class of virulence factors specifically expressed in particular gram-negative microbial infections

Keywords: Paraoxonase 2, paraoxonase 3, atherosclerosis, quorum quencher, lactonase, inflammation, oxidative stress

1.   INTRODUCTION

Infection, inflammation and oxidative stress correlate with mechanisms underlying a variety of diseases, including atherosclerosis, diabetes mellitus, systemic lupus erythematosus, rheumatoid arthritis and several ailments associated with age. A number of proteins have evolved for maintaining oxidative homeostasis in cells and tissues. *Understanding the biology and function of such proteins will pave way for the discovery of novel therapeutic agents in the fight against inflammatory diseases such as atherosclerosis.*

The paraoxonase (PON) gene family consists of three members, PON1, PON2 and PON3. All PON proteins have been implicated in the pathogenesis of several inflammatory diseases including atherosclerosis, Alzheimer's, Parkinson's, diabetes and cancer. A property shared by all PON proteins is a capacity to protect cells from oxidative stress, perhaps as cellular antioxidants; however their mechanism of action is currently unknown and has been the focus of a great deal of research in recent years. In this review, we will summarize some of the recent findings on the antioxidant properties, localization, and regulation of the PON proteins, with an emphasis on the athero-protective role of PON2 and PON3 in mouse models of atherosclerosis. In addition, we will describe a potential novel role for PON proteins in host defense against gram-negative microbial infection.

1.1.   The PON Gene Family

The PON gene family consists of three members, PON1, PON2, and PON3, located adjacent to each other on the long arm of chromosome 7 in human and chromosome 6 in mice (Primo-Parmo et al., 1996). The three human PON (hPON) genes share approximately 65% similarity at the amino acid level, and approximately 70% similarity at the nucleotide level. From a molecular evolutionary standpoint, PON2 appears to be the oldest member, followed by PON3 and PON1 (Draganov and La

Du, 2004). In humans, PON1 mRNA expression is limited to the liver. Similarly, hPON3 is expressed primarily in the liver, with expression also seen in the kidneys (Reddy et al., 2001). In contrast, hPON2 is more widely expressed and is found in a variety of tissues including the heart, kidney, liver, lung, placenta, small intestine, spleen, stomach, and testis (Ng et al., 2001; Ng et al., 2005; Primo-Parmo et al., 1996). Furthermore, unlike hPON1 and hPON3, hPON2 message is also detected in the cells of the artery wall, including endothelial cells, smooth muscle cells, and macrophages (Ng et al., 2001). In addition, while PON1 and PON3 associate with HDL in the circulation (Ng et al., 2005), by western blot, PON2 protein is undetectable in HDL, LDL, or the media of cultured cells, but appears to remain intracellular, associated with membrane fractions (Ng et al., 2001; Ng et al., 2005).

1.2.   Lactonase Activity is Common to All PON Proteins

Recent findings suggest that the name PON is in fact a misnomer, since PON2 and PON3 lack any significant paraoxonase activity (Draganov et al., 2000; Ng et al., 2001; Reddy et al., 2001). At the molecular level, PON1, PON2, and PON3 do however share an ability to hydrolyze aromatic and long-chain aliphatic lactones, and thus the term lactonase may be more appropriate (Draganov and La Du, 2004; Draganov et al., 2005). One class of lipo-lactones, acyl homoserine lactones (AHL), has recently been the focus of considerable attention due to their significant role in gram-negative microbial infection, thereby possibly associating the PON proteins with innate host immunity. Studies by Kobayashi et al. (Kobayashi et al., 1998) were the first to implicate this new enzyme activity for the PON proteins. Kobayashi et al. (Kobayashi et al., 1998) reported the cloning of a lactonehydrolase (lactonase) from *Fusarium oxysporum* AKU 3702 and discovered sequence similarities between the lactonase and human and rabbit PON1 gene. Jakuboski et al. (Jakubowski, 2000) demonstrated that PON1 indeed exhibits lactonase activity. Subsequently, the separation of rabbit serum PON3 from PON1 was facilitated both by the lactonase activity of PON3 as well as the lack of significant paraoxonase activity in PON3. Thus, the ability to hydrolyze lactones was proposed to be one of the common features shared by PON1 and PON3 (Draganov et al., 2000; Kobayashi et al., 1998). Recent studies have shown that PON1, PON2, and PON3 possess the ability to hydrolyze lactones with the three PON proteins exhibiting overlapping but also distinct substrate specificities (Draganov et al., 2005). More interestingly, PON proteins hydrolyze a number of AHL, molecules that mediate bacterial quorum-sensing signals which are important in regulating expression of virulence factors and in inducing a host inflammatory response (Draganov et al., 2005; Ozer et al., 2005; Yang et al., 2005).

2.   PON1

Initially characterized for its ability to hydrolyze organophosphates, the name, paraoxonase (PON1), reflects its ability to hydrolyze paraoxon, a metabolite of the insecticide parathion. PON1 is a calcium dependent esterase consisting of 354 amino acids with a molecular mass of approximately 45 kDa (Furlong et al., 1989; Hassett et al., 1991). In addition to paraoxon, PON1 has been shown to hydrolyze metabolites of a number of other insecticides and to also detoxify various nerve agents (Costa et al. 1999).

2.1. PON1 Gene Polymorphisms and Disease Association

To date, PON1 has two common coding polymorphisms, a methionine to leucine substitution at codon 55 (M/L55), and a glutamine to arginine substitution at position 192 (Q/R192). Both polymorphisms have been associated with a number of pathophysiological conditions. In addition to these two coding polymorphisms, at least five polymorphisms have been detected in the PON1 promoter region. Population studies and studies using reporter gene constructs in cells suggest that these polymorphisms affect expression levels of PON1 (Brophy et al., 2001; Leviev and James, 2000). The M/L55 coding polymorphism has been associated with stroke, coronary artery disease, Parkinson's disease, variations in plasma levels of total and LDL cholesterol, and in levels of PON1 message, protein, and activity towards paraoxon (Akhmedova et al., 2001; Fortunato et al., 2003; Kriska et al., 2007; Leviev and James, 2000; Mackness et al., 1997; Malin et al., 2001). The Q/R192 polymorphism has been found to affect the rates of hydrolysis towards various substrates including paraoxon, diazoxon, soman, and sarin (Costa et al., 2003; Davies et al., 1996; Furlong et al., 2002). In addition, the Q/R192 polymorphism has been associated with coronary artery disease, stroke, type 2 diabetes, and Parkinson's disease (Chen et al., 2003; Hegele et al., 1995; Hu et al., 2003; Kondo and Yamamoto, 1998; Odawara et al., 1997; Pfohl et al., 1999; Voetsch et al., 2002).

2.2. PON1 Enzymatic Activities and Antioxidant Properties

PON1 has been reported to have a number of other enzymatic activities including arylesterase, lactonase, peroxidase and phospholipase A2-like activities (Aviram et al., 1998; Rodrigo et al., 2001; Rozenberg et al., 2003; Teiber et al., 2003; Watson et al., 1995). One mechanism by which PON1 is believed to exert its anti-atherogenic effects is by hydrolyzing biologically active oxidized phospholipids and destroying lipid hydroperoxides. Whether PON1 indeed has peroxidase and phospholipase A2-like activities has been questioned (Marathe et al., 2003), with these activities being attributed to a contaminating protein(s) in PON1 preparations used in previous studies. Indeed, attempts to crystallize PON1 from 'purified preparations' led to the crystallization of a protein that co-purified with PON1 (Fokine et al., 2003). However, Harel and colleagues studied the crystal structure of a recombinant variant of PON1 obtained by directed evolution (Harel et al., 2004). In their manuscript, Harel and colleagues illustrate that PON1 is a six-bladed β-propeller, which structurally resembles the *Loligo vulgaris* DFPase (Harel et al., 2004). In addition, PON1's active site and catalytic mechanism is reminiscent of secreted phospholipase A2, supporting previous reports that PON1 has phospholipase A2-like activities.

2.3. PON1 and Atherosclerosis

Throughout the past decade, PON1 has been demonstrated to protect LDL against oxidation, to reverse the biological effects of oxidized LDL, and to preserve the function of HDL by inhibiting its oxidation (Aviram et al., 1998; Durrington et al., 2001; Mackness et al., 1993; Watson et al., 1995). PON1 null mice by gene targeting (Shih et al., 1998) and transgenic mice (Ng et al., 2005; Oda et al., 2002; Tward et al., 2002) corroborate the hypothesis that PON1 protects against atherogenesis and is an important contributor to HDL's antioxidant capacity. These *in vivo* studies underscore the potential of PON1 as a therapeutic agent to prevent atheroma.

3. PON2

PON2 is a widely expressed intracellular protein with a molecular mass of approximately 44 kDa (Ng et al., 2001; Primo-Parmo et al., 1996). While little is known about the physiological or pathophysiological role of this protein, PON2 has been reported to possess antioxidant properties. Stably transfected cells overexpressing PON2 exhibit significantly lower levels of intracellular oxidative stress when exposed to hydrogen peroxide or oxidized phospholipids (Ng et al., 2001). Rosenblat and colleagues demonstrated that purified recombinant PON2 can protect LDL against oxidation (Rosenblat et al., 2003). In addition, LDL incubated with stably transfected cells overexpressing PON2 has lower levels of lipid hydroperoxides, and is less able to induce monocyte transmigration through endothelial cells (Ng et al., 2001) Furthermore, PON2 is able to reduce the oxidative stress caused by preformed mildly-oxidized LDL and decrease its ability to induce inflammatory response in cells (Ng et al., 2001). Thus, one function of PON2 may be to act as a cellular antioxidant, protecting cells from oxidative stress. However, the mechanism by which PON2 participates as an antioxidant is unknown and warrants further investigation.

3.1. PON2 Gene Polymorphisms and Disease Association

Similar to PON1, PON2 has several polymorphisms that are associated with a number of pathophysiological conditions. Population studies have yielded a pair of amino acid substitutions with an alanine or glycine at codon 148 (A/G148), and either cysteine or serine at codon 311 (C/S311). The A/G148 polymorphism has been associated with variations in total and LDL cholesterol levels, fasting plasma glucose levels, and birth weight (Hegele et al., 1997; Hegele et al., 1998; Hegele, 1999). In addition, the polymorphism at position 311 has been associated with coronary artery disease, ischemic stroke in patients with type 2 diabetes mellitus, late-onset Alzheimer's disease, and reduced bone mass in post-menopausal women

(Chen et al., 2003; Yamada et al., 2003; Janka et al., 2002; Kao et al., 2002; Martinelli et al., 2004; Pan et al., 2002; Sanghera et al., 1998; Shi et al., 2004; Wang et al., 2003).

3.2. PON2 Enzymatic Activity and Antioxidant Properties

Similar to PON1, characterization of the antioxidant properties of PON2 has predominantly focused on principal mediators of atherogenesis, such as LDL oxidation and HDL oxidative protection. In 2001, Ng et al. (Ng et al., 2001) demonstrated that HeLa cells overexpressing human PON2 were less effective in promoting oxidative modification LDL compared to control cells, (Ng et al., 2001), and reduced oxidative stress caused by preformed mildly-oxidized LDL (Ng et al., 2001). This protective capacity against LDL mediated cellular oxidation was supported by later studies that demonstrated that addition of purified recombinant PON2 to apoE null derived peritoneal macrophages reduced cell-mediated LDL oxidation (Rosenblat et al., 2003). However, studies from several laboratories including our own, suggest that oxidized phospholipids or lipid peroxidation products are perhaps not the direct substrates for purified PON enzymes (Connelly et al., 2005; Kriska et al., 2007; Marathe et al., 2003). For instance, PON proteins purified from an insect system (kindly provided by D. Dragonov) do not confer phospholipase activity reminiscent of any of the known phospholipases that act on phospholipids (PLA2, PLC and PLD). Our laboratory has performed detailed profiling experiments following the action of PON proteins on biologically active oxidized phospholipids present in MM-LDL and we have determined that the PONs do not hydrolyze oxidized phospholipids such as Ox-PAPC as previously thought (Fig. 1).

Using an alternative approach, we analyzed the lipid profiles of cell supernatants and lysates from adenoviral-mediated overexpression of human PON proteins by electron spray ionization mass spectromentry (ESI-MS) but did not observe any differences in the pro-inflammatory oxidized lipids between the empty vector and any of the PON proteins in the distribution of phospholipids (positive ion ESI-MS) or fatty acids (negative ion ESI-MS) (data not shown). However, differences do exist in other uncharacterized lipid peaks that are currently being investigated suggesting that the substrates for PON enzymes are either different from the atherogenic oxidized phospholipids and/or changes in LDL oxidation are secondary to the action of PON proteins.

3.3. Tissue Distribution and Cellular Localization of Mouse and Human PON2

Characterizing the distribution and expression pattern of PON2 represents an important step in terms of identifying a physiological function and possibly identifying a substrate(s) for this orphan enzyme. Human PON2 is widely expressed and found in a variety of tissues including the heart, kidney, liver, lung, placenta,

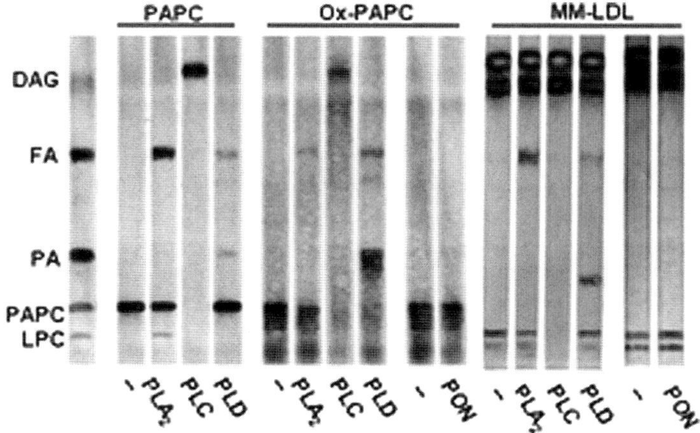

Figure 1. PON proteins do not have lipase activity. Ox-PAPC and MM-LDL were subjected to the respective treatments for 5 hours at 37 °C, extracted and resolved immediately by TLC, charred at 120 °C for 10 min. DAG; Diacyl glycerol, FA; Fatty Acid, PA; Phosphatidic Acid, PAPC; Palmitoyl Arachidonyl phosphatidyl choline, LPC; Lyso-PC, Ox-PAPC; MM-LDL; Minimally-Modified LDL, PLA$_2$; Phospholipase A$_2$, PLC; Phospholipase C, PLD; Phospholipase D, PON; Paraoxonase, –; buffer alone

small intestine, spleen, stomach, and testis (Ng et al., 2001; Ng et al., 2005; Primo-Parmo et al., 1996). Furthermore, unlike human PON1, human PON2 message is also detected in the cells of the artery wall, including endothelial cells, smooth muscle cells, and macrophages (Ng et al., 2001). In addition, while PON1 and PON3 associate with HDL in the circulation (Ng et al., 2005), PON2 protein is undetectable in HDL, LDL, or cell supernatants by western blot, but appears to remain intracellular, associated with membrane fractions (Ng et al., 2001; Ng et al., 2005). Recently developed PON2 deficient (PON2-def) mice carry an insertional mutation containing a lacZ gene under the control of the PON2 promoter and we were able to exploit this design to determine cell-type specific distribution of PON2 using X-gal staining of frozen sections.

In terms of tissue distribution, the expression pattern of mouse PON2 was essentially identical to that reported for human PON2 (Ng et al., 2001). However, further analysis revealed that PON2 expression was surprisingly not detected in all cell types, but found predominantly and specifically localized to epithelial cells and macrophages in all tissues tested. In mouse lung tissue (Fig. 2), lacZ expression was mainly localized in the epithelial cells of the bronchioles, in the cytoplasm of both apical and basal cells of this epithelium.

Also, PON2 was expressed in alveolar macrophages as determined by colocalization of β-gal activity with a macrophage specific marker, CD68. In mouse kidney tissue (Fig. 3), the distribution of X-Gal staining was noted in the mesangial cells and endothelial cells. In mouse heart tissue (Fig. 4), the staining of X-Gal was found in smooth muscle cells in arterial walls and focal staining in the myocardial cells. In

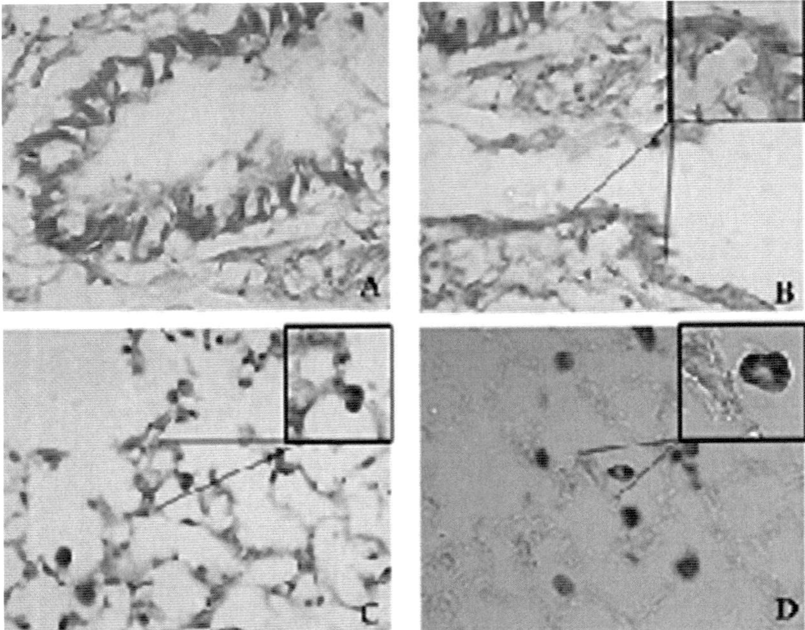

Figure 2. Localization of PON2 in cryosections of PON2-def mouse hung. (A-C) X-Gal counterstained with Nuclear Fast Red. PON2 is localized in the epithelial cells of the bronchiole (A, x40), in the smooth muscle layer of the arteries (B, x40) and in alveolar macrophages (C, x20). (D) Macrophages labeled with CD68 antibody (x20)

spleen tissue (Fig. 5) most of the blue staining was in the red pulp. In the white pulp, only the smooth muscle of the central arterioles show staining. Megakaryocytes are also PON2 positive. In mouse brain tissue (Fig. 6), the cell type that stains mainly for PON2 is the Purkinje cell of the cerebellum. Consistent staining is also found in the epithelium and endothelial cells of the choroid plexuses. The endothelium of a small number of capillaries and larger vessels in all areas of the brain also shows sporadic staining. In mouse liver tissue (Fig. 7), apart from hepatocytes a small proportion of Kupffer cells do stain as well as the endothelial lining of vascular structures. Current emphasis in our laboratory involves determining the specific membrane-associated subcellular localization of PON2 within these cell types.

3.4.   Regulation of PON2 Expression

The robust expression of PON2 in the resident macrophages of various tissues becomes extremely relevant when one considers i) the role of macrophages in inflammatory diseases and ii) the adverse effects of oxidative stress on macrophage function. Not only does PON2 have an anti-oxidant capacity, but we and others have shown (Ng et al., 2001; Rosenblat et al., 2003) that in response to an increase in

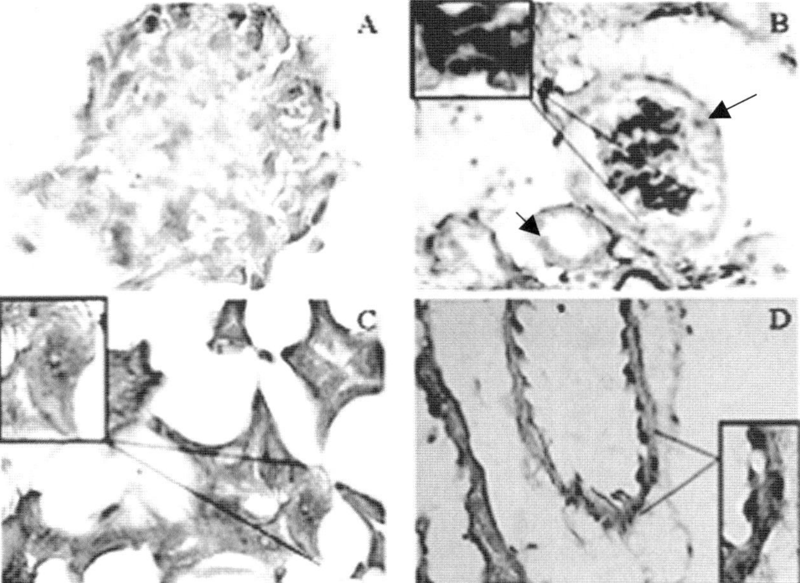

Figure 3. Localization of PON2 in cryosections of PON2-def mouse kidney. (A, C and D) X-Gal counterstained with Nuclear Fast Red. (A) Staining of glomerular mesangium (x100), (C) granular staining in the tubular epithelium (x100), and (D) staining of vascular smooth muscle cells (x100). (B) Double staining of X-Gal counterstained with Nuclear Fast Red and PECAM-1 immunostain, showing the localization of X-Gal staining (arrows) in the mesangium (x40)

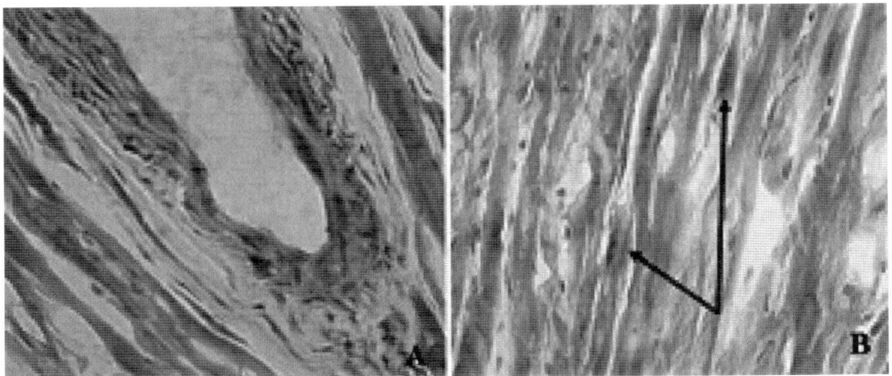

Figure 4. Localization of PON2 in cryosections of PON2-def mouse heart. X-Gal counterstained with Nuclear Fast Red. (A) PON2 is localized in the media layer of the arteries and (B) in the myocardial cells (x100, arrows)

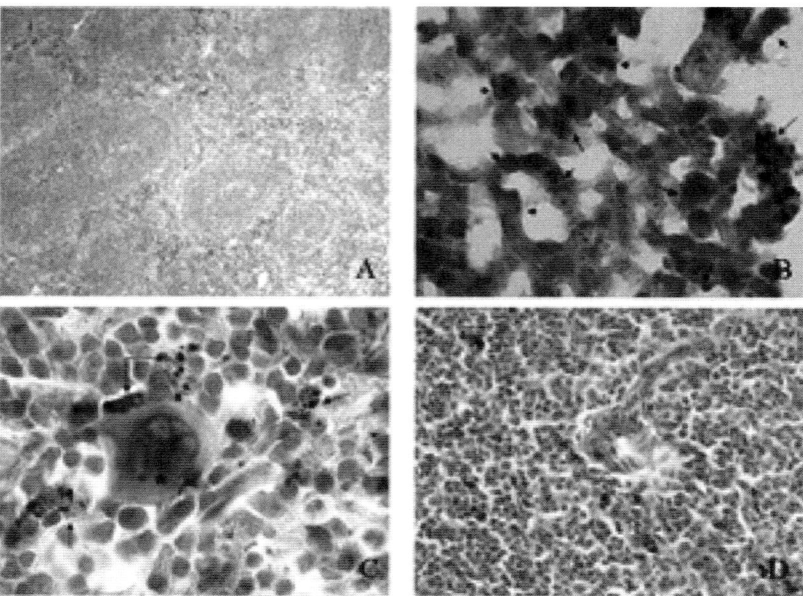

Figure 5. Localization of PON2 in cryosections of PON2-def mouse spleen. (A-D) X-Gal counterstained with Nuclear Fast Red. (A) Staining is primarily localized in the red pulp (x4). PON2 is typically localized in macrophages (B, x100, arrows) and megacaryocytes (C, x100, star). (D) In the white pulp, the staining is limited to the smooth muscle of the central arterioles (x40)

oxidative stress elicited by numerous agents, PON2 expression and lactonase activity correspondingly increase. Shiner et al., (Shiner et al., 2006) recently demonstrated that a moderate decrease in activity actually precedes the sharp increase in PON2 activity associated with increasing oxidative stress.

An increase in oxidative stress is also associated with monocyte differentiation and Rosenblat et al (Cathcart, 2004; Rosenblat et al., 2003; Shiner et al., 2004) have shown that PON2 expression increases in parallel with the production of NADPH-oxidase dependent macrophage superoxide anions. This mechanism further appears to play a role upstream of PON2 induction as peritoneal macrophages from P47$^{phox-/-}$ mice (inactive NADPH oxidase) have a significant reduction in PON2 expression. Superoxide anion production is involved in the regulation of multiple cellular functions, including cell-mediated oxidation by LDL. The oxidative stress inducible transcriptional activator, AP-1, is one such target of this system and the human PON2 promoter, which contains three putative AP-1 response elements, was further shown to be sensitive to direct AP-1 mediated activation. Aside from this valuable contribution, PON2 promoter regulation is an aspect that has not been effectively characterized and would lend considerable insight into PON2 function. To characterize oxidative stress induced PON gene expression *in vivo* and in the context of atherosclerosis, we quantified hepatic message levels in both C57Bl6/J

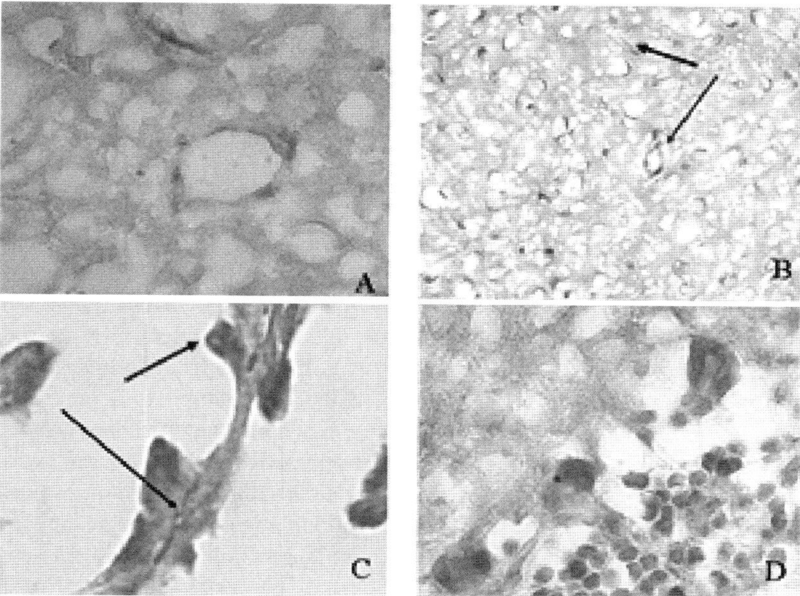

Figure 6. Localization of PON2 in cryosections of PON2-def mouse brain. (A-C and E) X-Gal counterstained with Nuclear Fast Red. (A and B) PON2 is localized in endothelial layers of vessels and capillaries (arrows) (x40 and x20), (C) epithelial cells of the choroids plexus and their capillary vessels (arrows) (x40) and (E) cerebellar Purkinje cells (x100)

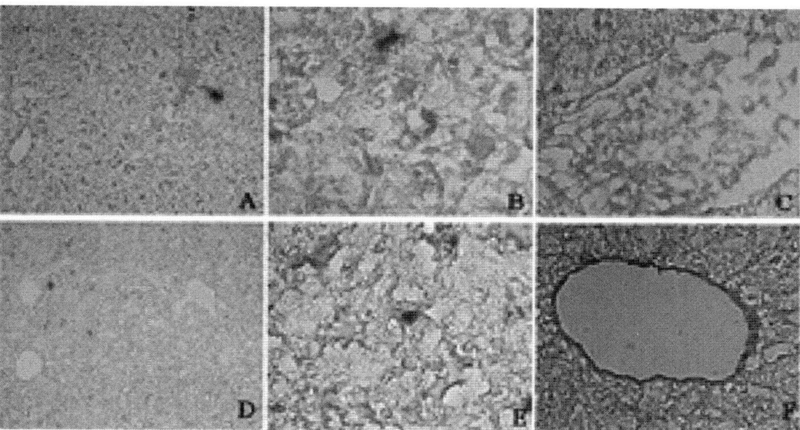

Figure 7. Localization of PON2 in cryosections of PON2-def mouse liver. (A-C) X-Gal counterstained with Nuclear Fast Red. (D, E) CD68 and (F) CD31 (PECAM-1) immunostain. (A, x20) and (B, x100) Staining is primarily localized in the Kuppfer cells and in the endothelial cell of the vessels (C, x40). (D, E) Immunoreactivity of Kuppfer cells with CD68. (F) Immunoreactivity of vascular endothelium with PECAM

and apoE null mice fed a high fat diet for 15 weeks or 8 weeks, respectively (Ng et al., 2005). Quantification of hepatic PON2 expression showed an approximate 1.7-fold increase in expression as a consequence of both diet and genetic background (Ng et al., 2005), confirming *in vitro* findings.

3.5.   PON2 and Atherosclerosis

Atherosclerosis is a disease of chronic inflammation and lipid accumulation. Increasing evidence suggests that oxidative stress plays a key role in the pathogenesis of this disease. More specifically, the oxidation of LDL has been shown to trigger a number of pro-inflammatory events that initiate and exacerbate the atherogenic process (Navab et al., 2002). Atherosclerosis arises from an interaction between the blood and the endothelial surface of the vasculature. It is currently believed that the important components in the blood that contribute to atherogenesis are the plasma lipids and monocytes. PON2 is prominently expressed in monocytes and influences lipoprotein properties and cellular oxidation *in vitro*. To assess the anti-atherogenic capacity of PON2 *in vivo*, PON2-def mice and their wild type C57Bl6/J counterparts were placed on a high fat, high cholesterol and cholate containing diet (Ng et al., 2006).

3.5.1.   PON2-def mice on an atherogenic diet have decreased VLDL/LDL

PON2-def mice developed significantly larger atheromatous lesions in the aortas than their wild type counterparts (Ng et al., 2006). Although high circulating levels of apoB containing cholesterol particles are an established risk factor for coronary heart disease, the PON2-def mice, however exhibited significantly lower levels of total (32%) and VLDL/LDL cholesterol (46%), with no significant differences in fasting levels of HDL cholesterol, triglycerides, free-fatty acids, or glucose. To underscore this disparity, we see that PON2-def mice develop 3.8 times more lesion area than their wild type counterparts when LDL-cholesterol is taken out of the equation (Fig. 8).

Serum cholesterol profiling did not reveal any significant differences in lipoprotein particle size between wild type and PON2-def mice, but western blot analysis demonstrated lower levels of serum apoB protein in the PON2-def mice. Hepatic synthesis of apoB is constitutive and several co- and post- translational mechanisms have been shown to modulate its secretion. The post-ER pre-secretory proteolysis pathway (PERPP) is one pathway that is particularly relevant to this mouse model as it is stimulated by oxidant stress (Fisher et al., 2001; Pan et al., 2004). Because PON2 is expressed in the liver and has also been shown to possess antioxidant capacities, we hypothesized that the lower levels of apoB in the serum of PON2-def mice may be due to an increase in apoB degradation via upregulation of the PERPP pathway by oxidative stress. Indeed, the livers of PON2 deficient mice exhibited higher levels of oxidative stress, containing significantly more lipid hydroperoxides relative to their control counterparts.

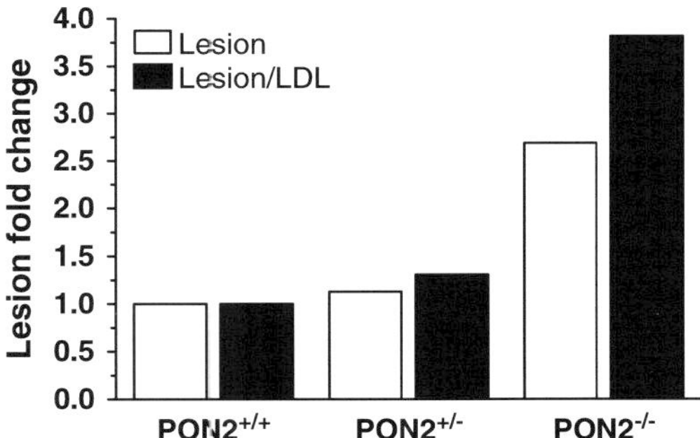

Figure 8. Lesion fold changes and lesion:LDL ratio in PON2 heterozygote and homozygote mice relative to wild type counterparts fed high fat, high cholesterol, and cholate-containing diet for 15 weeks

3.5.2.   *Adenovirus mediated expression of human PON2 protects against the development of atherosclerosis in apoE deficient mice*

An *in vivo* anti-atherogenic role for PON2 was also demonstrated by adenoviral-mediated transient overexpression of human PON2 (AdPON2) in six month old apoE null mice. We showed significant reduction in aortic lesion burden just three weeks following a single intravenous administration of recombinant adenovirus expressing human PON2 in apoE null mice. This suppression of atheroma formation in the aortic sinus was however not associated with significant differences in total, HDL, and non-HDL cholesterol levels (Ng et al., 2006). It remains to be determined if long-term overexpression of human PON2 will have an influence on lipoprotein metabolism, perhaps in terms of prolonged and increased protection against oxidative stress thereby preventing activation of the PERPP pathway.

3.6.   Lipoprotein Function in PON2 Mouse Models of Atherosclerosis

In all the animal models studied to date the effects of PONs on atherosclerosis have been associated with changes in LDL oxidation status as determined by LDL's lipid peroxidation and the presence or absence of atherogenic oxidized phospholipids. Increasing evidence suggests that it is not just the quantity but also the functional quality of the lipoproteins that is important in determining the risk of atherosclerosis (Ansell et al., 2003; Tsimikas et al., 2005).

The serum from AdPON2 treated mice exhibited an increase in antioxidant capacity, containing significantly lower levels of lipid hydroperoxides compared to mice treated with either PBS or the empty vector. In addition, serum from AdPON2 treated mice demonstrated an enhanced ability to efflux cholesterol from cholesterol-loaded macrophages, LDL isolated from AdPON2 treated mice was less susceptible to oxidation, inducing significantly less monocyte chemotactic activity and HDL isolated from AdPON2 treated mice exhibited enhanced anti-inflammatory properties, significantly inhibiting LDL-induced monocyte chemotactic activity (Ng et al., 2006). Although PON2 is not normally found in the circulation and was not detected in the serum of these mice, plasma proteins and lipoprotein particles may certainly interact with a number of tissues including the liver, which may modify these proteins and lipoproteins.

Results obtained from PON2-def mouse also strongly suggested that one manner in which PON2 protects against atherosclerosis is by modulating the properties of circulating lipoproteins, including the susceptibility of LDL to oxidation and the capacity of HDL to protect LDL against oxidation. LDL isolated from PON2-def mice was more susceptible to oxidation, inducing significantly more monocyte chemotaxis relative to LDL from wild type mice. HDL isolated from PON2 deficient mice was significantly more inflammatory and less able to protect against LDL induced monocyte chemotactic activity compared to HDL from wild type littermates. Unlike PON1, PON2 is a cell-associated enzyme (Ng et al., 2001) and thus its mechanism of action is likely dependent on its function in the cell. The ability of PON2 to alter the properties of LDL and HDL may be mediated through the interactions of these lipoproteins with a number of cell types including macrophages. Indeed, LDL incubated with peritoneal macrophages deficient in PON2 developed significantly higher levels of lipid hydroperoxides than LDL incubated with macrophages from wild type mice.

3.7.   Macrophages are Hypersensitive to Oxidative Stress in PON2-def mice

Unmistakably, there was enhanced macrophage trafficking into the artery wall in the PON2-def mice, as determined by staining macrophages in aortic sections using CD68 as the marker (Ng et al., 2006). Not only were there more macrophages in the artery wall of PON2-def mice, macrophages deficient in PON2 exhibited higher levels of

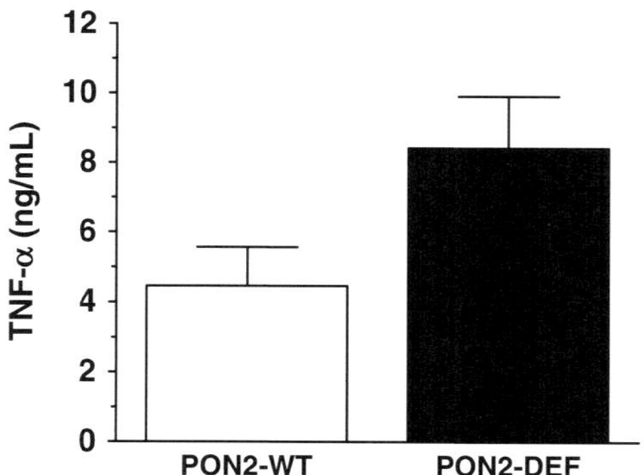

Figure 9. PON2 wild type and deficient mice ($n = 3$) were intraperitoneally injected with 5mg/kg LPS. Serum was collected 2 hours post-injection and TNFα measured by ELISA. $p < 0.05$

oxidative stress and enhanced pro-inflammatory properties (Ng et al., 2006). Peritoneal macrophages isolated from PON2-def mice exhibited higher levels of oxidative stress when treated with a number oxidative stress inducing agents. Moreover, when treated with LPS, PON2-def macrophages exhibited an exacerbated inflammatory response, inducing higher levels of the proinflammatory cytokines TNF-α and IL-1β than their wild type counterparts. The enhanced oxidative stress and proinflammatory state of PON2-def macrophages may help explain the higher levels of atheroma formation in PON2-def mice. An association between PON1 and endotoxemia was demonstrated by La Du et al. (La Du et al., 1999) in which purified PON1, co-administered together with a lethal dose of endotoxin, was shown to be protective. We therefore sought to extend our in vitro findings in an *in vivo* context. We subsequently observed that following intraperitoneal administration of LPS there was higher induction of TNF-α in PON2-def mice compared to wild type controls (Fig. 9).

4.   PON3

Human PON3 is an approximately 40kDa protein that associates with HDL in the circulation, albeit at much lower levels than PON1 (Draganov et al., 2000; Reddy et al., 2001). Of the three PON proteins, PON3 is the most recently identified family member and also the least characterized. Recently, a serine to threonine substitution at codon 311 (S/T311), and a glycine to aspartic acid substitution at codon 324 (G/D324) were detected in a population in southern Italy (Campo et al., 2004). However to date, the functional consequences of these two polymorphisms have not been reported.

4.1.   PON3 Enzymatic Activity and Antioxidant Properties

In contrast to PON1, PON3 has limited arylesterase and altogether lacks any paraoxonase activity. PON3 rapidly hydrolyzes numerous lactones including lovastatin, a statin prodrug, which has been used to assess PON3-selective enzymatic activity in both tissue and cell lysates (Draganov et al., 2005). Similar to PON1 and PON2, PON3 also possesses antioxidant properties (Reddy et al., 2001) although its physiological substrate is unknown. Draganov and colleagues reported that rabbit PON3 purified from serum was capable of inhibiting copper induced LDL oxidation *in vitro* (Draganov et al., 2000). In fact, in the same report, these authors demonstrated that on a per μg basis, rabbit PON3 was approximately 100 times more potent than rabbit PON1 in protecting LDL against oxidation. Reddy et al. (Reddy et al., 2001) showed that lipoproteins incubated with supernatants from cells overexpressing PON3 were less proinflammatory and contained less lipid hydroperoxides.

4.2.   PON3 Expression Pattern

Both human PON1 and PON3 are present on HDL. In contrast, by western blots, mouse PON3 is not detectable on HDL (Reddy et al., 2001). Using a similar immunoblotting approach, human PON3 was easily detected in 10 µg of human HDL, but mouse PON3 was undetectable with as much as 200 µg of mouse HDL (unpublished data). Human and mouse PON3 also exhibit significantly different tissue expression patterns. Northern analysis showed that hPON3 is predominantly expressed in the liver and also moderately expressed in the kidney (Reddy et al., 2001). In contrast, quantitative RT-PCR analysis of mouse PON3 in various tissues showed that mouse PON3 is expressed in a wide range of tissues, with highest expression found in the lungs (Shih et al., 2004). Rosenblat et al. (Rosenblat et al., 2003) have demonstrated the presence of PON3 in murine macrophages, whereas in human macrophages, paraoxonase expression appears to be limited to PON2. The fact that PON3 is detected in the macrophages of mice but not humans (Rosenblat et al., 2003) suggests that mouse PON3 may influence atherogenesis more directly through its expression in the artery wall cells. It remains to be determined how these differences in tissue expression pattern and association with HDL will comparatively affect the roles of mouse and human PON3 on atherogenesis and other biological processes.

4.3.   Regulation of PON3 Expression and Activity

Although mouse PON2 and PON3 proteins have overlapping tissue distribution, the differences in their gene expression and activity in response to environmental factors may result in very specific physiological functions. In the context of atherosclerosis, hepatic expression of murine PON3 does not appear to be significantly affected by genetic background or diet (Ng et al., 2005). The elevated oxidative stress associated with the apoE null background did not have a significant influence on hepatic message levels compared to the wild type C57Bl6/J mice, nor did administration of a high fat, high cholesterol diet with or without cholic acid alter hepatic message levels compared to wild type mice.

Work by Rosenblat et al (Rosenblat et al., 2003) confirmed these findings *ex vivo* in peritoneal macrophages harvested from progressively older apoE null mice. Although age was associated with increasing oxidative stress and the extent of atherosclerosis in this mouse model, there were no significant differences in PON3 expression. However, PON3 associated lovastatinase activity was almost two-fold reduced in four month old compared to two month old apoE null mice. Oxidative stress elicited by numerous agents shown to increase PON2 expression and associated activity in peritoneal macrophages isolated from C57Bl6/J mice all significantly reduced PON3 lovastatinase activity in the membrane enriched protein lysates, with no significant effect on PON3 expression levels (Ng et al., 2001; Rosenblat et al., 2003).

4.4.   PON3 and Mouse Models of Atherosclerosis and Obesity

There has recently been a considerable amount of work aimed at identifying the *in vivo* contribution of PON3 in atherogenesis. Using adenovirus as a vector, we demonstrated that transient expression of human PON3 (AdPON3) was sufficient to suppress the progression of atheromatous lesion formation in twenty-six week old female apoE null mice relative to mice administered the empty vector (Ng et al., 2004). Given the substantial degree of aortic lesions in these mature mice, the approximately 20% reduction in lesion formation was unforeseen.

Adenovirus mediated transgene expression was detected in the livers of mice throughout the course of the study, with expression also detected in lungs, kidneys, spleen and the aorta. Despite this broad tissue distribution, human PON3 was however not detected in serum. Similar to results obtained with transient expression of human PON2, serum lipid levels were not affected by the single administration and short length of the study, but exhibited lower levels of oxidized lipids, containing significantly less lipid hydroperoxides than their control counterparts (Ng et al., 2004).

Although the precise mechanism responsible for this suppression in lesion formation is not known, the reduction in atheroma may be mediated by the ability of PON3 to enhance the anti-atherogenic properties of plasma. In fact, LDL isolated from mice three weeks after injection of AdPON3 was significantly less susceptible to oxidative modification, while HDL demonstrated enhanced anti-inflammatory properties. Furthermore, serum exhibited an enhanced ability to efflux cholesterol from cholesterol-loaded macrophages. Whether PON3 is directly involved in the efflux of cellular cholesterol is unknown, but PON3 may affect the efflux capacity of serum by preserving the function of HDL and other serum factors that may be inhibited by oxidative stress. Indeed, oxidized HDL has been reported to be less effective in effluxing cholesterol from cholesterol-loaded macrophages relative to control HDL (Aviram et al., 1998). The enhancement in cholesterol efflux seen in AdPON3 treated mice may be attributable to other serum components whose functions may be attenuated by oxidative stress, but the mechanism by which PON3 reduces serum lipid hydroperoxides and modulates lipoprotein properties is unclear.

4.4.1. Human PON3 transgenic mice have decreased atherosclerosis lesion development

More recently, we generated two lines of human PON3 transgenic mice with high-level expression detected in several organs including the liver, kidney and lung (Shih et al., 2004). Liver extracts isolated from the human PON3 transgenic mice exhibited 2-fold higher activity in hydrolyzing lovastatin (a PON3-specific substrate), as compared to those of the non-transgenic littermates. Relative to littermate controls, no significant differences were observed in plasma total, HDL, and VLDL cholesterol or triglyceride levels, regardless of diet (normal or atherogenic diet). The atherosclerotic lesion area as measured by oil-red-o staining was significantly smaller in PON-3 transgenic mice (Fig. 10). The human PON3 transgene was also introduced onto the LDL receptor knockout (LDLRKO) mouse background and atherosclerotic lesions were shown to be significantly lower on both the C57Bl6/J and LDLRKO background (55% and 34% reduction, respectively), however this protection was observed only in male cohorts (Shih et al., 2007). Surprisingly,

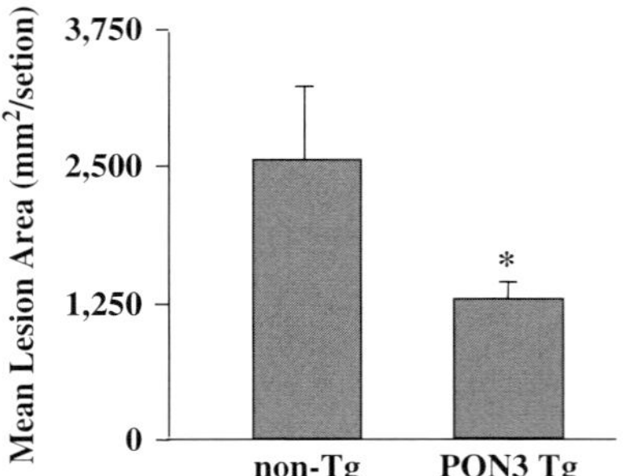

Figure 10. Decreased atherosclerosis in male human PON3 transgenic mice. Three months old male PON3 transgenic (Tg, $n = 17$) and non-Tg littermates ($n = 19$) were fed an atherogenic diet for 15 weeks before atherosclerotic lesion areas in the aortic root region were measured. Mean and standard error are shown. *: $p < 0.05$ vs. non-Tg

results from this study further associated elevated human PON3 expression with decreased obesity in male mice. These studies suggest a protective role for PON3 against atherosclerosis and obesity.

5. HOST INNATE IMMUNITY: QUORUM QUENCHING ACTIVITY OF PARAOXONASES

5.1. Quorum Sensing

Bacteria use small secreted chemicals or peptides as autoinducers to coordinately regulate gene expression in a process called quorum sensing. Quorum sensing (QS) controls several important functions in different bacterial species, including the production of virulence factors and biofilm formation. Quorum sensing was first characterized in *V. fisheri* and *Vibrio harveyi*, two bioluminescent marine symbionts of some fish and squids (Fuqua et al., 2001). Subsequently, QS systems have been described in a large and growing number of bacterial species that control multiple cellular functions (Miller and Bassler, 2001). Many gram-negative bacteria utilize AHLs that either diffuse or are transported across cell membranes and function as ligands for a family of inducible transcription factors (Fuqua and Greenberg, 2002). The mechanism of QS has been extensively characterized in several pathogenic bacteria including *P. aeruginosa* (Shiner et al., 2005). This bacterium causes catastrophic infections in immunocompromised hosts, such as individuals with cystic fibrosis, cancer and severe burns.

5.2. Quorum Quenching

Given that diverse bacterial species use QS to boost their competitive advantages, for example, production of antibiotics and virulence factors (Whitehead et al., 2001; Zhang and Dong, 2004), it is rational that hosts may also have evolved certain mechanisms to disarm the QS systems of microbes. Over the last few years, a range of quorum quenching enzymes and inhibitors have been identified from different sources, including both prokaryotic and eukaryotic organisms (Hentzer et al., 2002; Zhang, 2003; Zhang and Dong, 2004). These novel enzymes and inhibitors established the concept of quorum quenching, antipathogenic and signal interference (Hentzer et al., 2002; Zhang, 2003; Zhang and Dong, 2004). Interesting observations came from the study on mammalian cells in Greenberg's laboratory while testing the transgenic expression of the prokaryotic quorum quencher, *aiiA* gene in cell lines, they observed a high background of AHL-inactivation activity from the untransformed control cell lines (Dong and Zhang, 2005). Further study showed that the enzyme activity was associated with the cell membrane of differentiated human airway epithelia (Chun et al., 2004). The enzyme was able to inactivate the principal quorum sensor of *P. aeruginosa*, N-(3-oxododecanoyl)-L-homoserine lactone ($3OC_{12}$-HSL). The ability to inactivate $3OC_{12}$-HSL varied significantly among different cell types, with tissues likely to be exposed to pathogens showing the highest inactivation of the QS signal, such as A549 cells from human lungs and CaCo-2 cells from human colon. Subsequent studies showed that the $3OC_{12}$-HSL degradation activities are most likely due to the *PON* genes (Ozer et al., 2005). Indeed, the lactonase activity of the human paraoxonases has been demonstrated over 30 different non-AHL type lactones (Teiber et al., 2003). *Inactivation of QS signals has become a new index to the diverse spectrum of d biochemical functions assigned to PON enzymes.*

5.3.   Are PON Proteins Quorum Quenchers?

PON1, PON2, and PON3 all possess the ability to hydrolyze lactones (Draganov et al., 2005) and the purified recombinant enzymes have further all been shown to specifically hydrolyze acyl homoserine lactone (AHL) molecules, with PON2 exhibiting the highest specific activity (Draganov et al., 2005; Ozer et al., 2005) (Fig. 11). This substrate specificity has garnered a tremendous amount of attention recently because of the possibility of associating a paraoxonase function with host innate immunity against microbial infection. However, it is important to take into account that *in vivo* evidence is still lacking despite compelling *ex vivo* evidence.

AHLs are only one of a class of small secreted molecules, known as autoinducers, which coordinately regulate gene expression in a large and growing number of bacterial species in a process called quorum sensing. In addition to AHLs, gram-positive bacteria have been shown to quorum sense utilizing small secreted peptides and a third class, shared by both gram-positive and gram-negative bacteria, utilizes a

PON

$3OC_{12}$-HSL

Figure 11. Lactonase activity of PON proteins on acyl-homoserine lactones

structurally related group of signaling molecules initially referred to as autoinducer-2. Interestingly, although not yet tested, virtually all of the known QS molecules can be potential substrates for the enzyme activities that have been associated with PON proteins over the last two decades, namely arylesterase and lactonase.

Quorum sensing has been extensively characterized in several pathogenic bacteria, including the opportunistic pathogen, *P. aeruginosa*, responsible for life threatening infections in immunocompromised hosts, such as individuals with cystic fibrosis, cancer and severe burns (Shiner et al., 2005). Cell density dependent expression of its principal quorum sensor, $3OC_{12}$-HSL, has been shown to be crucial for the pathogenesis of both acute and chronic infections. $3OC_{12}$-HSL is involved in the regulation of a number of genes including the expression of key virulence factors and biofilm formation.

The pulmonary system is a primary site of infection for *P. aeruginosa*, thus we initially assessed the capacity of membrane-enriched lung homogenates from wild type and PON2-def mice to hydrolyze purified $3OC_{12}$-HSL, following a 2-hour incubation. PON2-def lung homogenates had the highest and essentially all of the $3OC_{12}$-HSL remaining, suggesting that the quorum-quenching system in these lysates is impaired (Fig. 12) (Bourquard et al., 2006). In a similar set of experiments using primary tracheal epithelial cells, we observed that membrane-enriched protein lysates from wild type mice hydrolyzed over 75% of the $3OC_{12}$-HSL whereas lysates from PON2-def mice failed to degrade essentially all of the $3OC_{12}$-HSL (Stoltz et al., 2006). Interestingly, the rate and extent of $3OC_{12}$-HSL hydrolysis was not significantly different with epithelial cell lysates obtained from PON1 and PON3 KO mice. Using a similar experimental approach using whole cells in lieu of tracheal epithelial cell lysates, the ability to hydrolyze $3OC_{12}$-HSL was surprisingly indistinguishable between wild type and PON2-def cells. However, this latter experiment was performed by assessing $3OC_{12}$-HSL remaining in the supernatant. Kravchenko et al (Kravchenko et al., 2006) recently showed that $3OC_{12}$-

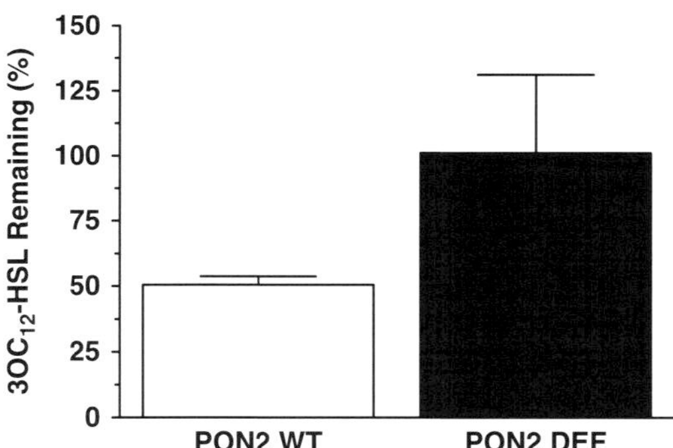

Figure 12. $3OC_{12}$-HSL bioassay. Membrane-enriched lung lysates (n = 3) were incubated with $3OC_{12}$-HSL for 2 hours and $3OC_{12}$-HSL remaining (%) was measured by quantitative bioassay. p < 0.05

HSL quite readily crosses the plasma membrane and can be detected in cellular fractions of macrophages, where it may possibly interact with the paraoxonases.

In an effort to associate paraoxonase function more directly with quorum sensing by gram-negative bacteria, airway epithelial cells from PON2-def and wild type mice were incubated with a strain of *P. aeruginosa* that carries a $3OC_{12}$-HSL responsive reporter gene. We observed a roughly two-fold increase in reporter gene expression in PON2-def cells incubated with *P. aeruginosa* relative to controls after 6 hours, but interestingly, this was not associated with any changes in bacterial density (Stoltz et al., 2006). This suggests that PON2 expression does not affect the growth of *P. aeruginosa* but degrades the bacterial quorum-sensing signal.

5.4. Can PON Proteins Modulate the Proinflammatory Pathways Activated by Quorum Sensing Lactones?

AHLs exhibit structural similarities to many eukaryotic hormones and a growing number of reports have documented apparent biological effects of AHLs on eukaryotic cells (Shiner et al., 2005). Accumulating evidence suggests that $3OC_{12}$-HSL, in particular, has substantial host immunomodulatory activity at low concentrations and reduces cellular viability at higher concentrations (Kravchenko et al., 2006; Schwartz et al., 2007; Telford et al., 1998). Although evidence suggests that $3OC_{12}$-HSL may be recognized by receptors of the host innate immune system, the host receptor is presently unknown and cell activation appears to involve a unique non-canonical toll-like receptor pathway (Kravchenko et al., 2006).

6. CONCLUSION

Epidemiological and biochemical studies have long associated PON2 and PON3 with inflammatory diseases, in particular atherosclerosis. Recently developed transgenic mouse models support a role for PON2 and PON3 genes in atherogenesis. These mouse models reinforce previous in *vitro* studies and further demonstrate that expression of PON2 and PON3 protect against cellular oxidative stress.

Ng et al. (Ng et al., 2006) demonstrated that PON2 plays a protective role in atherosclerosis. PON2 deficient mice on an atherogenic diet showed i) decreased apoB containing VLDL/LDL ii) enhanced LDL oxidation, iii) attenuated anti-atherogenic capacity of HDL, and iv) a heightened state of oxidative stress coupled with an exacerbated inflammatory response from PON2 deficient macrophages. Unlike PON1, PON2 is a cell-associated enzyme (Ng et al., 2001) and thus its mechanism of action is likely dependent on its function within the cell. Conversely, elevated levels of human PON2 in the context of established atherosclerosis effectively protected against lesion development (Ng et al., 2006). This was associated with a marked improvement in the antioxidant and functional properties of serum just three weeks following a single administration of the transgene.

Although PON2 and PON3 are highly conserved and share similar anti-oxidant capacities, differences in enzymatic activities, localization, and regulation suggest that their physiological function is far from redundant. Characterization of mouse PON3 suggests that it does not associate with HDL, as demonstrated for human PON3, but remains cell associated with an expression pattern that overlaps with that of PON2. Adenoviral-mediated expression of human PON3 in six month old female apoE null mice was also shown to protect against atherosclerosis (Ng et al., 2004). Liver targeting by adenovirus following systemic administration is rapid and is therefore an important consideration with this approach. Interestingly, male transgenic mice expressing hPON3 (Shih et al., 2004; Shih et al., 2007) developed significantly less atherosclerosis and decreased obesity, compared to control mice.

6.1. Does a Molecular Link Exist Between the Quorum Quenching Activity of PON2 and the Pathogenesis of Atherosclerosis?

Despite the animal models and the recent discovery that PON proteins share a common lactonase activity, the identity of the endogenous substrate(s) of the PON proteins still remains unclear. Interestingly, recent findings suggest that the PON gene family members may also be involved in host defense due to their ability to degrade quorum sensing molecules produced by Gram-negative bacteria such as *P. aeruginosa* (Ozer et al., 2005; Stoltz et al., 2006; Yang et al., 2005). Infections have been recognized as risk factors for atherosclerosis. The infection hypothesis also relates to current atherogenesis theories that accept the crucial role of inflammation in the development of the atherosclerotic plaque (Libby et al., 2002). A variety of agents including herpesvirus (such as cytomegalovirus), Chlamydia pneumonia, Helicobacter pylori, and periodontal pathogens (such as *Porphyromonas gingivalis*) have been associated with increased risk for atherosclerosis (Danesh et al., 1997; Epstein et al., 1999; Meurman et al., 2004). Chronic *P. aeruginosa* infection in the lung has been demonstrated to enhance atherosclerotic development in the aorta and coronary artery in rats that were fed a high cholesterol diet (Turkay et al., 2004). Since PON proteins inactivate the quorum sensors of Gram-negative bacteria, it is possible that under physiological conditions, PON proteins (depending on their location in the body) keep QS molecules under check and prevent bacterial colonization. Bacterial infection is associated with several pro-inflammatory factors including quorum sensing molecules and atherogenic lipids. PON proteins could directly inactivate AHL-like quorum sensing molecules and/or modulate the pro-inflammatory pathways mediated by bacterial products. Altogether, these putative dual roles of PONs in innate host defense against microbial infection and protection against atherosclerosis provide new opportunities and directions in paraoxonase research.

REFERENCES

Akhmedova, S. N., A. K. Yakimovsky, et al. 2001. Paraoxonase 1 Met–Leu 54 polymorphism is associated with Parkinson's disease. *J Neurol Sci* 184(2): 179–82.

Ansell, B. J., M. Navab, et al. 2003. Inflammatory/antiinflammatory properties of high-density lipoprotein distinguish patients from control subjects better than high-density lipoprotein cholesterol levels and are favorably affected by simvastatin treatment. *Circulation* 108(22): 2751–6.

Aviram, M., M. Rosenblat, et al. 1998. Paraoxonase inhibits high-density lipoprotein oxidation and preserves its functions. A possible peroxidative role for paraoxonase. *J Clin Invest* 101(8): 1581–90.

Bourquard, N., S. T. Reddy et al. 2006. Studies on the role of PON2 in inflammation and atherosclerosis. 2nd *international conference on the paraoxonases*. September 7–10, 2006. Hajduszoboszlo, Hungary.

Brophy, V. H., R. L. Jampsa, et al. 2001. Effects of 5' regulatory-region polymorphisms on paraoxonase-gene (PON1) expression. *Am J Hum Genet* 68(6): 1428–36.

Campo, S., A. M. Sardo, et al. 2004. Identification of paraoxonase 3 gene (PON3) missense mutations in a population of southern Italy. *Mutat Res* 546(1–2): 75–80.

Cathcart, M. K. 2004. Regulation of superoxide anion production by NADPH oxidase in monocytes/macrophages: contributions to atherosclerosis. *Arterioscler Thromb Vasc Biol* 24(1): 23–8.

Chen, Q., S. E. Reis, et al. 2003. Association between the severity of angiographic coronary artery disease and paraoxonase gene polymorphisms in the National Heart, Lung, and Blood Institute-sponsored Women's Ischemia Syndrome Evaluation (WISE) study. *Am J Hum Genet* 72(1): 13–22.

Chun, C. K., E. A. Ozer, et al. 2004. Inactivation of a Pseudomonas aeruginosa quorum-sensing signal by human airway epithelia. *Proc Natl Acad Sci U S A* 101(10): 3587–90.

Connelly, P. W., D. Draganov, et al. 2005. Paraoxonase-1 does not reduce or modify oxidation of phospholipids by peroxynitrite. *Free Radic Biol Med* 38(2): 164–74.

Costa, L. G., T. B. Cole, et al. 2003. Functional genomic of the paraoxonase (PON1) polymorphisms: effects on pesticide sensitivity, cardiovascular disease, and drug metabolism. *Annu Rev Med* 54: 371–92.

Costa, L. G., W. F. Li, et al. 1999. The role of paraoxonase (PON1) in the detoxication of organophosphates and its human polymorphism. *Chem Biol Interact* 119–120: 429–38.

Danesh, J., R. Collins, et al. 1997. Chronic infections and coronary heart disease: is there a link? *Lancet* 350(9075): 430–6.

Davies, H. G., R. J. Richter, et al. 1996. The effect of the human serum paraoxonase polymorphism is reversed with diazoxon, soman and sarin. *Nat Genet* 14(3): 334–6.

Dong, Y. H. and L. H. Zhang. 2005. Quorum sensing and quorum-quenching enzymes. *J Microbiol* 43 Spec No: 101–9.

Draganov, D. I. and B. N. La Du. 2004. Pharmacogenetics of paraoxonases: a brief review. *Naunyn Schmiedebergs Arch Pharmacol* 369(1): 78–88.

Draganov, D. I., P. L. Stetson, et al. 2000. Rabbit serum paraoxonase 3 (PON3) is a high density lipoprotein-associated lactonase and protects low density lipoprotein against oxidation. *J Biol Chem* 275(43): 33435–42.

Draganov, D. I., J. F. Teiber, et al. 2005. Human paraoxonases (PON1, PON2, and PON3) are lactonases with overlapping and distinct substrate specificities. *J Lipid Res* 46(6): 1239–47.

Durrington, P. N., B. Mackness, et al. 2001. Paraoxonase and atherosclerosis. *Arterioscler Thromb Vasc Biol* 21(4): 473–80.

Epstein, S. E., Y. F. Zhou, et al. 1999. Infection and atherosclerosis: emerging mechanistic paradigms. *Circulation* 100(4): e20–8.

Fisher, E. A., M. Pan, et al. 2001. The triple threat to nascent apolipoprotein B. Evidence for multiple, distinct degradative pathways. *J Biol Chem* 276(30): 27855–63.

Fokine, A., R. Morales, et al. 2003. Direct phasing at low resolution of a protein copurified with human paraoxonase (PON1). *Acta Crystallogr D Biol Crystallogr* 59(Pt 12): 2083–7.

Fortunato, G., P. Rubba, et al. 2003. A paraoxonase gene polymorphism, PON 1 (55), as an independent risk factor for increased carotid intima-media thickness in middle-aged women. *Atherosclerosis* 167(1): 141–8

Fuqua, C. and E. P. Greenberg. 2002. Listening in on bacteria: acyl-homoserine lactone signalling. *Nat Rev Mol Cell Biol* 3(9): 685–95.

Fuqua, C., M. R. Parsek, et al. 2001. Regulation of gene expression by cell-to-cell communication: acyl-homoserine lactone quorum sensing. *Annu Rev Genet* 35: 439–68.

Furlong, C. E., T. B. Cole, et al. 2002. Pharmacogenomic considerations of the paraoxonase polymorphisms. *Pharmacogenomics* 3(3): 341–8.

     BOURQUARD ET AL.

Furlong, C. E., R. J. Richter, et al. 1989. Spectrophotometric assays for the enzymatic hydrolysis of the active metabolites of chlorpyrifos and parathion by plasma paraoxonase/arylesterase. *Anal Biochem* 180(2): 242–7.

Harel, M., A. Aharoni, et al. 2004. Structure and evolution of the serum paraoxonase family of detoxifying and anti-atherosclerotic enzymes. *Nat Struct Mol Biol* 11(5): 412–9.

Hassett, C., R. J. Richter, et al. 1991. Characterization of cDNA clones encoding rabbit and human serum paraoxonase: the mature protein retains its signal sequence. *Biochemistry* 30(42): 10141–9.

Hegele, R. A. 1999. Paraoxonase genes and disease. *Ann Med* 31(3): 217–24.

Hegele, R. A., J. H. Brunt, et al. 1995. A polymorphism of the paraoxonase gene associated with variation in plasma lipoproteins in a genetic isolate. *Arterioscler Thromb Vasc Biol* 15(1): 89–95.

Hegele, R. A., P. W. Connelly, et al. 1997. Paraoxonase-2 gene (PON2) G148 variant associated with elevated fasting plasma glucose in noninsulin-dependent diabetes mellitus. *J Clin Endocrinol Metab* 82(10): 3373–7.

Hegele, R. A., S. B. Harris, et al. 1998. Genetic variation in paraoxonase-2 is associated with variation in plasma lipoproteins in Canadian Oji-Cree. *Clin Genet* 54(5): 394–9.

Hentzer, M., K. Riedel, et al. 2002. Inhibition of quorum sensing in Pseudomonas aeruginosa biofilm bacteria by a halogenated furanone compound. *Microbiology* 148(Pt 1): 87–102.

Hu, Y., H. Tian, et al. 2003. Gln-Arg192 polymorphism of paraoxonase 1 is associated with carotid intima-media thickness in patients of type 2 diabetes mellitus of Chinese. *Diabetes Res Clin Pract* 61(1): 21–7.

Jakubowski, H. 2000. Calcium-dependent human serum homocysteine thiolactone hydrolase. A protective mechanism against protein N-homocysteinylation. *J Biol Chem* 275(6): 3957–62.

Janka, Z., A. Juhasz, et al. 2002. Codon 311 (Cys –> Ser) polymorphism of paraoxonase-2 gene is associated with apolipoprotein E4 allele in both Alzheimer's and vascular dementias. *Mol Psychiatry* 7(1): 110–2.

Kao, Y., K. C. Donaghue, et al. 2002. Paraoxonase gene cluster is a genetic marker for early microvascular complications in type 1 diabetes. *Diabet Med* 19(3): 212–5.

Kobayashi, M., M. Shinohara, et al. 1998. Lactone-ring-cleaving enzyme: genetic analysis, novel RNA editing, and evolutionary implications. *Proc Natl Acad Sci U S A* 95(22): 12787–92.

Kondo, I. and M. Yamamoto. 1998. Genetic polymorphism of paraoxonase 1 (PON1) and susceptibility to Parkinson's disease. *Brain Res* 806(2): 271–3.

Kravchenko, V. V., G. F. Kaufmann, et al. 2006. N-(3-oxo-acyl)homoserine lactones signal cell activation through a mechanism distinct from the canonical pathogen-associated molecular pattern recognition receptor pathways. *J Biol Chem* 281(39): 28822–30.

Kriska, T., G. K. Marathe, et al. 2007. Phospholipase action of platelet-activating factor acetylhydrolase, but not paraoxonase-1, on long fatty acyl chain phospholipid hydroperoxides. *J Biol Chem* 282(1): 100–8.

La Du, B. N., M. Aviram, et al. 1999. On the physiological role(s) of the paraoxonases. *Chem Biol Interact* 119–120: 379–88.

Leviev, I. and R. W. James. 2000. Promoter polymorphisms of human paraoxonase PON1 gene and serum paraoxonase activities and concentrations. *Arterioscler Thromb Vasc Biol* 20(2): 516–21.

Libby, P., P. M. Ridker, et al. 2002. Inflammation and atherosclerosis. *Circulation* 105(9): 1135–43.

Mackness, B., M. I. Mackness, et al. 1997. Effect of the molecular polymorphisms of human paraoxonase (PON1) on the rate of hydrolysis of paraoxon. *Br J Pharmacol* 122(2): 265–8.

Mackness, M. I., S. Arrol, et al. 1993. Is paraoxonase related to atherosclerosis. *Chem Biol Interact* 87(1–3): 161–71.

Malin, R., A. Loimaala, et al. 2001. Relationship between high-density lipoprotein paraoxonase gene M/L55 polymorphism and carotid atherosclerosis differs in smoking and nonsmoking men. *Metabolism* 50(9): 1095–101.

Marathe, G. K., G. A. Zimmerman, et al. 2003. Platelet-activating factor acetylhydrolase, and not paraoxonase-1, is the oxidized phospholipid hydrolase of high density lipoprotein particles. *J Biol Chem* 278(6): 3937–47.

Martinelli, N., D. Girelli, et al. 2004. Interaction between smoking and PON2 Ser311Cys polymorphism as a determinant of the risk of myocardial infarction. *Eur J Clin Invest* 34(1): 14–20.

Meurman, J. H., M. Sanz, et al. 2004. Oral health, atherosclerosis, and cardiovascular disease. *Crit Rev Oral Biol Med* 15(6): 403–13.

Miller, M. B. and B. L. Bassler. 2001. Quorum sensing in bacteria. *Annu Rev Microbiol* 55: 165–99.

Navab, M., S. Y. Hama, et al. 2002. Oxidized lipids as mediators of coronary heart disease. *Curr Opin Lipidol* 13(4): 363–72.

Ng, C. J., N. Bourquard, et al. 2006. Paraoxonase-2 deficiency aggravates atherosclerosis in mice despite lower apolipoprotein-B-containing lipoproteins: anti-atherogenic role for paraoxonase-2. *J Biol Chem* 281(40): 29491–500.

Ng, C. J., S. Y. Hama, et al. 2006. Adenovirus mediated expression of human paraoxonase 2 protects against the development of atherosclerosis in apolipoprotein E-deficient mice. *Mol Genet Metab* 89(4): 368–73.

Ng, C. J., D. M. Shih, et al. 2005. The paraoxonase gene family and atherosclerosis. *Free Radic Biol Med* 38(2): 153–63.

Ng, C. J., D. J. Wadleigh, et al. 2001. Paraoxonase-2 is a ubiquitously expressed protein with antioxidant properties and is capable of preventing cell-mediated oxidative modification of low density lipoprotein. *J Biol Chem* 276(48): 44444–9.

Ng, C. J., S. Y. Hama, et al. 2004. Adenovirus-mediated overexpression of human PON3 enhances high-density lipoprotein function in apolipoprotein E deficient mice. *Circulation*. Supplement III. 110:III-36.

Oda, M. N., J. K. Bielicki, et al. 2002. Paraoxonase 1 overexpression in mice and its effect on high-density lipoproteins. *Biochem Biophys Res Commun* 290(3): 921–7.

Odawara, M., Y. Tachi, et al. 1997. Paraoxonase polymorphism (Gln192-Arg) is associated with coronary heart disease in Japanese noninsulin-dependent diabetes mellitus. *J Clin Endocrinol Metab* 82(7): 2257–60.

Ozer, E. A., A. Pezzulo, et al. 2005. Human and murine paraoxonase 1 are host modulators of Pseudomonas aeruginosa quorum-sensing. *FEMS Microbiol Lett* 253(1): 29–37.

Pan, J. P., S. T. Lai, et al. 2002. The risk of coronary artery disease in population of Taiwan is associated with Cys-Ser 311 polymorphism of human paraoxonase (PON)-2 gene. *Zhonghua Yi Xue Za Zhi (Taipei)* 65(9): 415–21.

Pan, M., A. I. Cederbaum, et al. 2004. Lipid peroxidation and oxidant stress regulate hepatic apolipoprotein B degradation and VLDL production. *J Clin Invest* 113(9): 1277–87.

Pfohl, M., M. Koch, et al. 1999. Paraoxonase 192 Gln/Arg gene polymorphism, coronary artery disease, and myocardial infarction in type 2 diabetes. *Diabetes* 48(3): 623–7.

Primo-Parmo, S. L., R. C. Sorenson, et al. 1996. The human serum paraoxonase/arylesterase gene (PON1) is one member of a multigene family. *Genomics* 33(3): 498–507.

Reddy, S. T., D. J. Wadleigh, et al. 2001. Human paraoxonase-3 is an HDL-associated enzyme with biological activity similar to paraoxonase-1 protein but is not regulated by oxidized lipids. *Arterioscler Thromb Vasc Biol* 21(4): 542–7.

Rodrigo, L., B. Mackness, et al. 2001. Hydrolysis of platelet-activating factor by human serum paraoxonase. *Biochem J* 354(Pt 1): 1–7.

Rosenblat, M., D. Draganov, et al. 2003. Mouse macrophage paraoxonase 2 activity is increased whereas cellular paraoxonase 3 activity is decreased under oxidative stress. *Arterioscler Thromb Vasc Biol* 23(3): 468–74.

Rozenberg, O., D. M. Shih, et al. 2003. Human serum paraoxonase 1 decreases macrophage cholesterol biosynthesis: possible role for its phospholipase-A2-like activity and lysophosphatidylcholine formation. *Arterioscler Thromb Vasc Biol* 23(3): 461–7.

Sanghera, D. K., C. E. Aston, et al. 1998. DNA polymorphisms in two paraoxonase genes (PON1 and PON2) are associated with the risk of coronary heart disease. *Am J Hum Genet* 62(1): 36–44.

Schwartz, T., S. Walter, et al. 2007. Use of quantitative real-time RT-PCR to analyse the expression of some quorum-sensing regulated genes in Pseudomonas aeruginosa. *Anal Bioanal Chem* 387(2): 513–21.

Shi, J., S. Zhang, et al. 2004. Possible association between Cys311Ser polymorphism of paraoxonase 2 gene and late-onset Alzheimer's disease in Chinese. *Brain Res Mol Brain Res* 120(2): 201–4.

Shih, D. M., L. Gu, et al. 1998. Mice lacking serum paraoxonase are susceptible to organophosphate toxicity and atherosclerosis. *Nature* 394(6690): 284–7.

Shih, D. M., Y-R Xia et al. 2004. Decreased atherosclerosis in human PON3 transgenic mice. *Circulation.* Supplement III. 110: III-290.

Shih, D. M., Y-R Xia et al. 2007. Decreased obesity and atherosclerosis in human paraoxonase 3 transgenic mice. *Circ. Res.* 100(8): 1200–7.

Shiner, E. K., K. P. Rumbaugh, et al. 2005. Inter-kingdom signaling: deciphering the language of acyl homoserine lactones. *FEMS Microbiol Rev* 29(5): 935–47.

Shiner, M., B. Fuhrman, et al. 2004. Paraoxonase 2 (PON2) expression is upregulated via a reduced-nicotinamide-adenine-dinucleotide-phosphate (NADPH)-oxidase-dependent mechanism during monocytes differentiation into macrophages. *Free Radic Biol Med* 37(12): 2052–63.

Shiner, M., B. Fuhrman, et al. 2006. A biphasic U-shape effect of cellular oxidative stress on the macrophage anti-oxidant paraoxonase 2 (PON2) enzymatic activity. *Biochem Biophys Res Commun* 349(3): 1094–9.

Stoltz, D. A., E. A. Ozer, et al. 2006. Paraoxonase-2 Deficiency Enhances Pseudomonas aeruginosa Quorum Sensing in Murine Tracheal Epithelia. *Am J Physiol Lung Cell Mol Physiol.*

Teiber, J. F., D. I. Draganov, et al. 2003. Lactonase and lactonizing activities of human serum paraoxonase (PON1) and rabbit serum PON3. *Biochem Pharmacol* 66(6): 887–96.

Telford, G., D. Wheeler, et al. 1998. The Pseudomonas aeruginosa quorum-sensing signal molecule N-(3-oxododecanoyl)-L-homoserine lactone has immunomodulatory activity. *Infect Immun* 66(1): 36–42.

Tsimikas, S., E. S. Brilakis, et al. 2005. Oxidized phospholipids, Lp(a) lipoprotein, and coronary artery disease. *N Engl J Med* 353(1): 46–57.

Turkay, C., R. Saba, et al. 2004. Effect of chronic Pseudomonas aeruginosa infection on the development of atherosclerosis in a rat model. *Clin Microbiol Infect* 10(8): 705–8.

Tward, A., Y. R. Xia, et al. 2002. Decreased atherosclerotic lesion formation in human serum paraoxonase transgenic mice. *Circulation* 106(4): 484–90.

Voetsch, B., K. S. Benke, et al. 2002. Paraoxonase 192 Gln–>Arg polymorphism: an independent risk factor for nonfatal arterial ischemic stroke among young adults. *Stroke* 33(6): 1459–64.

Wang, X. Y., Y. M. Xue, et al. 2003. [The association of paraoxonase 2 gene C311S variant with ischemic stroke in Chinese type 2 diabetes mellitus patients]. *Zhonghua Yi Xue Yi Chuan Xue Za Zhi* 20(3): 215–9.

Watson, A. D., J. A. Berliner, et al. 1995. Protective effect of high density lipoprotein associated paraoxonase. Inhibition of the biological activity of minimally oxidized low density lipoprotein. *J Clin Invest* 96(6): 2882–91.

Whitehead, N. A., A. M. Barnard, et al. 2001. Quorum-sensing in Gram-negative bacteria. *FEMS Microbiol Rev* 25(4): 365–404.

Yamada, Y., F. Ando, et al. 2003. Association of polymorphisms of paraoxonase 1 and 2 genes, alone or in combination, with bone mineral density in community-dwelling Japanese. *J Hum Genet* 48(9): 469–75.

Yang, F., L. H. Wang, et al. 2005. Quorum quenching enzyme activity is widely conserved in the sera of mammalian species. *FEBS Lett* 579(17): 3713–7.

Zhang, L. H. 2003. Quorum quenching and proactive host defense. *Trends Plant Sci* 8(5): 238–44.

Zhang, L. H. and Y. H. Dong. 2004. Quorum sensing and signal interference: diverse implications. *Mol Microbiol* 53(6): 1563–71.

CHAPTER 7

PARAOXONASE 1 AND POSTPRANDIAL LIPEMIA

A. ALIPOUR[1], B. COLL[2], A.P. RIETVELD[1], J. MARSILLACH[2], J. CAMPS[2], J. JOVEN[2], J.W.F. ELTE[1] AND M. CASTRO CABEZAS[1]

[1] *Department of Internal Medicine, Sint Franciscus Gasthuis, Rotterdam, The Netherlands*
[2] *Centre de Recerca Biomèdica and Department of Internal Medicine, Hospital Universitari de Sant Joan, Reus, Spain*

Abstract: Risk factors for coronary heart disease (CHD) include decreased insulin sensitivity and glucose intolerance, hypertension, increased body fat mass, unfavorable body fat distribution, the prothrombotic state and dyslipidemia. A clustering of these metabolic disturbances is also seen in Insulin Resistance and the metabolic syndrome. It has been shown that lipoproteins, triglycerides (TG), fatty acids and glucose can activate endothelial cells most likely due to the production of reactive oxygen species (ROS). Elevation of TG, fatty acids and glucose are common disturbances in the metabolic syndrome and are most prominent in the postprandial phase. Furthermore, postprandial lipemia has been associated with decreased HDL-C concentrations. The postprandial phase is therefore considered to be a pro-atherogenic and pro-oxidative condition. Paraoxonase 1 (PON1) is a HDL-associated enzyme with the ability to hydrolyze oxidized lipids, reducing oxidative stress in lipoproteins and in macrophages. Postprandial HDL-C decrease is associated with a decrease of PON1 activity. Subjects at risk for CHD, such as type 2 diabetics and patients with hypercholesterolemia have low PON1 activity. Furthermore, expression of PON1 inhibits the development of atherosclerosis in mice. Classic determinants of postprandial lipemia such as fasting TG, apolipoprotein B (apoB) and central obesity do not predict the postprandial PON1 activity and drugs like statins, glitazons

Corresponding author:
M. Castro Cabezas, MD, PhD
internist-endocrinologist/vascular specialist
St. Franciscus Gasthuis
Dpt. Of Internal Medicine
Centre for Diabetes and Vascular Medicine
PO Box 10900
3004 BA Rotterdam
the Netherlands
Tel: +3110-4617267
Fax: +3110-4612692
E-mail: m.castrocabezas@sfg.nl

B. Mackness et al. (eds.), The Paraoxonases: Their Role in Disease Development and Xenobiotic Metabolism, 129–138.
© 2008 *Springer.*

and metformin do not affect the postprandial PON1 changes, although some effects on fasting concentrations may be seen. Lifestyle interventions using the Mediterranean diet increase PON1 activity and may thereby decrease the risk fro atherosclerosis

Keywords: coronary heart disease, inflammation, oxidation, paraoxonase 1, postprandial

1.   INTRODUCTION

Coronary Heart Disease (CHD) is the major contributor of morbidity and mortality in Western societies (Braunwald, 1997). Important risk factors for CHD include decreased insulin sensitivity and glucose intolerance, hypertension, increased body fat mass, unfavorable body fat distribution, the prothrombotic state and dyslipidemia (Bonora et al., 1998; Despres et al., 1990; Despres et al., 1996; Ferrannini et al., 1991; Turner et al., 1998). These risk factors are strongly interrelated and all of them are part of metabolic syndrome (Reaven, 1988) and the Insulin Resistance (IR) syndrome to which type 2 diabetes belongs (Bonora et al., 1998; Despres et al., 1990; Turner et al., 1998). In type 2 diabetes, there is a two to four times increased risk for CHD mortality and the risk for a primary coronary event is as high as the risk for a second coronary event in non-diabetics (Haffner et al., 1998; Kannel and McGee, 1979; Nathan et al., 1997). These are other groups of patients like the HIV-infected patients using highly active antiretroviral therapy who have a clustering of cardiovascular risk factors and endothelial dysfunction showing striking similarities with the IR syndrome (van Wijk et al., 2006a). Macrovascular complications have only been reduced when treatment has been targeted to modifiable components of the metabolic syndrome. In this regard, dyslipidemia could be the most important component since lipid interventions reduce cardiovascular events in IR, as was shown in a post-hoc subgroup analysis in the Scandinavian Simvastatin Survival Study (4S) and the Cholesterol And Recurrent Events (CARE) trial (Goldberg et al., 1998; Pyörälä et al., 1997).

Lipid parameters, TG levels in particular but also HDL-C, are important predictors of CHD in type 2 diabetes and the metabolic syndrome, whereas total cholesterol is a less strong independent predictor (Fontbonne et al., 1989; Laakso et al., 1993; Syvänne and Taskinen, 1997). HDL-C and TG levels are inversely associated, since in hypertriglyceridemia cholesterol esters from HDL-C are readily exchanged for TG from triglyceride-rich particles by the action of Cholesteryl Ester Transfer Protein (CETP), resulting in relative cholesterol depleted small-dense HDL-C. Furthermore, a similar exchange occurs between LDL-C and triglyceride-rich particles, resulting in small dense LDL. After oxidation and glycation of these lipoproteins, small dense LDL particles become highly atherogenic. One of the factors, which protects serum lipids against oxidation is Paraoxonase1 (PON1) (Billecke et al., 2000). PON1 is an esterase enzyme that is synthesized by the liver and is associated with HDL in the blood (Mackness et al., 1996). It has been shown that PON1 is able to hydrolyze specific oxidized lipids (Aviram et al., 1998; Aviram et al., 2000) leading to a reduction of oxidative stress in serum lipoproteins, macrophages, and atherosclerotic lesions (Rozenberg et al., 2003). Low PON1 activity is present in subjects at high risk of CHD, including those with hypercholesterolemia (Mackness et al., 1991) and DM (Abbott et al., 1995). Furthermore, recently it has been shown that the expression of human PON1 inhibited the development of atherosclerosis in a mouse model of the metabolic syndrome, probably by reducing the amount of oxidized LDL-C in plasma and in the plaque, thereby preventing its proatherogenic effects (Mackness et al., 2006).

It is important to realize that triglyceride-rich particles are mainly produced in the postprandial state (Castro Cabezas et al., 2001; Zilversmit, 1979). Since people are non-fasting most part of the day, atherosclerosis is regarded as a postprandial phenomenon (Castro Cabezas, 2003; Castro Cabezas et al., 2001; Geluk et al., 2004; Halkes et al., 2001; Zilversmit, 1979). Indeed, in IR both fasting and postprandial hyperlipidemia have been described (Coppack, 1997; Lewis and Steiner, 1996; Lewis et al., 1991; Syvänne et al., 1994; van Oostrom et al., 2000; van Oostrom et al., 2007). In this chapter, we review the importance of postprandial lipemia and the role of PON1 in the process of atherosclerosis.

2. POSTPRANDIAL LIPEMIA AND ATHEROGENESIS

While LDL particles have an established role in atherogenesis, the contribution of remnant lipoproteins from intestinal and hepatic origin to plaque formation has been increasingly recognized (van Oostrom et al., 2004a; van Wijk et al., 2005; Verseyden et al., 2004). After a meal, dietary fat is transported in chylomicrons produced by the intestine. These chylomicrons, with apolipoprotein (apo) B48 as structural protein, become chylomicron remnants in the blood by the action of endothelial-bound lipoprotein lipase (LPL) (van Oostrom et al., 2004a). Chylomicrons compete with endogenous VLDL for the action of LPL (Brunzell et al., 1973), resulting in elevated remnant levels in situations of VLDL overproduction. Subsequently, chylomicron remnants are removed from the circulation by the concerted action of hepatic lipase (HL), apoE, the chylomicron remnant receptor, binding to heparan sulfate proteoglycans, and the LPL protein. Competition with VLDL remnants may occur again at this level. The role of the LDL-receptor in the clearance of chylomicron remnants is still debated (Castro Cabezas et al., 1998a; Eriksson et al., 1991; Rubinsztein et al., 1990). The remnants of chylomicrons and VLDL (intermediate density lipoproteins, IDL) are involved in particular in the development of atherosclerosis (Castro Cabezas, 1998b). Several investigators have demonstrated a positive association between delayed chylomicron remnant clearance and atherosclerosis (Castro Cabezas and Erkelens, 1998b; Castro Cabezas et al., 1998a; Halkes et al., 2001; Senti et al., 1992).

There are several potential mechanisms involved in the generation of atherosclerosis by enhanced postprandial lipemia and especially chylomicron remnant accumulation. Remnant particles may migrate into the sub-endothelial space and induce foam cell formation by direct interaction with monocytes/macrophages (Fig. 2, Castro Cabezas and Erkelens, 1998b). Remnants and fatty acids liberated by LPL from triglyceride rich lipoproteins may induce acute endothelial dysfunction and cause damage to the endothelial layer (van Oostrom et al., 2003). Postprandial hyperlipidemia is also associated with the generation of small dense LDL which can become easily modified and convert monocytes into foam cells (van Oostrom et al., 2004a). Furthermore, the postprandial period has been recognized as a situation of increased oxidative stress and inflammation by activation of leukocytes and the complement system (van Oostrom et al., 2003; van Oostrom et al., 2004a).

Finally, postprandial lipemia has been associated with decreased HDL-C concentrations by triglyceride enrichment of HDL during this phase and increased action by Hepatic Lipase which converts HDL2 into the less protective HDL3 (Castro Cabezas et al., 1993).

3. POSTPRANDIAL INFLAMMATORY AND OXIDATIVE CHANGES

Markers of inflammation like CRP, complement components and leukocytes have been identified as important predictors of the risk of atherosclerosis, but also as predictors of the response to treatment (Blake and Ridker, 2001). While CRP concentrations do not change postprandially (Coll et al., 2006), leukocytes and complement component 3 (C3) show a significant postprandial increase (Halkes et al., 2001; Halkes et al., 2003; Meijssen et al., 2002; Verseyden et al., 2003; van Oostrom et al., 2004b), especially after fat intake (van Oostrom et al., 2003., van Oostrom et al., 2004b), and become more activated during the postprandial phase (van Oostrom et al., 2004c). The factors involved are in part related to oxidative stress as has been suggested in vivo (van Oostrom et al., 2003). Interestingly, statins can modulate part of the postprandial inflammatory changes of leukocytes without affecting the postprandial leukocyte recruitment (van Oostrom et al., 2004c, van Oostrom et al., 2006). It has been postulated that one of the initiators of the postprandial leukocyte activation is opsonization of triglyceride rich lipoproteins in the blood stream (van Oostrom et al., 2004a). Atherosclerosis may therefore start in the bloodstream rather than at the endothelial cellular surface (van Oostrom et al., 2004a).

As postprandial leukocyte activation and oxidative stress are important steps in atherosclerosis, the opportunities are being explored to find ways to inhibit these factors. PON1 is definitely one of the important mediators in reducing oxidative stress (Aviram et al., 1998; Aviram et al., 2000).

4. ATHEROGENIC CHANGES OF HDL-C AND PON1 DURING THE POSTPRANDIAL PERIOD

During postprandial lipemia, HDL-C concentrations tend to decrease (Fig. 1). In theory, this may impair the reverse cholesterol transport in this highly pro-atherogenic situation, but HDL-C decrease has also been associated with a decrease

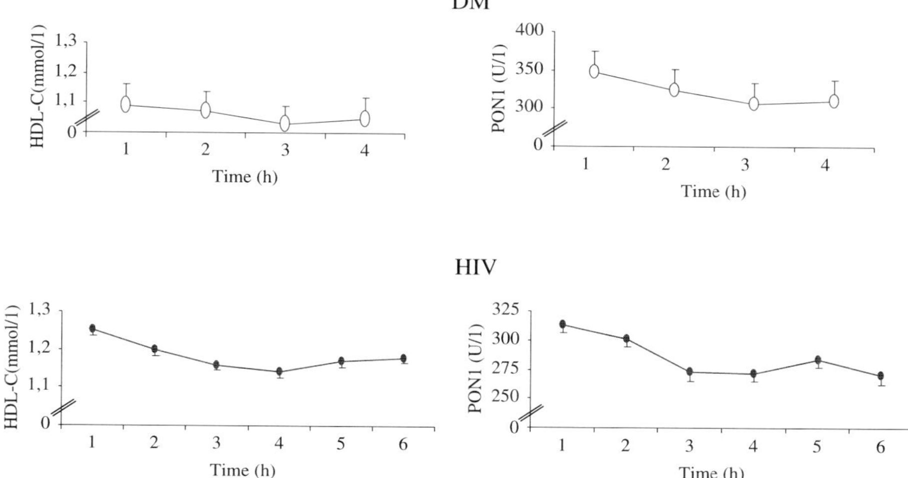

Figure 1. Examples of postprandial HDL-C and PON1 changes in non-diabetic: hyperlipidemic patients (upper panel) and HIV-infected patients (lower panel) during a four hours and a six hours oral fat loading test, respectively

of PON1 activity (Coll et al., 2006; Sutherland et al., 1999; van Wijk et al., 2006b). Therefore, the postprandial situation is a pro-oxidative condition (Fig. 2). This has already been suggested in previous studies showing that during postprandial lipemia there is an increase of oxidative stress (van Oostrom et al., 2003). Interestingly, addition of anti-oxidative vitamins to the diet reverses this and restores postprandial endothelial function to normal. However, there is no current evidence available of clinical trials suggesting that this anti-inflammatory cocktail may reduce atherosclerosis.

Efforts have been undertaken to explore the determinants of postprandial HDL-C and PON1 decrease (van Oostrom et al., 2003; van Wijk et al., 2006b). The classic determinants of postprandial lipemia such as fasting TG, apoB, and central obesity do not predict the postprandial PON1 decrease (van Oostrom et al., 2003, van Wijk et al., 2006b). PON1 is present on postprandial chylomicrons and VLDL and PON1 arises in part from serum HDL (Fuhrman et al., 2005). However, no changes in PON1 activity were observed in the latter study. It has also been shown that PON1 on chylomicrons possesses triacylglycerol lipase-like activity, although to a lesser extent in comparison with LPL (Fuhrman et al., 2006). This finding raises the possibility of a second pathway by which PON1 attenuates the increase of oxidative stress in serum.

It has been shown that *fasting* PON1 activity can be modulated by statins, TZD's and metformin in patients with premature CHD, DM and HIV. However, modulation of *postprandial* PON1 by these drugs has not been successful (van Oostrom et al., 2003; van Wijk et al., 2006b, data on file). For example,

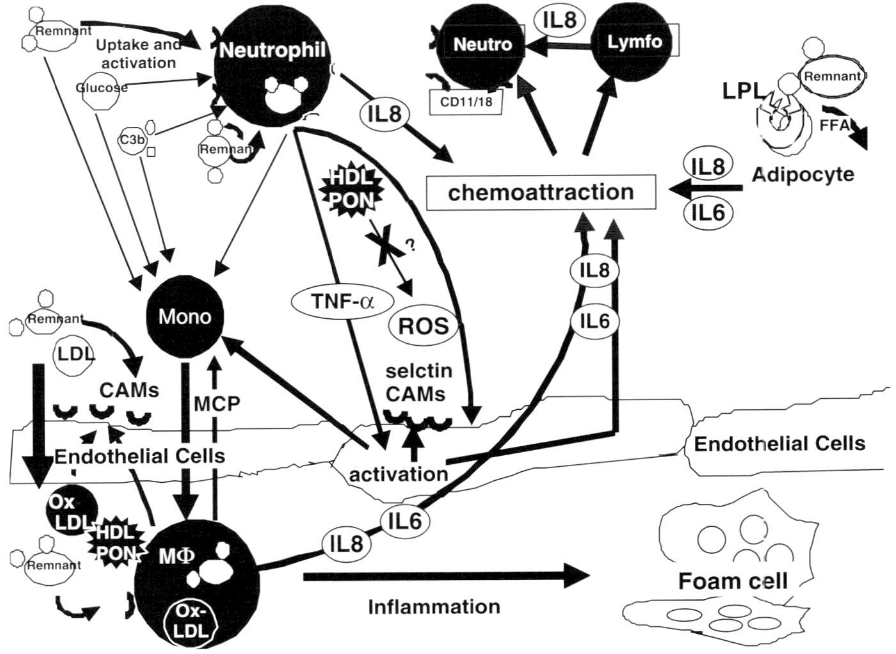

Figure 2. Concept of the initiation of atherosclerosis and the role of PON1: Postprandial increase of remnants, glucose and C3b induce leukocyte activation. This will lead to adherence of these cells to the endothelium as well as the production of chemo-attractants leading to the recruitment and activation of other leukocytes. Furthermore, these activated leukocytes can activate the endothelium facilitating adherence of all inflammatory cells. The activated leukocytes can also destabilize atherosclerotic lesions by the production of tumor necrosis factor α (TNFα) and reactive oxygen species (ROS). The postprandial phase is also associated with decrease of HDL-C and PON1 activity. PON1 hydrolyzes specific oxidized lipids and thus reduces the extent of oxidative stress in lipoproteins and in macrophages, resulting in higher ROS concentrations in the bloodstream. Monocytes and lymphocytes transmigrate across the endothelial wall. Monocytes residing in the arterial wall become activated as a result of pro-inflammatory cytokines and differentiate into macrophages. LDL and remnant particles can enter the vessel wall as well. Oxidative modification of LDL results in highly atherogenic particles that can easily be taken up by macrophages. The postprandial PON1 decrease results in increased oxidation of LDL and macrophages. Activated monocytes and macrophages in the vessel wall can activate endothelial cells resulting in the production of CAMs and cytokines and monocyte chemoattractant protein-1 (MCP-1). These effects combined will lead to the recruitment and activation of leukocytes to the activated endothelium and will eventually adhere firmly to the vessel wall. Monocytes and lymphocytes can then transmigrate through the endothelial barrier

Rosiglitazone only attenuated the postprandial fall of PON1 in type 2 diabetics (van Wijk et al., 2006b).

Consumption of a meal rich in thermally stressed olive oil significantly increased the postprandial serum PON1 levels (Wallace et al., 2001). Interestingly, this was only the case in women and not in men. Ingestion of different sorts of oils by mice showed increased postprandial lipid peroxidation (a marker for oxidative stress) and decreased postprandial PON1 activity (Fuhrman et al., 2006). The most striking finding was that fish oil, and its major fatty acid docosahexaenoic acid (DHA), resulted in a higher postprandial increase of oxidative stress and a higher postprandial PON1 decrease when compared to soy oil and olive oil, while long-term fish oil intake has been associated epidemiologically with cardiovascular protection (Kris-Etherton et al., 2003). However, a Mediterranean meal with monounsaturated fatty acids, increased the postprandial PON1 activity in 8 healthy males (Blum et al., 2006). Unheated, unused fat also increased postprandial PON1 activity, while used cooking fat showed a lower postprandial PON1 activity accompanied by a decrease of peroxide content of LDL, suggesting reduced postprandial protection of LDL against accumulation of peroxides and increased atherogenic oxidative modification (Sutherland et al., 1999).

There is need for a better understanding of the processes determining the postprandial PON1 activity and novel interventions that may modulate this response. Whether improvements of postprandial PON1 changes will reduce the risk for atherosclerosis in humans, remains to be determined.

REFERENCES

Abbott, C.A., Mackness, M.I., Kumar, S., Boulton, A.J., Durrington, P.N., 1995, Serum paraoxonase activity, concentration, and phenotype distribution in diabetes mellitus and its relationship to serum lipids and lipoproteins. *Arterioscler Thromb Vasc Biol*, 15:1812–1818.

Aviram, M., Hardak, E., Vaja, J., Mahmood, S., Milo, S., Hoffman, A., Billicke, S., Draganov, D., Rosenblat, M., 2000, Human serum paraoxonases (PON1) Q and R selectively decrease lipid peroxides in human coronary and carotid atherosclerotic lesions: PON1 esterase and peroxidase-like activities. *Circulation*, 101:2510–7.

Aviram, M., Rosenblat, M., Bisgaier, C.L., Newton, R.S., Primo-Parmo, S.L., La Du, B.N., 1998, Paraoxonase inhibits high-density lipoprotein oxidation and preserves its functions. A possible peroxidative role for paraoxonase. *J Clin Invest*, 101:1581–90.

Billicke, S., Draganov, D., Counsell, R., Stetson, P., Watson, C., Hsu, C., La Du, B.N., 2000, Human serum paraoxonase (PON1) isozymes Q and R hydrolyze lactones and cyclic carbonate esters. *Drug Metab Dispos*. 28:1335–42.

Blake, G.J., Ridker, P.M., 2001, Novel clinical markers of vascular wall inflammation. *Circ Res*, 89(9):763–771.

Blum, S., Aviram, M., Ben-Amotz, A., Levy, Y., 2006, Effect of Mediterranean meal on postprandial carotenoids, paraoxonase activity and C-reactive protein levels. *Ann Nutr Metab*, 50(1):20–4.

Bonora, E., Kiechl, S., Willeit, J., Oberhollenzer, F., Egger, G., Targher, G. et al., 1998, Prevalence of insulin resistance in metabolic disorders: the Bruneck Study. *Diabetes*, 47(10):1643–1649.

Braunwald, E., 1997, Shattuck lecture–cardiovascular medicine at the turn of the millennium: triumphs, concerns, and opportunities. *N Engl J Med*, 337(19):1360–1369.

Brunzell, J.D., Hazzard, W.R., Porte, D.J., Bierman, E.L., 1973, Evidence for a common, saturable, triglyceride removal mechanism for chylomicrons and very low density lipoproteins in man. *J Clin Invest*, 52, 1578–1585.

Castro Cabezas, M., 2003, Postprandial lipemia in familial combined hyperlipidemia. *Biochem Soc Trans*, 31:1090–1093.

Castro Cabezas, M., Erkelens, D.W., 1998b, The direct way from gut to vessel wall: atheroinitiation. *Eur J Clin Invest*, 28:504–505.

Castro Cabezas, M., De Bruin, T.W.A., Jansen, H., Kock, L.A.W., Kortlandt, W., Erkelens, D.W., 1993, Impaired chylomicron remnant clearance in familial combined hyperlipidemia. *Arterioscler Thromb*, 13:804–814.

Castro Cabezas, M., De Bruin, T.W.A., Westerveld, H.E., Meijer, E., Erkelens, D.W., 1998a, Delayed chylomicron remnant clearance in subjects with heterozygous familial hypercholesterolaemia. *J Intern Med*, 244:299–307.

Castro Cabezas, M., Van Heusden, G.P.H., de Bruin, T.W.A., van Beckhoven, J.R.C.M., Kock, L.A.W., Wirtz, K.W.A., Erkelens, D.W., 1993, Reverse cholesterol transport: relationship between free cholesterol uptake and HDL3 in normolipidemic and hyperlipidemic subjects. *Eur J Clin Invest*, 23:122–129.

Castro Cabezas, M., Van Oostrom, A.J.H.H.M., Erkelens, D.W., 2001, Gender differences in diurnal triglyceride profiles in healthy normolipidemic subjects. *Atherosclerosis*, 155:219–228.

Coll, B., Van Wijk, J.P.H., Parra, S., Castro Cabezas, M., Hoepelman, I.M., Alonos Vilaverde, C., De Koning, E.J.P., Camps, J., Ferre, N., Rabelink, T.J., Tous, M., Joven, J., 2006, Effects of rosiglitazone and metformin on postprandial paraoxonase-1 and monocyte chemoattractant protein-1 in human immunodeficiency virus-infected patients with lipodystrophy. *Eur J Pharmacol*, 544:104–110.

Coppack, S.W., 1997, Postprandial lipoproteins in non-insulin-dependent diabetes mellitus. *Diabet Med*, 14(Suppl 3):S67–S74.

Despres, J.P., Lamarche, B., Mauriege, P., Cantin, B., Dagenais, G.R., Moorjani, S. et al., 1996, Hyperinsulinemia as an independent risk factor for ischemic heart disease. *N Engl J Med*, 334(15):952–957.

Despres, J.P., Moorjani, S., Lupien, P.J., Tremblay, A., Nadeau, A., Bouchard, C., 1990, Regional distribution of body fat, plasma lipoproteins, and cardiovascular disease. *Arteriosclerosis*, 10(4): 497–511.

Eriksson, M., Angelin, B., Henriksson, P., Ericsson, S., Vitols, S., Berglund, L., 1991, Metabolism of lipoprotein remnants in humans. Studies during intestinal infusion of fat and cholesterol in subjects with varying expression of the low density lipoprotein receptor. *Arterioscler Thromb*, 11:827–837.

Ferrannini, E., Haffner, S.M., Mitchell, B.D., Stern, M.P., 1991, Hyperinsulinaemia: the key feature of a cardiovascular and metabolic syndrome. *Diabetologia*, 34(6):416–422.

Fontbonne, A., Eschwege, E., Cambien, F., Richard, J.L., Ducimetiere, P., Thibult, N. et al. 1989, Hypertriglyceridaemia as a risk factor of coronary heart disease mortality in subjects with impaired glucose tolerance or diabetes. Results from the 11-year follow-up of the Paris Prospective Study. *Diabetologia*, 32(5):300–304.

Fuhrman, B., Volkova, N., Aviram, M., 2005, Paraoxonase (PON1) is present in postprandial chylomicrons. *Atherosclerosis*, 180; 55–61.

Fuhrman, B., Volkova, N., Aviram, M., 2006, Postprandial serum triacylglycerols and oxidative stress in mice after consumption of fish oil, soy oil or olive oil: Possible role for paraoxonase-1 triacylglycerol lipase-like activity. *Nutrition*, 22:922–930.

Geluk, C.A., Halkes, C.J.M., de Jaegere, P.P.Th., Plokker, Th.W.M., Castro Cabezas, M. 2004, Daytime triglyceridemia in normocholesterolemic patients with premature atherosclerosis and in their first-degree relatives. *Metabolism*, 53:49–53.

Goldberg, R.B., Mellies, M.J., Sacks, F.M., Moye, L.A., Howard, B.V., Howard, W.J. et al. 1998, Cardiovascular events and their reduction with pravastatin in diabetic and glucose-intolerant myocardial infarction survivors with average cholesterol levels: subgroup analyses in the cholesterol and recurrent events (CARE) trial. The Care Investigators. *Circulation*, 98(23):2513–2519.

Haffner, S.M., Lehto, S., Ronnemaa, T., Pyorala, K., Laakso, M., 1998, Mortality from coronary heart disease in subjects with type 2 diabetes and in nondiabetic subjects with and without prior myocardial infarction. *N Engl J Med*, 339(4):229–234.

Halkes, C.J.M., Van Dijk, H., De Jaegere, P.P.Th., Plokker, H.W.M., Erkelens, D.W., Castro Cabezas, M., 2001, Postprandial increase of complement component 3 in normolipidemic patients with coronary artery disease. Effects of expanded dose simvastatin. *Arterioscl Thromb Vasc Biol*, 21:1526–1530.

Halkes, C.J.M., Van Dijk, H., De Jaegere, P.P.Th., Plokker, H.W.M., Erkelens, D.W., Castro Cabezas, M., 2003, Gender differences in postprandial ketone bodies in normolipidemic subjects and in untreated patients with familial combined hyperlipidemia. *Arterioscl Thromb Vasc Med*, 23:1875–1880.

Kannel, W.B., McGee, D.L., 1979, Diabetes and glucose tolerance as risk factors for cardiovascular disease: the Framingham study. *Diabetes Care*, 2(2):120–126.

Kris-Etherton, P.M., Harris, W.S., Appel, L.J., for the Nutrition Committee, 2003, Fish consumption, fish oil, omega-3 fatty acids, and cardiovascular disease. *Arterioscler Thromb Vasc Biol*, 23:e20–30.

Laakso, M., Lehto, S., Penttila, I., Pyorala, K., 1993, Lipids and lipoproteins predicting coronary heart disease mortality and morbidity in patients with non-insulin-dependent diabetes. *Circulation*, 88(4 Pt 1):1421–1430.

Lewis, G.F., O'Meara, N.M., Soltys, P.A., Blackman, J.D., Iverius, P.H., Pugh, W.L. et al. 1991, Fasting hypertriglyceridemia in noninsulin-dependent diabetes mellitus is an important predictor of postprandial lipid and lipoprotein abnormalities. *J Clin Endocrinol Metab*, 72(4):934–944.

Lewis, G.F., Steiner, G., 1996, Hypertriglyceridemia and its metabolic consequences as a risk factor for atherosclerotic cardiovascular disease in non-insulin-dependent diabetes mellitus. *Diabetes Metab Rev*, 12(1):37–56.

Mackness, B., Quarck, R., Verreth, W., Mackness, M., Holvoet, P., 2006, Human praoxonase-1 overexpression inhibits atherosclerosis in a mouse model of metabolic syndrome. *Arterioscl Thromb Vasc Biol*, 26:1545–1550.

Mackness, M.I., Harty, B., Bhatnagar, D., Wincour, P.H., Arrol, S., Ishola, M., Durrington, P.N., 1991, Serum Paraoxonase activity in familial hypercholesterolemia and insulin dependent diabetes mellitus. *Atherosclerosis*, 86:193–199.

Mackness, M.I., Mackness, B., Durrington, P.N., Connelly, P.W., Hegele, R.A., 1996, Paraoxonase: biochemistry, genetics and relationship to plasma lipoproteins. *Curr Opin Lipidol*, 7:69–76.

Meijssen, S., van Dijk, H., Verseyden, C., Erkelens, D.W., Castro Cabezas, M., 2002, Delayed and exaggerated postprandial complement 3 response in familial combined hyperlipidemia. *Arterioscl Thromb Vasc Biol*, 22:811–816.

Nathan, D.M., Meigs, J., Singer, D.E., 1997, The epidemiology of cardiovascular disease in type 2 diabetes mellitus: how sweet it is ... or is it? *Lancet*, 350(Suppl 1):SI4–SI9.

Patsch, J.R., Miesenböck, G., Hopferwieser, T., Mühlberger, V., Knapp, E., Dunn, J.K., Gotto, A.M., Patsch, W., 1991, Relation to triglyceride metabolism and coronary artery disease. Studies in the postprandial state. *Arterioscler Thromb*, 12:1336–1345.

Pyorala, K., Pedersen, T.R., Kjekshus, J., Faergeman, O., Olsson, A.G., Thorgeirsson, G., 1997, Cholesterol lowering with simvastatin improves prognosis of diabetic patients with coronary heart disease. A subgroup analysis of the Scandinavian Simvastatin Survival Study (4S). *Diabetes Care*, 20(4):614–620.

Reaven, G.M., 1988, Banting lecture. Role of insulin resistance in human diseae. *Diabetes*, 37:1595–1607.

Rozenberg, O., Rosenblat, M., Coleman, R., Shih, D.M., Aviram, M., 2003, Paraoxonase (PON1) deficiency is associated with increased macrophage oxidative stress: studies in PON1-knockout mice. *Free Radic Biol Med*, 34:774–84.

Rubinsztein, D.C., Cohen, J.C., Berger, M., van der Westhuyzen, D.R., Coetzee, G.A., Gevers, W., 1990, Chylomicron remnant clearance from the plasma is normal in familial hypercholesterolemic homozygotes with defined receptor defects. *J Clin Invest*, 86:1306–1312.

Senti, M., Nogues, X., Botet, J.P., Rubies-Prat, J., Vidal Barraquer, F., 1992, Lipoprotein profile in men with peripheral vascular disease: Role of intermediate density lipoproteins and apoprotein E phenotypes. *Circulation*, 85:30–36.

Sutherland, W.H.F., Walker, R.J., de Jong, S.A., van Rij, A.M., Phillips, V., Walker, H.L., 1999, Reduced postprandial serum paraoxonase activity after a meal rich in used cooking fat. *Arterioscler Thromb Vasc Biol*, 19:1340–1347.

Syvänne, M., Hilden, H., Taskinen, M.R., 1994, Abnormal metabolism of postprandial lipoproteins in patients with non- insulin-dependent diabetes mellitus is not related to coronary artery disease. *J Lipid Res*, 35(1):15–26.

Syvänne, M., Taskinen, M.R., 1997, Lipids and lipoproteins as coronary risk factors in non-insulin-dependent diabetes mellitus. *Lancet*, 350(Suppl 1):SI20–SI23.

Turner, R.C., Millns, H., Neil, H.A., Stratton, I.M., Manley, S.E., Matthews, D.R. et al., 1998, Risk factors for coronary artery disease in non-insulin dependent diabetes mellitus: United Kingdom Prospective Diabetes Study (UKPDS: 23). *BMJ*, 316(7134):823–828.

Van Oostrom, A.J.H.H.M., Alipour, A., Plokker, T.W., Sniderman, A.D., Castro Cabezas, M., 2007, The metabolic syndrome in relation to complement component 3 and postprandial lipemia in patients from an outpatient lipid clinic and healthy volunteers. *Atherosclerosis*, 190:167–173.

Van Oostrom, A.J.H.H.M., Castro Cabezas, M., Ribalta, J., Masana, L.L., Twickler, Th.B., Remijnse, T.A., Erkelens, D.W., 2000, Diurnal triglyceride profiles in healthy normolipidemic male subjects are related to insulin sensitivity, body composition and diet. *Eur J Clin Invest*, 30:964–971.

Van Oostrom, A.J.H.H.M., Plokker, H.W.M., Van Asbeck, B.S., Rabelink, T.J., Van Kessel, K.P.M., Jansen, A.H.J.M., Stehouwer, C.D.A., Castro Cabezas, M., 2006, Effects of rosuvastatin on postprandial leukocytes in mildly hyperlipidemic patients with premature coronary sclerosis. *Atherosclerosis*, 185:331–339.

Van Oostrom, A.J.H.H.M., Rabelink, T.J., Verseyden, C., Sijmonsma, T.P., Plokker, H.W.M., De Jaegere, P.P.Th., Castro Cabezas, M., 2004c, Activation of leukocytes by postprandial lipemia in healthy volunteers. *Atherosclerosis*, 177:175–182.

Van Oostrom, A.J.H.H.M., Sijmonsma, T.P., Verseyden, C., Jansen, E.H.J.M., de Koning, E.J.P., Rabelink, T.P., Castro Cabezas, M., 2003, Postprandial recruitment of neutrophils is associated with endothelial dysfunction. *J Lipid Res*, 44:576–583.

Van Oostrom, A.J.H.H.M., Van Wijk, J.P.H., Castro Cabezas, M., 2004a, Lipaemia, Inflammation and Atherosclerosis: Novel Opportunities in the Understanding and Treatment of Atherosclerosis. *Drugs*, 64(Suppl 2):19–41.

Van Oostrom, A.J.H.H.M., Van Dijk, H., Verseyden, C., Sniderman, A.D., Cianflone, K., Rabelink, T.J., Castro Cabezas, M., 2004b, A mixed meal prevents the fat-mediated postprandial rise of complement component 3 by enhanced insulin-mediated free fatty acid trapping. *Am J Clin Nutr*, 79:510–5.

Van Wijk, J.P.H., Buirma, R., Van Tol, A., Halkes, C.J.M., De Jaegere, P.P.Th., Plokker, H.W.M., Van der Helm, Y.J.M., Castro Cabezas, M., 2005, Effects of increasing doses of simvastatin on fasting lipoprotein subfractions, and the effect of high-dose simvastatin on postprandial chylomicron remnant clearance in normotriglyceridemic patients with premature coronary sclerosis. *Atherosclerosis*, 178:147–155.

Van Wijk, J.P.H., Coll, B., Castro Cabezas, M., Camps, J., Rabelink, T.J., Mackness, B., Joven, J., 2006b, Rosiglitazone modulates fasting and postprandial paraoxonase activity in type 2 diabetic patients. *CEPP*, 33:1134–1137.

Van Wijk, J.P.H., De Koning, E.J.P., Castro Cabezas, M., Joven, J., Op 't Roodt, J., Rabelink, T.J., Hoepelman, I.M., 2006a, Functional and structural markers of atherosclerosis in HIV-infected patients. *J Am Coll Cardiol*, 47(6):1117–23.

Verseyden, C., Meijssen, S., Castro Cabezas, M., 2004, Effects of Atorvastatin on Fasting Plasma and Marginated Apolipoproteins B48 and B100 in Large, Triglyceride-Rich Lipoproteins in Familial Combined Hyperlipidemia. *J Clin Endocrinol Metab*, 89:5021–5029.

Verseyden, C., Meijssen, S., van Dijk, H., Jansen, H., Castro Cabezas, M., 2003, Effects of atorvastatin on fasting and postprandial complement component 3 response in familial combined hyperlipidemia. *J Lipid Res*, 44:2100–2108.

Wallace, A.J., Sutherland, W.H., Mann, J.I., Williams, S.M., 2001, The effect of meals rich in thermally stressed olive and safflower oils on postprandial serum paraoxonase activity in patients with diabetes. *Eur J Clin Nutr*, 55:951–8.

Zilversmit, D.B., 1979, Atherogenesis: a postprandial phenomenon. *Circulation*, 60(3):473–485.

CHAPTER 8

PON1 GENOTYPES AND CORONARY HEART DISEASE

M. ROEST AND H.A.M. VOORBIJ

Laboratory for clinical Chemistry and Haematology, UMC Utrecht, The Netherlands

Abstract: Paraoxonase type 1 (PON1) have emerged as predictor of coronary heart disease (CHD). To date it is not known if PON1 is a causal determinant of atherosclerosis. In mouse models and in vitro it has been shown that PON1 is functionally involved in atherosclerosis, as it inhibits both LDL-oxidation and atherosclerosis progression. In humans, it turned out to be more difficult to find evidence for causality, because epidemiological studies on PON1 levels and/or activity are sensitive methodological bias and confounding variables and do not distinguish between cause or consequence. PON1 genotype is fixed at conception and therefore not subject to most forms of bias, not affected by confounding variables and not a consequence of CHD. Therefore, a relationship between PON1 genotype and CHD would support causal involvement of PON1 in CHD.

In the past decades there have been a large number of studies on PON1 Q192R, L55M and C-107T genotype and CHD. A meta analysis of those studies showed no convincing relationship, indicating that there is currently no strong genetic support for causal involvement of PON1 in CHD

Keywords: PON1, genotype, Coronary Heart Disease

The last decades a multitude of risk factors for coronary heart disease (CHD) have been established. Among these risk factors, low levels of high density lipoprotein (HDL) have emerged as one of the strongest (Gordon et al., 1989). There does not appear to be a single explanation for this protective role of HDL in CHD, but it is obvious that the proteins on HDL, which determine the function of the particle, play a major role in the protection against CHD. The cardioprotective effects of HDL have been attributed to enhanced cholesterol efflux and reversed cholesterol transport (RCT) from peripheral cells to the liver (Lewis and Rader, 2005). More recently, it has become apparent that HDL attenuates the bioavailability of a number of pro-oxidant species that have been implicated in the propagation of atherogenesis and inhibits the oxidative modification of LDL and therefore preventing the detrimental effects of this lipoprotein on the arterial wall (Navab et al., 1996). Paraoxonase type 1 (PON1), which is physically associated with the high-density lipoprotein (HDL) particle, plays a prominent role in those anti-oxidative and anti-inflammatory properties of HDL (Mackness et al., 1991).

B. Mackness et al. (eds.), The Paraoxonases: Their Role in Disease Development
and Xenobiotic Metabolism, 139–147.
© 2008 *Springer.*

The human paraoxonase family consists of three members; PON1, PON2 and PON3, which are aligned next to each other on chromosome 7 (Primo-Parmo et al., 1996). PON1 is the best characterized member of this family, whereas the main functions and properties of PON2 and PON3 remain to be elucidated. The current chapter will focus on PON1 genotypes and CHD.

PON1 is primarily a lactonase that hydrolyses aromatic and long chain aliphatic lactones, but is also capable of hydrolysing a variety of other substrates, including pesticides, nerve agents and phenylacetate (Costa et al., 1999). Furthermore, it has been suggested that PON1 may be able to prevent or limit oxidation of low-density lipoprotein (LDL) and therefore protect against atherosclerosis (Mackness et al., 1991). This latter hypothesis was proposed for the first time by Mackness in the early 1990s and resulted in an exponential gain in the number of studies to the role of PON-1 in CHD. More recently, it has been shown that PON-1 is also capable to hydrolyse the oxidised lipid derivates hydroy-docosahexaenoic acid (5-HETEL) and 4-hydroxy-docosahexaeonic acid (4-HDoHE), which are derivates of the oxidised fatty acids arachidonic acid and docosahexaeonoic acid, respectively (Draganov et al., 2005; Khersonsky and Tawfik, 2005). Those oxidised lipid derivates are potent triggers of an inflammatory response and therefore determinants of atherosclerotic disease.

1.   PON1 AND CHD: MOUSE MODELS

The most convincing evidence for causal involvement of PON1 in CHD is derived from studies in PON-1 null mice. Those PON1 null mice exhibited no detectable plasma paraoxonase activity, whereas their heterozygous counterparts exhibited 50% lower plasma paraoxonase activity compared with wild-type mice (Shih et al., 2000). PON1 null mice were unable to protect LDL against oxidation in a culture model of the artery wall and when fed a high-fat, high-cholesterol, cholate-containing diet, PON1 null mice developed significantly larger lesions than their wild-type and heterozygous littermates (Shih et al., 2000). Furthermore, when PON1 null mice were crossed with apoE null mice, PON1/apoE null mice developed significantly larger lesions than apoE null mice (Shih et al., 2000). LDL freshly isolated from PON1/apoE null mice had higher levels of biologically active phospholipids relative to LDL from apoE null mice, suggesting higher levels of oxidative stress in the double knockout mice. Similar to PON1 null mice, HDL from PON1/apoE null mice failed to protect LDL against oxidation. Taken together, these loss of function studies corroborate the hypothesis that PON1 protects against atherogenesis and is an important contributor to HDL's antioxidant capacity.

In a model of PON-1 over-expressing mice it has been shown that their HDL is more resistant to lipid peroxidation than HDL from their control littermates (Oda et al., 2002). In a PON-1 over-expressing mouse model it was shown that PON1 transgenic mice exhibited enhanced abilities to protect LDL against oxidation and developed significantly smaller lesions than their non-transgenic littermates (Oda et al., 2002).

2.   PON1 PHENOTYPE AND CHD IN HUMANS

There have been several epidemiological studies that have studied the relation between PON1 status and CHD. PON status can be distinguished into PON1 activity towards paraoxon and PON1 concentration, which is mainly determined in serum with ELISA or estimated with phenylacetate hydrolytic activity. The first study on the relation between PON1 activity and CHD was conducted in 1985 (McElveen et al., 1986). The outcomes of this study indicated that lower PON1 activity and lower PON1 levels predict a higher risk of CHD. Subsequently Navab and colleagues showed that patients with high HDL but low PON1 were more susceptible to CHD than patients with low HDL but high PON1 (Navab et al., 1997). Indicating that PON1 may be an important HDL protein for the protection against CHD. In support, three other studies, investigating the relationship of PON1 status and CHD found that low PON1 activity or levels were associated with an increased risk of CHD (Ayub et al., 1999; Jarvik et al., 2000; Mackness et al., 2001). The only prospective study of the relationship of PON1 status and CHD to report so far is the Caerphilly Prospective Study (Mackness et al., 2003). PON1 activity toward paraoxon was 30% lower in men who developed a new coronary event (n = 163) than in those who did not (n = 1175; p = 0.04). However, there were no differences in PON1 activity toward diazoxon or phenylacetate, PON1 concentration, or Apo J concentration. There was a graded decrease in CHD incidence by quintile of paraoxon hydrolysis; men in the highest quintile of PON1 activity were 60% less likely to have a coronary event than those in the lowest quintile. PON1 activity predicted coronary events independent of HDL-C.

3.   PON1 GENOTYPE AND CHD IN HUMANS

The findings that low PON1 activity and low PON1 levels predict a higher risk of CHD was an important step forward in understanding the involvement of PON1 in CHD in humans, but it does not prove causal involvement. Observational studies on acquired PON1 status and CHD are often affected by bias and confounding and they cannot be used to distinguish between cause or consequence. The discussion about cause or consequence, confounding and bias can be solved by the outcomes of appropriate, sufficiently powered genetic studies on the relationship between PON1 and CHD. Genes are fixed at conception and therefore neither a symptom of CHD, nor a reflection of confounding factors (Ridker and Stampfer, 1999). For the same reason, genotype is less sensitive to the major forms of bias (Ridker and Stampfer, 1999).

The PON1 gene contains at least three allelic variants (Q192R, L55M and C-107T), which affect PON1 protein expression or activity (Ng et al., 2005). This makes PON1 an appropriate gene to study the relationship between functional genetic variations in the PON1 gene and risk of CHD, to establish potential causal involvement of PON1 in CHD. The research of the different genetic variants in the PON1 gene to CHD is outlined below.

PON1 Q192R is the most studied allelic variant in the PON1 gene. The Q192R variant determines PON1's substrate specific activity, in particular toward

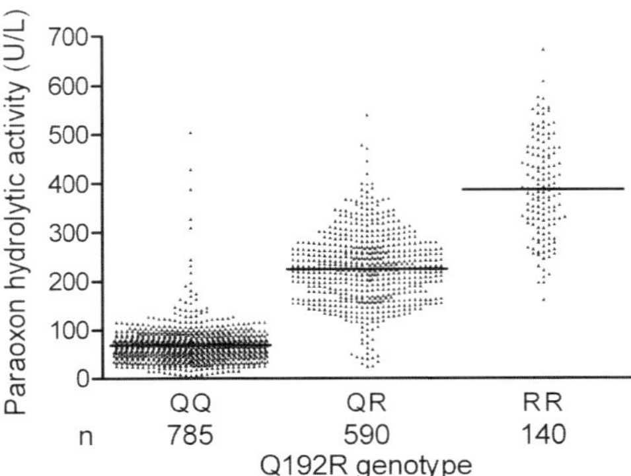

Figure 1. Serum paraoxon hydrolytic activity with regard to PON1 Q192R genotype. Illustrating that the 192R variant codes for increased PON1 activity towards paraoxon (Roest et al., 2007)

organophosphates. The R allele exhibits several-fold higher activity toward paraoxon, whereas the arylesterase activity and PON1 antigen concentration are similar in both isozymes. Recently, it has been shown that PON1 position 192 is involved in HDL binding. The PON1-192Q binds HDL with a 3-fold lower affinity than the R isozyme and consequently exhibits significantly reduced stability, lipolactonase activity, and macrophage cholesterol efflux (Gaidukov et al., 2006). In the same study, it was shown that the affinity and stability PON1 on HDL was lower in sera of individuals with the Q192 variant than in individuals with the 192R variant. In a recent cross sectional evaluation among 1527 post menopausal women we have shown that the R isoenzyme was indeed the major determinant of PON1 activity towards paraoxon ((Roest et al., 2007); Fig. 1). The Q192R allelic variant was also associated with serum arylesterase activity, but this turned out to be a reflection of allelic linkage to the C-107T variant in the PON1 gene (not shown).

4.   Q192R AND CHD

The Q192R polymorphism may play a role in CHD etiology because this genotype is associated with LDL oxidation; the PON1-192 R isoform is less effective at hydrolysing lipid peroxides than the Q isoform (Aviram et al., 1998; Mackness et al., 1999). In addition to its effects on LDL oxidation, the Q192R genotype may affect serum HDL levels (van Himbergen et al., 2005). Recent findings by our group indicate that the 192R variant was associated with higher HDL cholesterol levels in FH patients (Fig. 2). This was independently confirmed by others (Blatter Garin et al., 2006).

From 1990 to present, there have been a large number of studies on the relation between PON1 Q192R genotype and CHD. The majority of those studies lacked the power to draw firm conclusions about the relationships between PON1 Q192R genotype and CHD. Therefore, a very elegant meta analysis has been performed on the relation between PON1 Q192R genotype and CHD (Wheeler et al., 2004). This meta analysis included 35 studies, which contained 45 separate comparisons,

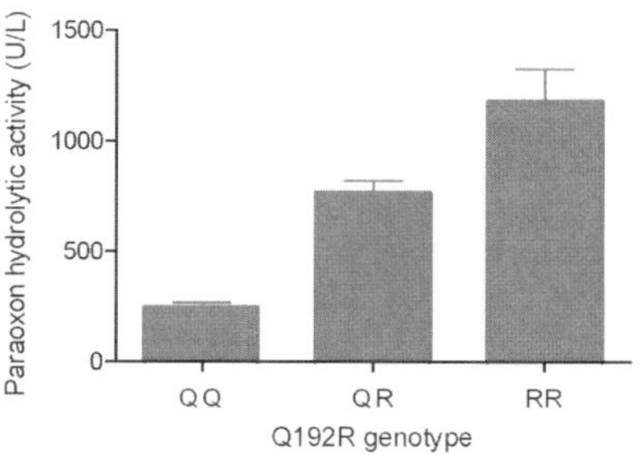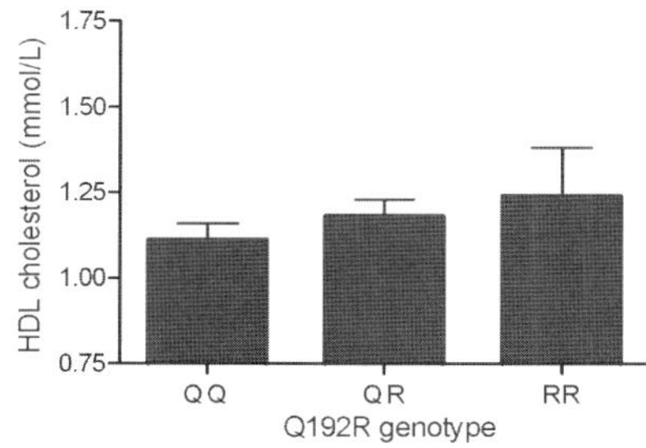

Figure 2. PON1 activity towards paraoxon and HDL cholesterol concentrations with regard to Q192R genotype. Illustrating that the 192R variant is associated with both higher PON1 activity towards paraoxon and HDL cholesterol concentration

when studies with different types of CHD and with different ethnic groups were considered separately. Among those, 19 studies involved a total of 5723 cases of MI and 8063 controls, while 26 involved a total of 4383 cases of coronary stenosis and 7545 controls. Studies included both males and females and were conducted in a wide range of geographical settings. Most studies had a retrospective design. The outcomes of the meta analysis suggested that the R192 varant was associated with an increased risk of CHD: the relative risk of the R192 variant for total CHD was 1.12 (95% CI 1.07–1.16; Fig. 1) per allele, with corresponding results under dominant and recessive genetic models of 1.15 (1.09–1.22) and 1.15 (1.05–1.25), respectively. Subsidiary analysis of specific CHD endpoints yielded a per-allele relative risk for MI of the R192 variant of 1.08 (1.02–1.14) and for coronary stenosis of 1.16 (1.09–1.24). This relationship may be affected by preferential publication of strikingly positive findings in smaller studies. To overcome such bias, the investigators restricted the attention to the five largest studies (each of which included more than 500 cases). In aggregate, these five studies yielded a per-allele relative risk of 1.05 (0.98–1.13).

In summary, if anything at al, the 192R variant of the Q192R genotype may be weakly associated with an increased risk of CHD. The interpretation of this relationship is complicated, because the 192R variant corresponds with high PON1 activity towards paraoxon, while there is general consensus that serum PON1 activity towards paraoxon is inversely associated with CHD risk. In other words, a meta analysis of genetic studies to the Q192R genotype and CHD do not support causal involvement of PON1 activity in CHD.

5. PON1 L55M AND C-107T GENOTYPE

The PON1 L55M and C-107T variants in the PON1 gene are both associated with variation in PON1 levels (Roest et al., 2007). Among those, the L55M genotype was the first to be discovered. Still, little is known about its functionality. It has been suggested that the PON1 55M variant may have a longer half life in the circulation and therefore correspond with higher PON1 concentrations (Leviev et al., 2001).

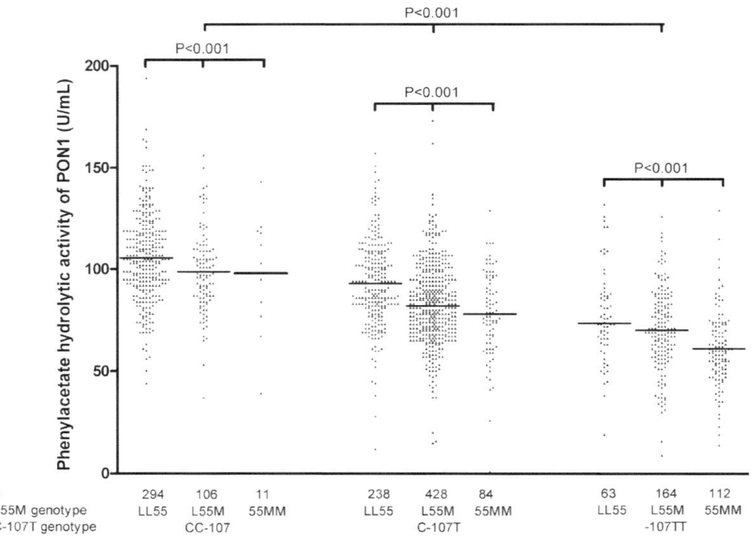

Figure 3. PON1 activity towards phenylacetate with regard to subgroups of the C-107T and L55M genotypes. Showing that the C-107T variant is a major determinant of PON1 levels and that there is also an independent effect of L55M genotype (Roest et al., 2007)

The C-107T genotype in the promoter sequence of PON1 has been described for the first time by two independent groups (Leviev and James, 2000; Brophy et al., 2001). The -107T variant is consistent with the disruption of a Sp1 recognition sequence and therefore a reduced PON1 expression (Deakin et al., 2003). It is now clearly shown that C-107T genotype is the strongest determinant of serum PON1 levels, while the L55M variant has also an independent effect on PON1 levels ((Roest et al., 2007); Fig. 3).

In addition, the C-107T variants are also associated with PON1 activity towards paraoxon, which is partly independent of linkage to the Q192R genotype (Fig. 4). The same was found for L55M genotype (data not shown).

There is no firm evidence that PON1 C-107T and L55M genotype affect LDL-oxidation *in vivo*. On the other hand, our group has shown that both the C/107T and L55M genotypes affect HDL content in FH patients (Fig. 5).

6. PON1 L55M AND C-107T GENOTYPES AND CHD

There have been several studies on the relation between PON1 L55M and C-107T genotypes and CHD, but on individual level those studies lacked the power to draw firm conclusions. The previously described meta analysis of Wheeler indicate that PON1 L55M and C-107T genotypes are not associated with CHD risk (Wheeler et al., 2004). In this meta analysis the relation between the L55M polymorphism and CHD was analysed in 21 studies (based on 19 reports), while the relation between the C-107T polymorphism and CHD has been analysed in four reports.

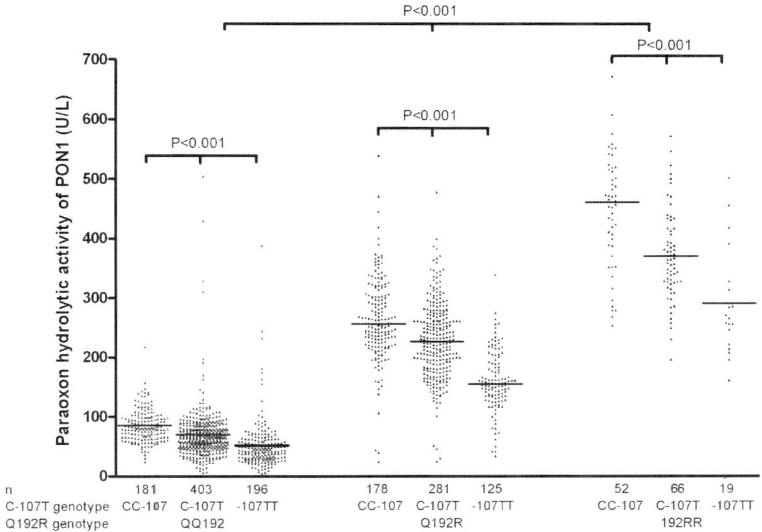

Figure 4. PON1 activity towards paraoxon with regard to subgroups of the Q192R and C-107T. Showing that PON1 activity towards paraoxon is not only affected by the Q192R genotype, but also by C-107T genotype, independent of Q192R (Roest et al., 2007)

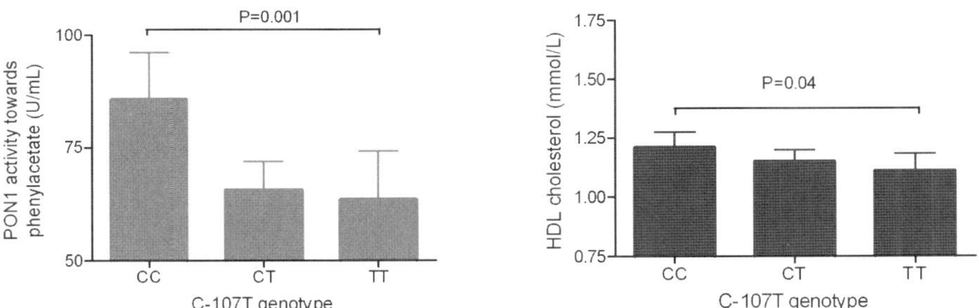

Figure 5. PON1 activity towards phenylacetate and HDL cholesterol concentrations with regard to PON1 L55M genotype. Illustrating that the high expressor C-107 variant is associated with higher HDL cholesterol concentration

7.   SUMMARY AND CONCLUSIONS

PON1 Q192R, C-107T and L55M genotypes are the major determinants of variation in both PON1 activity and PON1 expression. There is no consistent evidence for major involvement of PON1 genotype in CHD. If anything, the 192R variant may be associated with a slightly increased risk of CHD. The other functional genotypes L55M and C-107T were not associated with CHD. Although this argues against causal involvement of PON1 in CHD, it does not completely close the book because the interpretation of a meta analyses is very complex when the different studies of the analysis lack uniformity. This was the case in the meta analysis of studies on PON1 and CHD, which showed a strong variation in age, gender, ethnical background and CHD outcome definitions among studies. This heterogeneity leads to an attenuation of the relationship between PON1 genotypes and specific CHD outcomes. A further complicating issue of this meta analysis is that the ratio between cases and controls is highly variable among studies, which leads to unbalanced distribution of the different ethnical backgrounds in the pooled cases and controls. The definitive answer about the involvement of PON1 genotype in CHD should come from ultra large studies (either case-control or cohort studies) using homogeneous case definitions, and a balanced distribution of cases and controls regarding to age gender and ethnical background.

REFERENCES

Aviram, M. et al., 1998, Paraoxonase active site required for protection against LDL oxidation involves its free sulfhydryl group and is different from that required for its arylesterase/paraoxonase activities: selective action of human paraoxonase allozymes Q and R: Arterioscler. Thromb.Vasc. Biol., v. 18, no. 10, pp. 1617–1624.

Ayub, A., M. I. Mackness, S. Arrol, B. Mackness, J. Patel, and P. N. Durrington, 1999, Serum paraoxonase after myocardial infarction: Arterioscler. Thromb.Vasc. Biol., v. 19, no. 2, pp. 330–335.

Blatter Garin, M. C., X. Moren, and R. W. James, 2006, Paraoxonase-1 and serum concentrations of HDL-cholesterol and apoA-I: J.Lipid Res., v. 47, no. 3, pp. 515–520.

Brophy, V. H., M. D. Hastings, J. B. Clendenning, R. J. Richter, G. P. Jarvik, and C. E. Furlong, 2001, Polymorphisms in the human paraoxonase (PON1) promoter: Pharmacogenetics, v. 11, no. 1, pp. 77–84.

Costa, L. G., W. F. Li, R. J. Richter, D. M. Shih, A. Lusis, and C. E. Furlong, 1999, The role of paraoxonase (PON1) in the detoxication of organophosphates and its human polymorphism: Chem. Biol. Interact., v. 119–120, pp. 429–438.

Deakin, S., I. Leviev, M. C. Brulhart-Meynet, and R. W. James, 2003, Paraoxonase-1 promoter haplotypes and serum paraoxonase: a predominant role for polymorphic position - 107, implicating the Sp1 transcription factor: Biochem. J., v. 372, no. Pt 2, pp. 643–649.

Draganov, D. I., J. F. Teiber, A. Speelman, Y. Osawa, R. Sunahara, and B. N. La Du, 2005, Human paraoxonases (PON1, PON2, and PON3) are lactonases with overlapping and distinct substrate specificities: J. Lipid Res., v. 46, no. 6, pp. 1239–1247.

Gaidukov, L., M. Rosenblat, M. Aviram, and D. S. Tawfik, 2006, The 192R/Q polymorphs of serum paraoxonase PON1 differ in HDL binding, lipolactonase stimulation, and cholesterol efflux: J. Lipid Res., v. 47, no. 11, pp. 2492–2502.

Gordon, D. J., J. L. Probstfield, R. J. Garrison, J. D. Neaton, W. P. Castelli, J. D. Knoke, D. R. Jacobs, Jr., S. Bangdiwala, and H. A. Tyroler, 1989, High-density lipoprotein cholesterol and cardiovascular disease. Four prospective American studies: Circulation, v. 79, no. 1, pp. 8–15.

Jarvik, G. P., L. S. Rozek, V. H. Brophy, T. S. Hatsukami, R. J. Richter, G. D. Schellenberg, and C. E. Furlong, 2000, Paraoxonase (PON1) phenotype is a better predictor of vascular disease than is PON1(192) or PON1(55) genotype: Arterioscler. Thromb.Vasc. Biol., v. 20, no. 11, pp. 2441–2447.

Khersonsky, O., and D. S. Tawfik, 2005, Structure-reactivity studies of serum paraoxonase PON1 suggest that its native activity is lactonase: Biochemistry., v. 44, no. 16, pp. 6371–6382.

Leviev, I., S. Deakin, and R. W. James, 2001, Decreased stability of the M54 isoform of paraoxonase as a contributory factor to variations in human serum paraoxonase concentrations: J. Lipid Res., v. 42, no. 4, pp. 528–535.

Leviev, I., and R. W. James, 2000, Promoter polymorphisms of human paraoxonase PON1 gene and serum paraoxcnase activities and concentrations: Arterioscler. Thromb.Vasc. Biol., v. 20, no. 2, pp. 516–521.

Lewis, G. F., and D. J. Rader, 2005, New insights into the regulation of HDL metabolism and reverse cholesterol transport: Circ. Res., v. 96, no. 12, pp. 1221–1232.

Mackness, B., G. K. Davies, W. Turkie, E. Lee, D. H. Roberts, E. Hill, C. Roberts, P. N. Durrington, and M. I. Mackness, 2001, Paraoxonase status in coronary heart disease: are activity and concentration more importan: than genotype?: Arterioscler. Thromb. Vasc. Biol., v. 21, no. 9, pp. 1451–1457.

Mackness, B., P. Durrington, P. McElduff, J. Yarnell, N. Azam, M. Watt, and M. Mackness, 2003, Low paraoxonase activity predicts coronary events in the Caerphilly Prospective Study: Circulation, v. 107, no. 22, pp. 2775–2779.

Mackness, B., P. N. Durrington, and M. I. Mackness, 1999, Polymorphisms of paraoxonase genes and low-density lipoprotein lipid peroxidation: Lancet, v. 353, no. 9151, pp. 468–469.

Mackness, M. I., S. Arrol, and P. N. Durrington, 1991, Paraoxonase prevents accumulation of lipoperoxides in low-density lipoprotein: FEBS Lett., v. 286, no. 1–2, pp. 152–154.

McElveen, J., M. I. Mackness, C. M. Colley, T. Peard, S. Warner, and C. H. Walker, 1986, Distribution of paraoxon hydrolytic activity in the serum of patients after myocardial infarction: Clin. Chem., v. 32, no. 4, pp. 671–673.

Navab, M. et al., 1996, The Yin and Yang of oxidation in the development of the fatty streak. A review based on the 1994 George Lyman Duff Memorial Lecture: Arterioscler. Thromb.Vasc. Biol., v. 16, no. 7, pp. 831–842.

Navab, M. et al., 1997, Mildly oxidized LDL induces an increased apolipoprotein J/paraoxonase ratio: J. Clin. Invest, v. 99, no. 8, pp. 2005–2019.

Ng, C. J., D. M. Shih, S. Y. Hama, N. Villa, M. Navab, and S. T. Reddy, 2005, The paraoxonase gene family and atherosclerosis: Free Radic. Biol. Med., v. 38, no. 2, pp. 153–163.

Oda, M. N., J. K. Bielicki, T. T. Ho, T. Berger, E. M. Rubin, and T. M. Forte, 2002, Paraoxonase 1 overexpression in mice and its effect on high-density lipoproteins: Biochem. Biophys. Res. Commun., v. 290, no. 3, pp. 921–927.

Primo-Parmo, S. L., R. C. Sorenson, J. Teiber, and B. N. La Du, 1996, The human serum paraoxonase/arylesterase gene (PON1) is one member of a multigene family: Genomics, v. 33, no. 3, pp. 498–507.

Ridker, P. M., ard M. J. Stampfer, 1999, Assessment of genetic markers for coronary thrombosis: promise and precaution: Lancet, v. 353, no. 9154, pp. 687–688.

Roest, M., T. M. Himbergen, A. Barendrecht, P. H. Peeters, Y. T. van der Schouw, H. A. M. Voorbij, 2007, Genetic and environmental determinants of the PON-1 phenotype: Eur. J. Clin Invest, 2007, in press.

Shih, D. M. et al., 2000, Combined serum paraoxonase knockout/apolipoprotein E knockout mice exhibit increased lipoprotein oxidation and atherosclerosis: J. Biol. Chem., v. 275, no. 23, pp. 17527–17535.

van Himbergen, T. M., M. Roest, J. de Graaf, E. H. Jansen, H. Hattori, J. J. Kastelein, H. A. Voorbij, A. F. Stalenhoef, and L. J. van Tits, 2005, Indications that paraoxonase-1 contributes to plasma high density lipoprotein levels in familial hypercholesterolemia: J. Lipid Res., v. 46, no. 3, pp. 445–451.

Wheeler, J. G., B. D. Keavney, H. Watkins, R. Collins, and J. Danesh, 2004, Four paraoxonase gene polymorphisms in 11212 cases of coronary heart disease and 12786 controls: meta-analysis of 43 studies: Lancet, v. 363, no. 9410, pp. 689–695.

PART 3

PON1 STRUCTURE

CHAPTER 9

HUMAN PARAOXONASE I: A POTENTIAL BIOSCAVENGER OF ORGANOPHOSPHORUS NERVE AGENTS

DAVID T. YEUNG, DAVID E. LENZ AND DOUGLAS M. CERASOLI*

Research Division, U.S. Army Medical Research Institute of Chemical Defense, 3100 Ricketts Point Rd., Aberdeen Proving Ground, MD 21010-5400

Abstract: Human serum paraoxonase (HuPON1, EC 3.1.8.1) is a Ca^{2+}-dependent enzyme that hydrolyzes esters including organophosphorus (OP) nerve agents. Efforts to elucidate the putative roles of active site amino acid residues have been hampered by the lack of three-dimensional structural information of this enzyme. The advent of an homology model for HuPON1 (folded onto the six-fold β-propeller structure of squid diisopropylfluorophosphatase) followed by a confirmatory crystal structure of a closely related hybrid PON molecule has led to the design and expression of site-directed mutants of HuPON1. These mutants were analyzed for enzymatic activity against a variety of substrates, including OP nerve agents. Substitution of residues predicted from the model to be important for substrate binding, Ca^{2+} ion coordination, and catalysis resulted in enzyme inactivation, supporting the validity of the proposed structural model. Mutants with altered specificities for substrate were identified; some recognized OP substrates as competitive inhibitors. The OP nerve agent soman (GD) has two chiral centers; the four stereoisomers of GD vary in toxicity by several orders of magnitude. A novel GC/MS-based assay was developed to examine the stereospecificity of wild-type and mutant HuPON1 enzymes for GD. The wild-type HuPON1 catalyzed hydrolysis of all four GD isomers with modest stereoselectivity for the less toxic $(C \pm P+)$ isomers. Two of the mutants tested (S193A and S193G) demonstrated altered stereospecificity and kinetics, resulting in three to four-fold increased rate of hydrolysis for one of the toxic P- GD isomers. The capacity of recombinant HuPON1 to hydrolyze structurally isomeric OP nerve agents (VX and VR) was examined and found to differ in both affinity and rate of catalysis. Unlike VX, VR was not hydrolyzed by the H115W mutant of HuPON1, but instead was recognized as a competitive inhibitor. Given the structural similarity between VX and VR, these results suggest that residue H115 in wild-type HuPON1 is critical for determining the substrate specificity of HuPON1 for some classes of OPs. Together, the results presented have expanded our understanding of the amino acid

---

* To whom all correspondence should be addressed. The opinions or assertions contained herein are the private views of the authors, and are not to be construed as reflecting the view of the Department of the Army or the Department of Defense.

151

B. Mackness et al. (eds.), The Paraoxonases: Their Role in Disease Development and Xenobiotic Metabolism, 151–170.
© 2008 *Springer.*

residues critical for HuPON1 OP hydrolase activity. This information is being used to design mutants of PON1 with enhanced anti-OP activity and stereoselectivity towards the more toxic OP stereoisomers

Keywords: Nerve agents, bioscavengers, cholinesterase, organophosphorus poisons, chemical warfare agents, prophylaxis

1.   INTRODUCTION

Human paraoxonase-1 (PON1) belongs to the family of serum paraoxonases consisting of PON1, PON2, and PON3 (Aharoni et al., 2004; Draganov et al., 2000; Furlong et al., 2002; Harel et al., 2004b; van Himbergen et al., 2006; La Du et al., 1999; Primo-Parmo et al., 1996; Teiber et al., 2003). The 3 human PON genes are located adjacent to each other at bands q21-q22 on chromosome 7 (Furlong et al., 2002). Of the 3 PON enzymes, PON1 is the most studied and best understood (Clendenning et al., 1996; Furlong et al., 2000a; Furlong et al., 2000b). It was initially identified by Abraham Mazur in 1946 when he reported the discovery of an OP hydrolyzing enzyme in animal tissue (van Himbergen et al., 2006; Mazur, 1946). Although the enzyme was named after its capacity to hydrolyze paraoxon, it was later discovered that it actually exhibits a broad spectrum of activities (Aldridge 1953a; Aldridge 1953b). Among the diverse substrates recognized by PON1 are esters, lactones, phospholipids, and organophosphorus compounds (Billecke et al., 2000; Broomfield et al., 2000; Davies et al., 1996; Furlong et al., 1998; Furlong et al., 2000a; Furlong et al., 2000b; Gaidukov et al., 2006; Jakubowski et al., 2001; Josse et al., 1999a; Josse et al., 1999c; Josse et al., 2002; Khersonsky and Tawfik, 2005; Lacinski et al., 2004; Li et al., 1995; Li et al., 2000; Rodrigo et al., 2001; Teiber et al., 2003; Yeung et al., 2004). The enzyme is synthesized in the liver and bound to high density lipoprotein (HDL), likely in association with human phosphate binding protein (HPBP; (Morales et al., 2006)) before it is secreted into circulation (Blatter et al., 1993; Gaidukov et al., 2006; Harel et al., 2004b; James and Deakin 2004; Rosenblat et al., 2006; Sorenson et al., 1999). The concentration of PON1 in human plasma is $\sim 50\,mg/L$, but can vary by as much as 13-fold from one individual to another (Furlong et al., 2000b; Gaidukov and Tawfik 2007; Rochu et al., 2007). Of note, the level of serum PON1 in the circulation of small non-primate mammals has been inversely correlated with susceptibility to OP intoxication (Costa et al., 1999; Costa et al., 2003a; Costa et al., 2005; Furlong et al., 1998; Furlong et al., 2000b; Li et al., 1993). More importantly, the exogenous administration of purified rabbit PON1 has been shown to protect against OP toxicity in both rats and mice (Li et al., 1995).

In humans, several polymorphisms of HuPON1 are known to exist, where the most prominent is the Q192R allozyme (Table 1), which can have a substantial impact on PON1 activities (Billecke et al., 2000; Costa et al., 2003b; Furlong et al., 1998; Furlong et al., 2000b; Furlong et al., 2002; Gaidukov et al., 2006; Li et al., 2000; Yeung

Table 1. Kinetic properties of plasma purified HuPON1 (Masson et al., 1998)

Substrate	Bimolecular Rate Constant ($k_{cat}/K_m[M^{-1} \bullet min^{-1}]$)	
	Q192[b]	R192[b]
GD	2.80×10^6	2.10×10^6
GB	9.10×10^5	6.80×10^4
DFP	3.70×10^4	N.D.[a]
Paraoxon	6.80×10^5	2.40×10^6

[a] Not Determined.
[b] Two most commonly occurring allelic variants of PON in the human population

et al., 2004). Although the *in vivo* substrate(s) of HuPON1 is presently unknown, there is increasing evidence that the enzyme acts as a lactonase (Jakubowski et al., 2001; Khersonsky and Tawfik, 2005; Khersonsky and Tawfik, 2006; Lacinski et al., 2004; Teiber et al., 2003), and as such, appears to have anti-atherosclerotic properties by preventing the accumulation of oxidized low-density lipoproteins (LDL) (Aviram et al., 1998; Brites et al., 2004; van Himbergen et al., 2006; Shih et al., 1998).

The mature HuPON1 enzyme retains its N-terminus signal peptide and consists of 354 amino acid residues. Depending on glycosylation state, its molecular weight can range from 38–45 kDa (Clendenning et al., 1996; Furlong et al., 2000a; Hassett et al., 1991). It is known that the PON1 enzyme is a serum OP-hydrolase with an absolute requirement for two Ca^{2+} ions (Kuo and La Du, 1998; Sorenson et al., 1995), wherein one calcium is required to maintain its overall structural integrity while the other is essential for enzymatic activities (Jakubowski, 2000; Josse et al., 1999c; Kuo and La Du, 1998; Kuo and La Du, 1998; Sorenson et al., 1995). Additionally, through chemical modification and site-directed mutagenesis studies, a number of residues have been identified as essential for activity (Aharoni et al., 2004; Amitai et al., 2006; Doorn et al., 1999; Harel et al., 2004b; Josse et al., 1999a; Josse et al., 1999b; Josse et al., 1999c; Josse et al., 2001; Josse et al., 2002; Khersonsky and Tawfik 2006; Rosenblat et al., 2006; Yeung et al., 2004). At present, the exact structure and catalytic mechanism of the HuPON1 enzyme is still unknown, although a recently published gene-shuffled bacterially-expressed chimeric PON1 (crePON1) variant suggests that HuPON1 is a six-fold beta-propeller protein and that catalysis may be mediated by a functionally unique dyad of histidine residues located at positions 115 and 134 (Harel et al., 2004b).

2. HUPON1 AS A POTENTIAL BIOSCAVENGER OF OP NERVE AGENTS

Organophosphorus (OP) chemical weapons are among the most toxic substances that have been identified (Ballantyne and Marrs, 1992; Brown and Brix, 1998; Dacre, 1984; Heath and Meredith, 1992; Holstege et al., 1997; Reutter 1999; Sidell, 1997;

Taylor, 2001). Originally, these OP poisons were developed for use as insecticides, (Ballantyne and Marrs, 1992; Brown and Brix, 1998; Holstege et al., 1997) but their extreme toxicity toward higher vertebrates has led to their adoption as weapons of warfare (Holstege et al., 1997; Maynard and Beswick, 1992; Sidell, 1997). The OP compounds most commonly utilized as chemical weapons (usually referred to as nerve agents) that might be encountered are tabun (GA), sarin (GB), soman (GD), cyclosarin (GF), VX and Russian VX (VR); GB, GD, and GF all have fluoride leaving groups bound to the phosphorus atom, while GA has cyanide and VX and VR have thiol leaving groups (Holstege et al., 1997; Lee, 1997). The potential for the use of these poisons as weapons of terrorism in civilian settings continues to grow, as exemplified by the use of sarin by the Aum Shinrikyo cult in the 1995 Tokyo subway attack (Masuda et al., 1995; Nagao et al., 1997). It is therefore becoming increasingly imperative that effective medical countermeasures against these terrorist threats be developed.

The molecular weights of nerve agents range from 140 to 267 Daltons (Da), and under standard conditions they are all liquids, but differ in their degrees of volatility (Somani et al., 1992). They have median lethal dose (LD_{50}) values in mammals, including estimates for humans, in the μg/kg dose range for all routes of exposure except dermal, where LD_{50} doses are in the mg/kg range (Maynard and Beswick, 1992). Although pharmacological treatments such as anticonvulsants, acetylcholine receptor antagonists and oximes therapies are in place to counteract the immediate effects of OP nerve agent intoxication (Ballantyne and Marrs, 1992; Heath and Meredith, 1992; Taylor, 2001), these therapies do not directly detoxify these poisons *in vivo*. However, the mechanism of OP toxicity suggests that a broad spectrum prophylactic approach based on a reduction in the concentration of OP toxicant in the blood before it can reach its site of action (synaptic endplates) should be particularly effective; potentially incapacitating or even toxic exposures could be mitigated to mild symptoms, while lower level exposures could be rendered inconsequential. This type of rapid onset therapy that specifically reduces the concentration of the nerve agent poison in circulation would be advantageous because it would avoid most or all of the behavioral incapacitation associated with conventional nerve agent therapy (Cerasoli and Lenz, 2002; Lenz et al., 2001).

The potential advantages of the inactivation of OP nerve agents by catalytically active enzymes (scavengers) have resulted in consideration of this approach as a possible therapeutic strategy. The use of catalytic scavengers would be advantageous because small amounts of enzyme, meaning lower concentrations in circulation, would be sufficient to detoxify large amounts of nerve agent (as in an acute exposure), assuming that the rate of reaction of the scavenger with the nerve agent was sufficiently fast (Masson et al., 1998). Additionally, catalytic scavengers would have an advantage over stoichiometric binders of nerve agents (Lenz et al., 2006), in that they would not be consumed in the process of detoxifying the nerve agent, and thus would be available to protect against multiple exposures of either high or low doses of OPs. In its wild-type form, HuPON1 has the capacity to detoxify many

OP nerve agents albeit at a rate too slow to protect against a substantial exposure to a nerve agent (Broomfield et al., 2000; Josse et al., 2002).

Despite its naturally low catalytic efficiency against OP poisons, it is this capacity to hydrolyze OPs that makes HuPON1 a potential bioscavenger candidate (Josse et al., 2002; Masson et al., 1998; Rochu et al., 2007). Since the enzyme is of human origin, the risk of an adverse immune response upon subsequent administration is low when compared with nonhuman proteins (Rochu et al., 2007). Additionally, injection of purified rabbit PON1 into rats and mice has previously been shown to protect against OP-induced intoxication (Li et al., 1995). Most importantly, it has been postulated that if the catalytic efficiency of PON1 could be enhanced by only one or two orders of magnitude through genetic engineering efforts, then such a PON1 variant would afford *in vivo* protection against OP intoxication by substantial (multiple LD_{50}) OP nerve agent exposures (Josse et al., 2001; Josse et al., 2002; Rochu et al., 2007).

Consequently, recently efforts have focused on examining the relative roles of different amino acid residues in and around the active site of HuPON1, to better understand the parameters that mediate OP binding and hydrolysis by this enzyme. Likewise, the capacity of HuPON1 to distinguish between different structurally related OP substrates, such as nerve agent structural and stereo isomers, has been examined.. The overall goal of these investigations is to develop a sufficiently complete body of information about the HuPON1 enzyme, and in particular its active site residues and catalytic mechanism, to enable the creation of rationally designed variants with enhanced activities against OP nerve agents.

3.   STRUCTURE-ACTIVITY ANALYSES OF ACTIVE SITE AMINO ACID RESIDUES

Initial efforts directed at evaluating HuPON1 as a catalytic scavenger were hampered by the fact that relatively little was known about either the structure of the HuPON1 enzyme or its OP hydrolyzing activities. It is very difficult to purify the enzyme, either from serum or *in vitro* mammalian expression systems, and efforts to crystallize the native enzyme were largely unsuccessful (Morales et al., 2006). Despite the lack of structural information, prior experiments utilizing group-specific reagents followed by site-directed mutagenesis identified many residues essential for enzyme activity (Josse et al., 1999a; Josse et al., 1999b; Josse et al., 1999c; Josse et al., 2001). From the results of these experiments and subsequent sequence-alignment studies, it was postulated that the enzyme was structurally homologous to the squid DFPase enzyme (Fig. 1; (Josse et al., 2002)). Based on this model, HuPON1 was postulated to adopt a six-fold beta-propeller structure in which the beta-pleated sheets are radially arranged into a torus-like shape with the two calcium ions located in the center (Josse et al., 2002; Yeung et al., 2004).

To test the validity of this hypothetical structural model (PDB: 1XHR), site-directed mutants were generated at amino acid residues predicted to be important for enzymatic activity based on their physical locations in the model. Although many of these amino acid substitutions resulted in mutant enzymes with no detectable

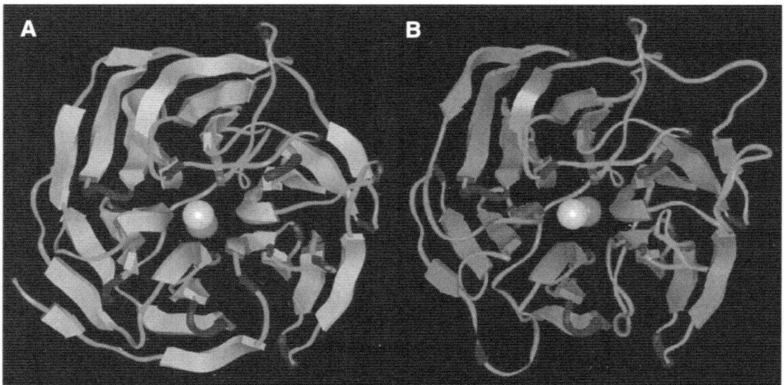

Figure 1. Crystal structure of DFPase used as a template to model a proposed secondary structure of HuPON1 (Yeung et al. 2004; Josse et al. 2002). (A) View of a ribbon diagram of the DFPase structure as determined from x-ray crystallography experiments (Scharff et al. 2001). (B) Ribbon diagram representation of the proposed secondary structure of HuPON1 (PDB: 1XHR) viewed along the axis with the catalytic calcium (white) above the structural calcium (pink). The model displays a slightly distorted six-bladed β-propeller shown in the same relative orientation as (A). HuPON1 was simulated using explicitly solvated dynamics with Insight/Discover software from Accelrys Inc. Computational resources were provided by the National Cancer Institute's Advanced Biomedical Computing Center in Frederick, Maryland

activity against either phenyl acetate or paraoxon, all of the mutants were readily detected by western blot from the supernatants of transiently transfected tissue culture cells, indicating that each mutant was structurally stable and capable of being secreted. The results of these mutagenesis experiments were consistent with the proposed homology model, as substitution at residues expected to be important for catalysis, active site structure, or coordination of the calcium ions led to loss of detectable enzymatic activity (Table 2; (Yeung et al., 2004)).

Several insights into the mechanics of the HuPON1 active site were gained from this study, specifically with respect to the importance of the histidine residue at position 115 (H115) and the phenylalanine at 222 (F222). Substitution of histidine with tryptophan (H115W) resulted in a mutant that was able to hydrolyze paraoxon but not phenyl acetate. (Yeung et al., 2004; Yeung et al., 2005). Further, phenyl acetate was found to competitively inhibit the hydrolysis of paraoxon by the H155W mutant (Table 3; (Yeung et al., 2005)). Together, these results indicated that the substitution of tryptophan for histidine at position 115 did not disrupt the OP hydrolytic machinery of the enzyme, but rather altered the capacity of some substrates (such as phenyl acetate) to occupy the active site in the correct position to be hydrolyzed.., The ability of phenyl acetate to competitively inhibit paraoxon hydrolysis strongly suggests that both compounds bind to HuPON1 at the same site. The reverse result was observed with the F222Y mutant, which hydrolyzed phenyl acetate but not paraoxon (Yeung et al., 2004). Interestingly, the F222Y mutant did not recognize diisopropyl-fluorophosphonate (DFP), a similar OP, as an inhibitor (Table 3; (Yeung et al., 2005)). This finding suggested that residue F222

Table 2. Amino acid residues predicted to play a functional role in HuPON1 substrate binding and catalysis (Yeung et al., 2004)

Predicted Role	reHuPON1	Phenyl Acetate		Paraoxon	
		K_m*mM	K_m Relative to wt (a)	K_m*(mM)	K_m Relative to wt (a)
N-Terminus Signaling Peptide	G11A	0.74 ± 0.06	~ 0.70	0.18 ± 0.02	~ 1.13
	G11C	0.74 ± 0.08	~ 0.70	0.19 ± 0.02	~ 1.19
	G11S	0.64 ± 0.09	~ 0.63	0.18 ± 0.01	~ 1.13
Catalytic Calcium Binding	N168E	nd	-	nd	-
	N224A	nd	-	nd	-
	D269E	nd	-	nd	-
Structural Calcium Binding	D54N	nd	-	nd	-
Substrate Binding Site	L69F	nd	-	nd	-
	H115W	nd	-	0.42 ± 0.08	~ 2.62
	N133S	0.59 ± 0.02	~ 0.58	0.16 ± 0.01	~ 1.06
	H115W/N133S	nd	-	0.15 ± 0.03	~ 0.83
	H134W	nd	-	nd	-
	H134Y	nd	-	nd	-
	D183E/H/K/N/Q	nd	-	nd	-
	H184A/D/Y	nd	-	nd	-
	S193A	1.00 ± 0.10	~ 0.94	0.29 ± 0.07^b	$\sim 0.87^b$
	S193G	0.88 ± 0.09	~ 0.83	0.34 ± 0.06^b	$\sim 1.02^b$
	F222D	nd	-	nd	-
	F222Y	0.77 ± 0.31	~ 0.75	nd	-
	N224A	nd	-	nd	-
	C284D	nd	-	nd	-
Catalytic Site	H285D	nd	-	nd	-
	H285Y	nd	-	nd	-
Surface Residue	R214Q	0.54 ± 0.01	~ 0.51	0.64 ± 0.13^b	$\sim 1.92^b$
	E313A	0.51 ± 0.04	~ 0.50	0.14 ± 0.01	~ 0.88
	E314A	0.66 ± 0.07	~ 0.65	0.19 ± 0.01	~ 1.19
	V304A	nd	-	nd	-

(a) Relative to recombinant wild-type K_m values for Phenyl Acetate ($K_m = 1.06 \pm 0.12$ mM at pH 8.0) and Paraoxon ($K_m = 0.15 \pm 0.04$ mM at pH 10.5) as substrates, respectively

(b) Data was determined at pH 7.4 (Recombinant wild-type paraoxonase K_m value was 0.33 ± 0.02 mM)

nd = not detectable at 3.3 mM and 2.6 mM substrate concentrations for Phenyl Acetate and Paraoxon, respectively

* K_m values are averages of at least three experiments with supernatants from independent transfections

was also important for defining the substrate specificity of the enzyme, but not for the arylesterase hydrolytic activity.

To directly test if the arylesterase and OP hydrolase activities of HuPON1 are mediated by the same active site, the potential for phenyl acetate, paraoxon and

Table 3. Kinetic parameters of recombinant HuPON1 mutants for the hydrolysis of phenyl acetate and paraoxon in the presence of competitive inhibitors (Yeung et al., 2005)

Mutants	Substrate	K_m (mM)	Inhibitor	K_i (mM)
H115W[1]	Paraoxon	0.43 ± 0.07	Phenyl acetate	1.95 ± 0.91
F222Y[2]	Phenyl acetate	0.81 ± 0.48	DFP	Nd

[1] phenyl acetate or

[2] paraoxon hydrolysis was not detectable (Yeung et al., 2004). Paraoxonase activities were assayed in 50 mM Glycine/NaOH, 1 mM $CaCl_2$, at pH 10.5 while arylesterase activities were assayed in 50 mM Tris-HCl, 1 mM $CaCl_2$, pH 8.0.

Kinetic parameters were derived from the average of at least three independent experiments, with values shown ± standard deviation. nd = inhibition not detectable at 1.30 mM DFP for F222Y.

DFP (used as an alternative OP substrate when tracking phenyl acetate hydrolysis) to act as mutually competitive substrates for wild-type HuPON1 was evaluated (Table 4; (Yeung et al., 2004)). The results indicate that each of these compounds is capable of competitively inhibiting the hydrolysis of the others, suggesting that a single active site on HuPON1 is responsible for hydrolysis of both phenyl acetate and paraoxon. When the pH dependence of HuPON1 activity with these substrates was examined, the Km values for phenyl acetate at pH 8.0 and 10.5 were found to differ (0.61 ± 0.09 mM and 1.09 ± 0.45 mM, respectively), while the Km values for paraoxon at pH 8.5 and at 10.5 were very similar (Table 4). This finding suggests that an amino acid residue at or near the active site deprotonates between pH 8.5 and 10.5, decreasing the affinity of phenyl acetate but not of paraoxon for the active site of the wild-type enzyme. Candidate amino acid residues with pKa values in this pH range include cysteine, tyrosine, and lysine. While multiple lysine and tyrosine residues (e.g., K70, K81, K192, Y190, and Y294) are located near the proposed active site of wild-type HuPON1, C284 is of particular interest because mutations at this residue have been shown to alter both arylesterase and paraoxonase activity (Sorenson et al., 1995; Yeung et al., 2004).

Table 4. K_i values for inhibition of the catalytic activity of recombinant HuPON1 by DFP or phenyl acetate (Yeung et al., 2005)

pH	Substrate	K_m (mM)	Inhibitor	K_i (mM)
8.0	Phenyl acetate	0.61 ± 0.09	DFP	0.52 ± 0.16
8.5	Paraoxon	0.31 ± 0.06	DFP	0.44 ± 0.13
8.5	Paraoxon	0.31 ± 0.06	Phenyl acetate	0.47 ± 0.17
10.5	Paraoxon	0.25 ± 0.08	Phenyl acetate	0.91 ± 0.19

Arylesterase assays were performed in 50 mM Tris-HCl, 1 mM $CaCl_2$, pH 8.0, and paraoxonase activities were assayed in 50 mM Glycine/NaOH, 1 mM $CaCl_2$, at pH 8.5 or pH 10.5, as indicated. Kinetic parameters were derived from the average of at least three independent experiments, with values shown ± standard deviation.

Concurrent with the mutagenesis studies described above, a different group took an alternative approach to examining the PON1 enzyme. Although HuPON1 was not functionally expressed by bacteria, the rabbit version of this protein was functionally produced at low levels in bacteria. Using the rabbit PON1 gene as a scaffold, Aharoni et al., generated a chimeric PON1 gene by randomly shuffling human, rabbit, mouse, and rat PON1 cDNA constructs (Aharoni et al., 2004). This chimeric recombinant PON1 (crePON1) enzyme, which has 59 amino acid substitutions relative to wild-type HuPON1, was capable of being functionally produced in bacteria, enabling large-scale production of this enzyme for crystallization studies. The X-ray analysis of the crystal structure of the crePON1 enzyme revealed a six-fold beta-propeller protein centrally arranged around two calcium ions (PDB: 1V04 (Harel et al., 2004b)), similar to the DFPase-based homology model (Josse et al., 2002; Yeung et al., 2004). Based on the derived structure, it was postulated that enzymatic catalysis of all substrates by PON1 was mediated by a unique H115–H134 dyad, wherein the actual substrate hydrolysis is driven by an "activated" water molecule (Harel et al., 2004b). This conclusion was supported by the finding that an H115Q mutant of crePON1 had dramatically reduced activity against paraoxon, as compared to the baseline crePON1. This result, in apparent contrast to the mutagenesis studies described above, was later modified when subsequent analysis revealed a second mutation at residue 177 that ablated paraoxonase activity; the H115Q mutation alone had very little effect on hydrolysis of paraoxon by crePON1 (Harel et al., 2004a). While amino acid residue H115 is still believed to be important for the lactonase activity of PON1 (Khersonsky and Tawfik, 2006), to date, the identity and precise functional roles of the amino acid residues that are responsible for the OP hydrolase activity of PON1 remain undefined.

While substantial efforts have focused on identifying amino acid residues important for HuPON1 enzymatic activity (Aharoni et al., 2004; Harel et al., 2004b; Josse et al., 1999c; Josse et al., 2002; Khersonsky and Tawfik, 2006; Yeung et al., 2004), relatively little attention has been paid to the more subtle question of substrate stereospecificity of this enzyme until recently (Amitai et al., 2006; Yeung et al., 2007). This question is important because OP nerve agents all have at least one chiral center (at the phosphorus molecule) and the toxicity of different nerve agent stereoisomers can vary dramatically. For example, the nerve agent GD, which in addition to the phosphorus chiral center contains a second center at one of the alkyl side chain carbon atoms, exists as four stereoisomers: $C+P+$, $C+P-$, $C-P+$, and $C-P-$ (Fig. 2; (Benschop, 1975; Benschop et al., 1981b; Benschop and de Jong, 1998; Benschop et al., 1984; Broomfield et al., 1986; Johnson et al., 2005; de Jong et al., 1988; Keijer and Wolring, 1969; Lenz et al., 1990; Sidell, 1997)). Two of these isomers ($C\pm P-$) are much more toxic and readily inhibit AChE, while the inhibition bimolecular rate constants for AChE with the $C\pm P+$ isomers are ~ 1000-fold lower than those with the $C\pm P-$ isomers (Benschop et al., 1981a; Benschop et al., 1984; Keijer and Wolring, 1969; Little et al., 1989). Knowledge of enzyme stereoselectivity for OP nerve agents is critical to understanding substrate

C+P+

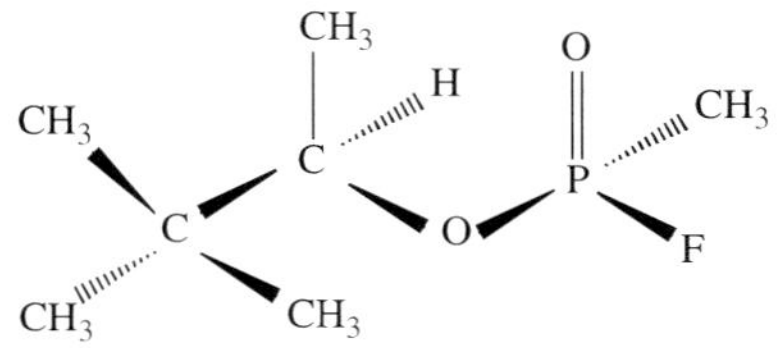

C+P−

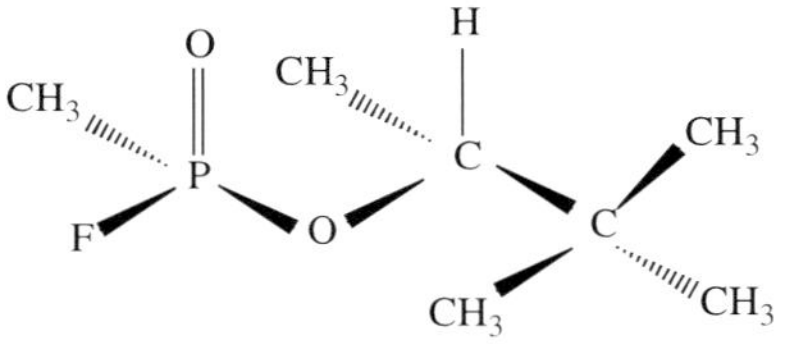

C−P+

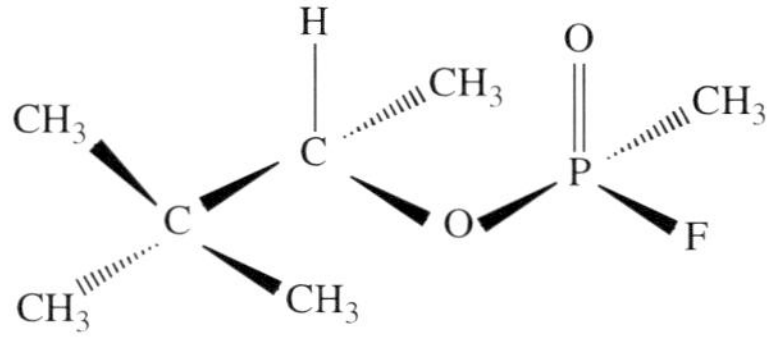

C−P−

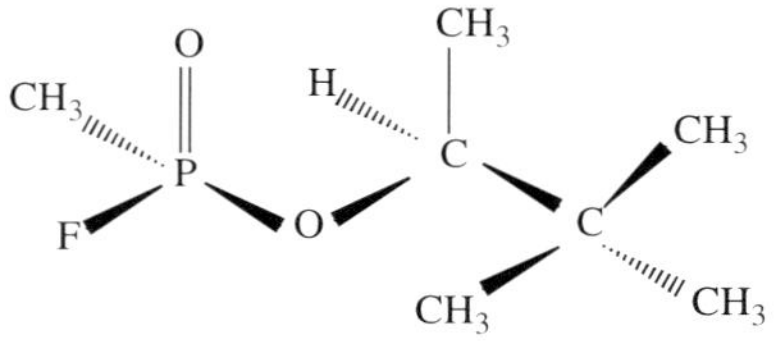

Figure 2. GD Stereoisomers GD was discovered in 1944 by Richard Kuhn. It is a member of the fluorinated organophosphorus family with the chemical formula $C_7H_{16}FO_2P$ (Molecular Weight, 182.2)

orientation and for the rational design of mutants with enhanced activity towards the more toxic isomers such as the $C \pm P-$ isomers of GD.

This issue was examined recently for crePON1-mediated hydrolysis of the OP nerve agents GD and GF, where the stereoselectivity of enzymatic hydrolysis was determined by simultaneously measuring the amount of the OP (GD or GF) and the inhibitory capacity of the same OP after incubation with the crePON1 enzyme for different time intervals (Amitai et al., 2006). While the results obtained suggested preferential degradation of the less toxic isomers, the approach could not distinguish between hydrolysis of the C+ and C− isomers. Additionally, attempts to obtain K_m and k_{cat} values for the degradation of individual isomers were largely unsuccessful (Amitai et al., 2006). As an alternative approach, a novel gas chromatography/mass spectrometry (GC/MS) based assay was used to demonstrate that recombinant wild-type HuPON1 exhibits modest, but distinct stereoselectivity in the catalytic hydrolysis of the four GD stereoisomers (Yeung et al., 2007). While the C + P+

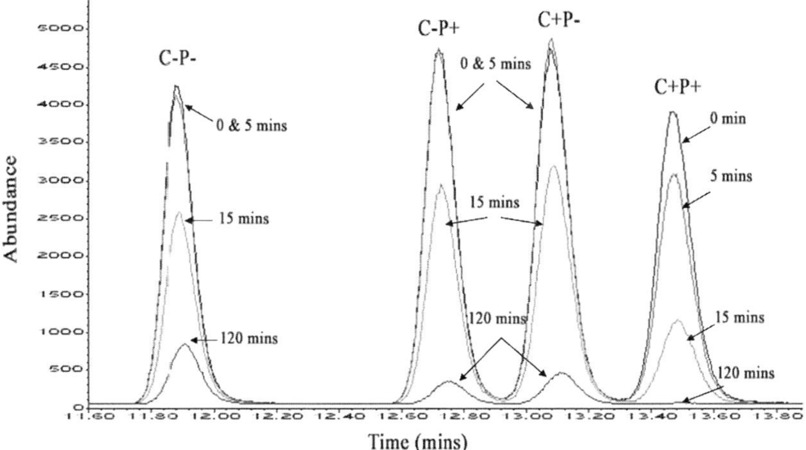

Figure 3. Overlay of reconstructed ion chromatograms (m/z 126) of GD hydrolysis by HuPON1 as assessed by gas chromatography/mass spectrometry. Typical ion chromatograms indicating the relative abundance of the four GD stereoisomers (0.75 mM racemic GD) after different incubation periods (0, 5.0, 15.0, and 120 minutes, as indicated) with wild-type HuPON1 enzyme

isomer was preferentially hydrolyzed by wild-type HuPON1 (Fig. 3, Table 5), the k_{cat} values for each of the C±P-isomers was similar to that for C-P+and were only half that for the C+P+ isomer (Yeung et al., 2007).

Using the GC/MS based assay of GD hydrolysis, the capacity of several mutants of HuPON1 to degrade GD was determined. The mutant H115W, previously found to have selective reactivity with different OP substrates (Harel et al., 2004a; Josse et al., 2002; Khersonsky and Tawfik, 2005; Khersonsky and Tawfik, 2006; Yeung et al., 2004; Yeung et al., 2005), displayed no detectable activity against GD (Table 6). This finding adds further support to the conclusion that residue H115 is one of several residues within the active site that dictates the specificity of OP substrate binding. In contrast, mutating S193 to A or G altered the kinetics of

Table 5. Kinetic parameters for the enzymatic hydrolysis of the various GD stereoisomers by recombinant wild-type HuPON1 (Yeung et al., 2007)

GD isomer	K_m (mM)	k_{cat} (min^{-1})	K_{cat}/K_m (mM^{-1} ∗min^{-1})
C(−)P(−)	0.91 ± 0.34	501 ± 45	625 ± 241
C(−)P(+)	0.58 ± 0.23	593 ± 54	1160 ± 469
C(+)P(−)	0.71 ± 0.49	553 ± 163	1040 ± 465
C(+)P(+)	0.27 ± 0.08	1030 ± 94	4130 ± 1090

HuPON1 catalyzed reactions GD hydrolysis was assayed in the presence of at least 1.0 mM. Kinetic results presented for each isomer were determined from at least eight independent kinetic experiments (n = 8).

Table 6. Catalytic parameters of selected PON1 mutants expressed as experimentally determined and as approximate fold increase/decrease (parentheses) relative to wild-type HuPON1 for the hydrolysis of GD stereoisomers

PON1 Mutants	C − P −		C − P +		C + P −		C + P +	
	K_m (mM)	k_{cat}/K_m (mM^{-1} ∗ min^{-1})	K_m (mM)	k_{cat}/K_m (mM^{-1} ∗ min^{-1})	K_m (mM)	k_{cat}/K_m (mM^{-1} ∗ min^{-1})	K_m (mM)	k_{cat}/K_m (mM^{-1} ∗ min^{-1})
S193A	2.34 ± 0.84 (2.57)	749.96 ± 243.87 (1.19)	3.21 ± 0.99 (5.53)	931.84 ± 227.27 (0.80)	1.69 ± 1.50 (2.38)	2199.96 ± 1289.9 (2.12)	0.53 ± 0.15 (1.96)	7446.75 ± 2051.83 (1.81)
S193G	5.79 ± 1.42 (6.36)	297.50 ± 69.17 (0.48)	6.86 ± 1.29 (11.83)	427.78 ± 83.33 (0.37)	3.94 ± 2.46 (5.55)	914.29 ± 630.29 (0.88)	1.09 ± 0.79 (4.03)	4498.75 ± 2462.30 (1.09)
R214Q	0.78 ± 0.16 (0.86)	410.02 ± 45.41 (0.66)	0.82 ± 0.33 (1.41)	580.20 ± 97.47 (0.49)	0.47 ± 0.27 (0.66)	1030.00 ± 471.17 (0.99)	0.11 ± 0.06 (0.41)	4500.06 ± 1903.94 (1.09)
S193A/ R214Q	2.92 ± 0.65 (3.21)	575.00 ± 169.93 (0.92)	4.26 ± 0.72 (7.34)	672.73 ± 112.61 (0.58)	2.22 ± 1.19 (3.13)	1259.09 ± 568.25 (1.21)	0.81 ± 0.40 (3.00)	5318.18 ± 1940.01 (1.29)
H115W	nd	nd	nd	nd	nd	nd	nd	nd

nd: Somanase activity not detectable from 0.50–3.0 mM racemic GD. Each indicated value corresponds to the average of at least six different determinations. The K_m (mM) and k_{cat}/K_m (mM^{-1} min^{-1}) values for wild-type HuPON1 are presented in Table 5.

hydrolysis of all four GD stereoisomers. Interestingly, changes in the kinetics of GD hydrolysis were also detected in the R214Q mutant. In both the crystal structure of crePON1 and the DFPase-based homology model, residue R214 is a solvent exposed amino acid located at the end of a propeller blade. Residue R214 appears to interact with other amino acid residues on an adjacent blade, particularly W254. Molecular dynamics and energy minimization modeling suggest that the R214Q mutation induces a global shift in the configuration of the active site relative to wild-type HuPON1, resulting in enhanced substrate access to the catalytic machinery.

Overall, the GC/MS data from both wild-type and HuPON1 mutant enzymes suggest that the observed variations in catalytic efficiency for GD can be attributed largely to differences in the K_m values of the enzyme for the various stereoisomers. While the stereochemistry of the substrates may be important for binding, the results suggest that once bound, the catalytic machinery is not overly sensitive to the chirality of the groups around the phosphorus atom. Therefore, small changes (via site directed mutagenesis) that reduce the K_m for the more toxic isomers might be singularly sufficient to make the enzyme a viable bioscavenger for detoxification of OP nerve agents *in vivo*. For example, a reduction in K_m by ten-fold with no change in the V_{max} value would enhance catalytic turnover of the more toxic stereoisomers of GD such that they would be preferentially hydrolyzed by several fold (Josse et al., 2001; Josse et al., 2002; Yeung et al., 2004).

The capacity of wild-type and mutant HuPON1 enzymes to hydrolyze the structurally isomeric nerve agents VX and VR, which both contain thiol leaving groups, was examined (Yeung et al., submitted) using a modified Ellman assay (Ellman et al., 1961). An approximate five-fold difference in binding affinity (K_m) is observed between the two V-agents, with an accompanying seven- to eight-fold difference in the rates of turnover (Fig. 4, Tables 7 and 8). Though the wild-type HuPON1 enzyme binds VX with less affinity than VR, the k_{cat} value suggests that

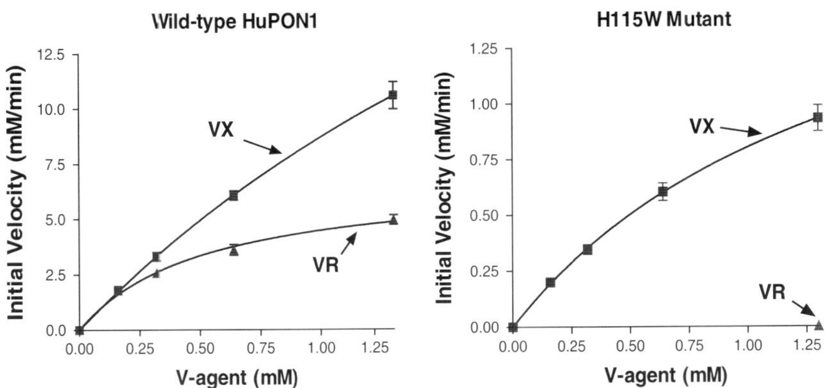

Figure 4. Kinetics of V-agents hydrolysis. Michaelis-Menten plots for the hydrolysis of VX (■) and VR (▲) by wild-type (A) and mutant H115W (B) recombinant HuPON1. Data shown are representative of three independent experiments. The kinetic parameter K_m derived from the fits is presented in Tables 7 and 8

Table 7. Kinetic parameters for the enzymatic hydrolysis of VX by recombinant HuPON1 enzymes

Samples	K_m (mM)	k_{cat} (min^{-1})	k_{cat}/K_m (mM^{-1}min^{-1})
Wild-type	2.74 ± 1.51	107.91 ± 31.13	44.99 ± 12.51
H115W	2.58 ± 0.56	53.91 ± 7.80	21.37 ± 8.06
S193G	2.14 ± 0.45	152.26 ± 65.97	69.41 ± 16.28
S193A	3.33 ± 0.24	393.25 ± 0.76	118.29 ± 16.28
R214Q	2.08 ± 0.56	284.48 ± 47.03	138.98 ± 15.06
S193A/R214Q	3.32 ± 1.79	125.30 ± 30.19	41.89 ± 11.05

Results shown are the average and standard deviation of at least three independent (n ≥ 3) transfections.

once bound, VX is quickly hydrolyzed. Interestingly, the opposite appears to be the case for VR hydrolysis, where the substrate is easily bound, but the rate of turnover is much lower. Despite the differences in both binding affinity and turnover rates, the catalytic efficiency (K_m/k_{cat}) of wild-type HuPON1 for the two V-agents differs by less than two-fold.

Supporting the previous discussion (*vide supra*) highlighting the importance of residue S193 in the hydrolysis of GD, mutating S193 to A or G also affected the kinetics of hydrolysis for VX and VR with a more pronounced effect observed in the S193A mutant. A three- to four-fold enhancement relative to wild-type HuPON1 in both turnover rate and catalytic efficiency were observed for VX and VR hydrolysis by the S193A mutant. Likewise, enhanced turnover of both V-agents was also detected in R214Q mutant. In contrast, the double mutant S193A/R214Q did not display enhanced hydrolysis of either VX or VR; rather, the k_{cat} and k_{cat}/K_m values for this double mutant were restored to wild-type levels. These results suggest that multiple mutations which individually result in enhanced activity against one or more OP nerve agents may not exhibit additive or synergistic effects when they are simultaneously present on the same molecule.

In comparison to wild-type HuPON1, the H115W mutant has reduced activity against VX, and no detectable activity against VR (Fig. 4). Since H115W-mediated turnover of VR was not observed, a competitive assay experiment using VX as

Table 8. Kinetics of VR hydrolysis mediated by recombinant HuPON1 enzymes

Samples	K_m (mM)	k_{cat} (min^{-1})	k_{cat}/K_m (mM^{-1}min^{-1})
Wild-type	0.57 ± 0.08	14.37 ± 1.03	25.49 ± 5.23
H115W	ND	ND	ND
S193G	0.31 ± 0.02	11.99 ± 3.72	39.71 ± 15.62
S193A	0.39 ± 0.12	40.98 ± 12.05	107.61 ± 17.80
R214Q	0.48 ± 0.14	35.93 ± 6.32	75.93 ± 9.18
S193A/R214Q	0.43 ± 0.05	9.47 ± 0.63	22.34 ± 2.19

Results (averages and standard deviations) shown were determined from at least three independent (n ≥ 3) transfections. ND: Catalysis was not detectable at VR concentrations ≤ 1.80 mM.

the reporting substrate was carried out to determine whether either of the two non-hydrolyzed nerve agents, VR or GD, was able to bind to the enzyme. Both VR and GD were found to competitively inhibit VX hydrolysis by the H115W mutant, with K_i values of 0.27 ± 0.09 mM and 0.90 ± 0.38 mM, respectively. These findings suggest that altering a single residue in the active site, i.e. H115, can dictate the specificity of OP substrate binding and hydrolysis for a variety of OP nerve agent substrates. Based on the results of these studies, residue H115 appears to be important for substrate docking in the active site of HuPON1, but is most likely not directly involved in the catalytic hydrolysis of at least OP substrates. Since all of the substrates tested with H115W can bind the active site (as judged by either substrate hydrolysis or competitive inhibition), it is likely that residue H115 is essential for proper orientation of the substrate(s) to promote activated water-mediated hydrolysis.

In light of these conclusions, the role of residue H115 in the activity of HuPON1 against other classes of substrates (such as phospholipids and lactones) needs to be reconsidered. While it is possible HuPON1 possesses more than one set of catalytic amino acid residues, each capable of promoting hydrolysis of different types of substrates, identifying multiple mechanisms in a single enzyme active site is basically unprecedented. Alternative mechanisms to describe the activity of HuPON1 which do not directly involve histidine residue 115 should be considered.

4.   SUMMARY

The work described herein offers valuable insights for the detailed analysis of the mechanism(s) involved in HuPON1 catalysis. The results support a structural model of HuPON1 as a six-fold β -propeller protein, and have identified several active site amino acid residues which can influence both the stereoselectivity and specificity of substrate hydrolysis by HuPON1. Most importantly, the results will inform future genetic engineering efforts to enhance the OP hydrolase activity of HuPON1 against highly toxic OP nerve agents. Finally, it appears based on the relatively limited panel of HuPON1 mutants analyzed to date that if PON1 is to be used as an effective bioscavengers against OP nerve agent intoxication, the best approach may be the utilization of a mixture of PON1 mutants, each with specificity for a different nerve agent.

REFERENCE

Aharoni, A., Gaidukov, L., Yagur, S., Toker, L., Silman, I., and Tawfik, D.S. 2004. Directed evolution of mammalian paraoxonases PON1 and PON3 for bacterial expression and catalytic specialization. *Proc. Natl. Acad. Sci. U. S. A.* **101**:482–7.

Aldridge, W.N. 1953a. Serum esterases. I. Two types of esterase (A and B) hydrolysing p-nitrophenyl acetate, propionate and butyrate, and a method for their determination. *Biochem. J.* **53**:110–117.

Aldridge, W.N. 1953b. Serum esterases. II. An enzyme hydrolysing diethyl p-nitrophenyl phosphate (E600) and its identity with the A-esterase of mammalian sera. *Biochem. J.* **53**:117–124.

Amitai, G., Gaidukov, L., Adani, R., Yishay, S., Yacov, G., Kushnir, M., Teitlboim, S., Lindenbaum, M., Bel, P., Khersonsky, O., Tawfik, D.S., and Meshulam, H. 2006. Enhanced stereoselective hydrolysis of toxic organophosphates by directly evolved variants of mammalian serum paraoxonase. *FEBS J.* **273**:1906–1919.

Aviram, M., Billecke, S., Sorenson, R., Bisgaier, C., Newton, R., Rosenblat, M., Erogul, J., Hsu, C., Dunlop, C., and La Du, B. 1998. Paraoxonase active site required for protection against LDL oxidation involves its free sulfhydryl group and is different from that required for its arylesterase/paraoxonase activities: selective action of human paraoxonase allozymes Q and R. *Arterioscler Thromb. Vasc. Biol.* **18**:1617–24.

Ballantyne, B. and Marrs, T.C. 1992. Overview of the biological and clinical aspects of organophosphates and carbamates. In *Clinical and Experimental Toxicology of Organophosphates and carbamates.* (eds. B Ballantyne and TC Marrs), p 1. Butterworth: Oxford, London.

Benschop, H.P. 1975. The absolute configuration of chiral organophosphorus anticholinesterase poisoning. *Pesticide Biochemistry and Physiology* **5**:348–349.

Benschop, H.P., Berends, F., and de Jong, L.P. 1981b. GLC-analysis and pharmacokinetics of the four stereoisomers of Soman. *Fundam. Appl. Toxicol.* **1**:177–182.

Benschop, H.P., Berends, F., and de Jong, L.P. 1981a. GLC-analysis and pharmacokinetics of the four stereoisomers of Soman. *Fundam. Appl. Toxicol.* **1**:177–182.

Benschop, H.P. and de Jong, L.P. 1998. Nerve agent stereoisomers: analysis, isolation, and toxicology. *Accounts of Chemical Research* **21**:368–374.

Benschop, H.P., Konings, C.A., van Genderen, J., and de Jong, L.P. 1984. Isolation, anticholinesterase properties, and acute toxicity in mice of the four stereoisomers of the nerve agent soman. *Toxicol. Appl. Pharmacol.* **72**:61–74.

Billecke, S., Draganov, D., Counsell, R., Stetson, P., Watson, C., Hsu, C., and La Du, B.N. 2000. Human serum paraoxonase (PON1) isozymes Q and R hydrolyze lactones and cyclic carbonate esters. *Drug Metab. Dispos.* **28**:1335–42.

Blatter, M.C., James, R.W., Messmer, S., Barja, F., and Pometta, D. 1993. Identification of a distinct human high-density lipoprotein subspecies defined by a lipoprotein-associated protein, K-45. Identity of K-45 with paraoxonase. *Eur. J. Biochem.* **211**:871–9.

Brites, F.D., Verona, J., Schreier, L.E., Fruchart, J.C., Castro, G.R., and Wikinski, R.L. 2004. Paraoxonase 1 and platelet-activating factor acetylhydrolase activities in patients with low hdl-cholesterol levels with or without primary hypertriglyceridemia. *Arch. Med. Res.* **35**:235–240.

Broomfield, C.A., Lenz, D.E., and MacIver, B. 1986. The stability of soman and its stereoisomers in aqueous solution: toxicological considerations. *Arch. Toxicol.* **59**:261–265.

Broomfield, C.A., Morris, B.C., Anderson, R., Josse, D., and Masson, P. Kinetics of nerve agent hydrolysis by a human plasma enzyme. 2000. Aberdeen Proving Ground, Maryland, U.S. Army Medical Research Institute of Chemical Defense. Journal of Medical Chemical Defense. 5-7-2000. Ref Type: Conference Proceeding

Brown, M.A. and Brix, K.A. 1998. Review of health consequences from high-, intermediate- and low-level exposure to organophosphorus nerve agents. *J. Appl. Toxicol.* **18**:393–408.

Cerasoli, D.M. and Lenz, D.E. 2002. Nerve agent bioscavengers: protection with reduced behavioral effects. *Military Psychology* **14**:121–143.

Clendenning, J.B., Humbert, R., Green, E.D., Wood, C., Traver, D., and Furlong, C.E. 1996. Structural organization of the human PON1 gene. *Genomics* **35**:586–9.

Costa, L.G., Cole, T.B., Vitalone, A., and Furlong, C.E. 2005. Measurement of paraoxonase (PON1) status as a potential biomarker of susceptibility to organophosphate toxicity. *Clin. Chim. Acta* **352**:37–47.

Costa, L.G., Li, W.F., Richter, R.J., Shih, D.M., Lusis, A., and Furlong, C.E. 1999. The role of paraoxonase (PON1) in the detoxication of organophosphates and its human polymorphism. *Chem Biol Interact* **119–120**:429–38.

Costa, L.G., Richter, R.J., Li, W.F., Cole, T., Guizzetti, M., and Furlong, C.E. 2003a. Paraoxonase (PON 1) as a biomarker of susceptibility for organophosphate toxicity. *Biomarkers* **8**:1–12.

Costa, L.G., Cole, T.B., Jarvik, G.P., and Furlong, C.E. 2003b. Functional genomics of the paraoxonase(PON1) polymorphisms: effects on pesticide sensitivity, cardiovascular disease, and drug metabolism. *Annu. Rev. Med.* **54**:371–92.

Dacre, J.C. 1984. Toxicology of some anticholinesterases used as a chemical warefare agents – a review. In *Cholinesterases, Fundamental and Applied Aspects*. (eds. M Brzin, EA Barnard, and D Sket), p 415. de Gruyter: Berlin, Germany.

Davies, H.G., Richter, R.J., Keifer, M., Broomfield, C.A., Sowalla, J., and Furlong, C.E. 1996. The effect of the human serum paraoxonase polymorphism is reversed with diazoxon, soman and sarin. *Nat Genet* **14**:334–6.

de Jong, L.P., van Dijk, C., and Benschop, H.P. 1988. Hydrolysis of the four stereoisomers of soman catalyzed by liver homogenate and plasma from rat, guinea pig and marmoset, and by human plasma. *Biochem. Pharmacol.* **37**:2939–2948.

Doorn, J.A., Sorenson, R.C., Billecke, S.S., Hsu, C., and La Du, B.N. 1999. Evidence that several conserved histidine residues are required for hydrolytic activity of human paraoxonase/arylesterase. *Chem. Biol. Interact.* **119–120**:235–241.

Draganov, D.I., Stetson, P.L., Watson, C.E., Billecke, S.S., and La Du, B.N. 2000. Rabbit serum paraoxonase 3 (PON3) is a high density lipoprotein-associated lactonase and protects low density lipoprotein against oxidation. *J Biol Chem* **275**:33435–42.

Ellman, G.L., Courtney, K.D., Andres, V., Jr., and Feather-Stone, R.M. 1961. A new and rapid colorimetric determination of acetylcholinesterase activity. *Biochem. Pharmacol.* **7**:88–95.

Furlong, C.E., Cole, T.B., Jarvik, G.P., and Costa, L.G. 2002. Pharmacogenomic considerations of the paraoxonase polymorphisms. *Pharmacogenomics.* **3**:341–348.

Furlong, C.E., Li, W.F., Brophy, V.H., Jarvik, G.P., Richter, R.J., Shih, D.M., Lusis, A.J., and Costa, L.G. 2000a. The PON1 gene and detoxication. *Neurotoxicology* **21**:581–7.

Furlong, C.E., Li, W.F., Costa, L.G., Richter, R.J., Shih, D.M., and Lusis, A.J. 1998. Genetically determined susceptibility to organophosphorus insecticides and nerve agents: developing a mouse model for the human PON1 polymorphism. *Neurotoxicology* **19**:645–50.

Furlong, C.E., Li, W.F., Richter, R.J., Shih, D.M., Lusis, A.J., Alleva, E., and Costa, L.G. 2000b. Genetic and temporal determinants of pesticide sensitivity: role of paraoxonase (PON1). *Neurotoxicology* **21**:91–100.

Gaidukov, L., Rosenblat, M., Aviram, M., and Tawfik, D.S. 2006. The 192R/Q polymorphs of serum paraoxonase PON1 differ in HDL binding, stimulation of lipo-lactonase, and macrophage cholesterol efflux. *J. Lipid Res.*

Gaidukov, L. and Tawfik, D.S. 2007. The development of human sera tests for HDL-bound serum paraoxonase PON1 and its lipo-lactonase activity. *J. Lipid Res.*

Harel, M., Aharoni, A., Gaidukov, L., Brumshtein, B., Khersonsky, O., Meged, R., Dvir, H., Ravelli, R.B., McCarthy, A., Toker, L., Silman, I., Sussman, J.L., and Tawfik, D.S. 2004a. Corrigendum: Structure and evolution of the serum paraoxonase family of detoxifying and anti-atherosclerotic enzymes. *Nat. Struct. Mol. Biol.* **11**:1253.

Harel, M., Aharoni, A., Gaidukov, L., Brumshtein, B., Khersonsky, O., Meged, R., Dvir, H., Ravelli, R.B., McCarthy, A., Toker, L., Silman, I., Sussman, J.L., and Tawfik, D.S. 2004b. Structure and evolution of the serum paraoxonase family of detoxifying and anti-atherosclerotic enzymes. *Nat. Struct. Mol. Biol.* **11**:412–419.

Hassett, C., Richter, R.J., Humbert, R., Chapline, C., Crabb, J.W., Omiecinski, C.J., and Furlong, C.E. 1991. Characterization of cDNA clones encoding rabbit and human serum paraoxonase: the mature protein retains its signal sequence. *Biochemistry* **30**:10141–10149.

Heath, A.J.W. and Meredith, T. 1992. Atropine in the management of anticholinesterase poisoning. In *Clinical and Experimental Toxicology of Organophosphates and Cabamates*. (eds. B Ballantyne and TC Marrs), p 543. Butterworth: Oxford, London.

Holstege, C.P., Kirk, M., and Sidell, F.R. 1997. Chemical warfare. Nerve agent poisoning. *Crit Care Clin.* **13**:923–942.

Jakubowski, H. 2000. Calcium-dependent human serum homocysteine thiolactone hydrolase. A protective mechanism against protein N-homocysteinylation. *J. Biol. Chem.* **275**:3957–3962.

Jakubowski, H., Ambrosius, W.T., and Pratt, J.H. 2001. Genetic determinants of homocysteine thiolactonase activity in humans: implications for atherosclerosis. *FEBS Lett.* **491**:35–39.

James, R.W. and Deakin, S.P. 2004. The importance of high-density lipoproteins for paraoxonase-1 secretion, stability, and activity. *Free Radic. Biol. Med.* **37**:1986–1994.

Johnson, J.K., Cerasoli, D.M., and Lenz, D.E. 2005. Role of immunogen design in induction of soman-specific monoclonal antibodies. *Immunol. Lett.* **96**:121–127.

Josse, D., Broomfield, C.A., Cerasoli, D., Kirby, S., Nicholson, J., Bahnson, B., and Lenz, D.E. Engineering of HuPON1 For a Use as a Catalytic Bioscavenger in Organophosphate Poisoning. 2002. Aberdeen Proving Ground, MD, U.S. Army Medical Research Institute of Chemical Defense. U.S. Army Medical Defense Bioscience Review. Ref Type: Conference Proceeding

Josse, D., Lockridge, O., Xie, W., Bartels, C.F., Schopfer, L.M., and Masson, P. 2001. The active site of human paraoxonase (PON1). *J Appl Toxicol* **21(Suppl 1)**:7–11.

Josse, D., Xie, W., Masson, P., and Lockridge, O. 1999a. Human serum paraoxonase (PON1): identification of essential amino acid residues by group-selective labelling and site-directed mutagenesis. *Chem Biol Interact* **119–120**:71–8.

Josse, D., Xie, W., Masson, P., Schopfer, L.M., and Lockridge, O. 1999b. Tryptophan residue(s) as major components of the human serum paraoxonase active site. *Chem Biol Interact* **119–120**: 79–84.

Josse, D., Xie, W., Renault, F., Rochu, D., Schopfer, L.M., Masson, P., and Lockridge, O. 1999c. Identification of residues essential for human paraoxonase (PON1) arylesterase/organophosphatase activities. *Biochemistry* **38**:2816–25.

Keijer, J.H. and Wolring, G.Z. 1969. Stereospecific aging of phosphonylated cholinesterases. *Biochim. Biophys. Acta* **185**:465–468.

Khersonsky, O. and Tawfik, D.S. 2005. Structure-reactivity studies of serum paraoxonase PON1 suggest that Its native activity is lactonase. *Biochemistry* **44**:6371–6382.

Khersonsky, O. and Tawfik, D.S. 2006. The histidine 115-histidine 134 dyad mediates the lactonase activity of mammalian serum paraoxonases. *J. Biol. Chem.* **281**:7649–7656.

Kuo, C.L. and La Du, B.N. 1998. Calcium binding by human and rabbit serum paraoxonases. Structural stability and enzymatic activity. *Drug Metab Dispos* **26**:653–60.

La Du, B.N., Furlong, C.E., and Reiner, E. 1999. Recommended nomenclature system for the paraoxonases. *Chem. Biol. Interact.* **119–120**:599–601.

Lacinski, M., Skorupski, W., Cieslinski, A., Sokolowska, J., Trzeciak, W.H., and Jakubowski, H. 2004. Determinants of homocysteine-thiolactonase activity of the paraoxonase-1 (PON1) protein in humans. *Cell Mol. Biol. (Noisy. -le-grand)* **50**:885–893.

Lee, E.J. 1997. Pharmacology and toxicology of chemical warfare agents. *Ann. Acad. Med. Singapore* **26**:104–107.

Lenz, D.E., Broomfield, C.A., Maxwell, D.M., and Cerasoli, D.M. 2001. "Nerve Agent Bioscavengers: Protection against High- and Low-Dose Organophosphorus Exposure". In *Chemical Warfare Agents: Toxicity at Low Levels.* (ed. JR S.Somani), pp. 215–243. CRC Press: Boca Raton.

Lenz, D.E., Little, J.S., Broomfield, C.A., and Ray, R. 1990. Catalytic properties of nonspecific diisopropylfluorophosphatases. In *Chirality and Biological Activity* pp. 169–175. Alan R. Liss, Inc..

Lenz, D.E., Yeung, D., Smith, J.R., Sweeney, R.E., Lumley, L.A., and Cerasoli, D.M. 2006. Stoichiometric and catalytic scavengers as protection against nerve agent toxicity: A mini review. *Toxicology* **233**:31–39.

Li, W.F., Costa, L.G., and Furlong, C.E. 1993. Serum paraoxonase status: a major factor in determining resistance to organophosphates. *J Toxicol Environ Health* **40**:337–46.

Li, W.F., Costa, L.G., Richter, R.J., Hagen, T., Shih, D.M., Tward, A., Lusis, A.J., and Furlong, C.E. 2000. Catalytic efficiency determines the in-vivo efficacy of PON1 for detoxifying organophosphorus compounds. *Pharmacogenetics* **10**:767–79.

Li, W.F., Furlong, C.E., and Costa, L.G. 1995. Paraoxonase protects against chlorpyrifos toxicity in mice. *Toxicol Lett* **76**:219–26.

Little, J.S., Broomfield, C.A., Fox-Talbot, M.K., Boucher, L.J., MacIver, B., and Lenz, D.E. 1989. Partial characterization of an enzyme that hydrolyzes sarin, soman, tabun, and diisopropyl phosphorofluoridate (DFP). *Biochem. Pharmacol.* **38**:23–29.

Masson, P., Josse, D., Lockridge, O., Viguie, N., Taupin, C., and Buhler, C. 1998. Enzymes hydrolyzing organophosphates as potential catalytic scavengers against organophosphate poisoning. *J. Physiol Paris* **92**:357–362.

Masuda, N., Takatsu, M., Morinari, H., and Ozawa, T. 1995. Sarin poisoning in Tokyo subway. *Lancet* **345**:1446.

Maynard, R.L. and Beswick, F.W. 1992. Organophosphorus compounds as chemical warfare agents. In *Clinical and Experimental Toxicology of Organophosphates and Carbamates*. (eds. B Ballantyne and TC Marrs), p. 373. Butterworth: Oxford, London.

Mazur, A. 1946. An enzyme in animal tissues capable of hydrolyzing the phosphorus-fluorine bond of alkyl fluorophosphates. *Journal of Biological Chemistry* **164**:271–289.

Morales, R., Beswick, F.W., Carpentier, P., Contreras-Martel, C., Renault, F., Nicodeme, M., Chesne-Seck, M., Bernier, F., Depuy, J., Schaeffer, C., Diemer, H., Van-Dorsselaer, A., Fontecilla-Camps, J.C., Masson, P., Rochu, D., and Chabrieres, E. 2006. Serendipitous discovery and x-ray structure of a human phosphate binding apoliprotein. *Structure* **14**:1–9.

Nagao, M., Takatori, T., Matsuda, Y., Nakajima, M., Iwase, H., and Iwadate, K. 1997. Definitive evidence for the acute sarin poisoning diagnosis in the Tokyo subway. *Toxicol. Appl. Pharmacol.* **144**:198–203.

Primo-Parmo, S.L., Sorenson, R.C., Teiber, J., and La Du, B.N. 1996. The human serum paraoxonase/arylesterase gene (PON1) is one member of a multigene family. *Genomics* **33**: 498–507.

Reutter, S. 1999. Hazards of chemical weapons release during war: new perspectives. *Environ. Health Perspect.* **107**:985–990.

Rochu, D., Chabrieres, E., and Masson, P. 2007. Human paraoxonase: A promising approach for pre-treatment and therapy of organophosphorus poisoning. *Toxicology* **233**:47–59.

Rodrigo, L., Mackness, B., Durrington, P.N., Hernandez, A., and Mackness, M.I. 2001. Hydrolysis of platelet-activating factor by human serum paraoxonase. *Biochem J* **354**:1–7.

Rosenblat, M., Gaidukov, L., Khersonsky, O., Vaya, J., Oren, R., Tawfik, D.S., and Aviram, M. 2006. The catalytic histidine dyad of high density lipoprotein-associated serum paraoxonase-1 (PON1) is essential for PON1-mediated inhibition of low density lipoprotein oxidation and stimulation of macrophage cholesterol efflux. *J. Biol. Chem.* **281**:7657–7665.

Scharff, E.I., Lucke, C., Fritzsch, G., Koepke, J., Hartleib, J., Dierl, S., and Ruterjans, H. 2001. Crystallization and preliminary X-ray crystallographic analysis of DFPase from Loligo vulgaris. *Acta Crystallogr D Biol Crystallogr* **57**:148–9.

Shih, D.M., Gu, L., Xia, Y.R., Navab, M., Li, W.F., Hama, S., Castellani, L.W., Furlong, C.E., Costa, L.G., Fogelman, A.M., and Lusis, A.J. 1998. Mice lacking serum paraoxonase are susceptible to organophosphate toxicity and atherosclerosis. *Nature* **394**:284–7.

Sidell, F.R. 1997. Nerve Agents. In *Medical Aspects of Chemical and Biological Warfare*. (ed. R Zajtchuk), pp.129–179. Office of the Surgeon General at TMM Publication: Washington DC.

Somani, S.M., Solana, R.P., and Dube, S.N. 1992. Toxiodynamics of nerve agents. In *Chemical Warfare Agents*. (ed. SM Somani), p. 68. Academic Press: San Diego, CA.

Sorenson, R.C., Bisgaier, C.L., Aviram, M., Hsu, C., Billecke, S., and La Du, B.N. 1999. Human serum Paraoxonase/Arylesterase's retained hydrophobic N-terminal leader sequence associates with HDLs by binding phospholipids : apolipoprotein A-I stabilizes activity. *Arterioscler Thromb Vasc Biol* **19**:2214–25.

Sorenson, R.C., Primo-Parmo, S.L., Kuo, C.L., Adkins, S., Lockridge, O., and La Du, B.N. 1995. Reconsideration of the catalytic center and mechanism of mammalian paraoxonase/arylesterase. *Proc Natl Acad Sci U S A* **92**:7187–91.

Taylor, P. 2001. Anticholinesterase agents. In *Goodman & Gillman's The Pharmacological Basis of Therapeutics*, 10 ed. (eds. JG Hardman, LE Limbird, and AG Gilman), pp. 175–192. McGraw-Hill: New York.

Teiber, J.F., Draganov, D.I., and La Du, B.N. 2003. Lactonase and lactonizing activities of human serum paraoxonase (PON1) and rabbit serum PON3. *Biochem Pharmacol* **66**:887–96.

van Himbergen, T.M., van Tits, L.J., Roest, M., and Stalenhoef, A.F. 2006. The story of PON1: how an organophosphate-hydrolysing enzyme is becoming a player in cardiovascular medicine. *Neth. J. Med.* **64**:34–38.

Yeung, D.T., Josse, D., Nicholson, J.D., Khanal, A., McAndrew, C.W., Bahnson, B.J., Lenz, D.E., and Cerasoli, D.M. 2004. Structure/function analyses of human serum paraoxonase (HuPON1) mutants designed from a DFPase-like homology model. *Biochim. Biophys. Acta* **1702**:67–77.

Yeung, D.T., Lenz, D.E., and Cerasoli, D.M. 2005. Analysis of active-site amino-acid residues of human serum paraoxonase using competitive substrates. *FEBS J.* **272**:2225–2230.

Yeung, D.T., Smith, J.R., Sweeney, R.E., Lenz, D.E., and Cerasoli, D.M. 2007. Direct detection of stereospecific soman hydrolysis by wild-type human serum paraoxonase. *FEBS J.* **274**:1183–1191.

CHAPTER 10

STABILISATION OF ACTIVE FORM OF NATURAL HUMAN PON1 REQUIRES HPBP

D. ROCHU[1,2], E. CHABRIERE[3], M. ELIAS[3], F. RENAULT[1],
C. CLERY-BARRAUD[1] AND P. MASSON[1]

[1] *Département de Toxicologie, Centre de Recherches du Service de Santé des Armées, BP 87, 38702 La Tronche cedex, France*
[2] *Bundeswehr Institute of Pharmacology and Toxicology, 80937 Munich, Germany*
[3] *Laboratoire de Cristallographie et Modélisation des Matériaux Minéraux et Biologiques, CNRS-Université Henri Poincaré, BP 239, 54506 Vandoeuvre-lès-Nancy, France*

Abstract: Human PON1 displays functional promiscuity correlated to its natural biological milieu. Devoid of its physiological HDL environment, the natural human enzyme is unstable. The serendipitous discovery of HPBP, a PON1 partner protein, and the contamination of current PON1 preparations causing misinterpretation of PON1's functions are central new data. Besides, both activities and stability of PON1 are completely dependent on the HDL components' molecular surrounding. Because of the variability of the HDL environment of PON1, identification of partner lipoprotein(s) or/and hydrophobic cofactor(s) capable of acting as reliable surrogate stabilizer(s) of the enzyme's functional conformation is crucial. Attempts at characterizing PON1's functional state(s), determining its thermal stability and describing factors involved in its storage stability demonstrated that HPBP helps to stabilize the active form(s) of natural human PON1. Together with the depiction of other HDL-associated components in the modulation of PON1 stability, these data shed light on the contribution of the HDL environment to the interactions of PON1 with its natural and/or multiples substrates

Keywords: PON1 (paraoxonase), protein purity, stability, functional state, HPBP (human phosphate binding protein)

1. INTRODUCTION

> *Alone: adj., In bad company*
> (Ambrose Bierce, The Devil's Dictionary, 1911)

A new episode in the human PON1 story was opened as the protein, associated with HDL particles (Blatter et al., 1993) and named for its ability to hydrolyse

B. Mackness et al. (eds.), The Paraoxonases: Their Role in Disease Development and Xenobiotic Metabolism, 171–183.

the organophosphate (OP) paraoxon, became a player in cardiovascular physiology (Mackness et al., 1991). Besides its phosphotriesterase activity, PON1 was shown to hydrolyse aromatic carboxylic acid esters such as phenyl acetate, and thus to be involved in degradation of aromatic drug esters, and lactones. Anti-oxidative, phospholipids-binding properties and anti-atherogenic properties of PON1 were also described (Shih et al., 1998; Watson et al., 1995). Albeit its physiological function is likely to be a lipophilic lactonase (Jakubowski et al., 2001; Khersonsky and Tawfik, 2005; Khersonsky and Tawfik, 2006; Lacinski et al., 2004; Teiber et al., 2003), PON1 is an enzyme with promiscuous activities (Aharoni et al., 2005). Hence, certain of the activities attributed to PON1 are for the moment largely debated.

The 3D structure and the catalytic mechanism of human PON1 are still unclear. It is a clue that the thorough characterization of a protein cannot be truthfully initiated until it has been completely purified. But for PON1, purity is a challenge. Protein purity is usually checked by SDS-PAGE, although proteomics investigators estimate that only 20–30% of expressed proteins are detectable by standard methods to date (Righetti et al., 2005). A vast part of the "hidden serum proteome" includes proteins with very low and high pI that are poorly represented using 2D electrophoresis. In addition, hydrophobic proteins cannot be properly solubilized in current buffers. The particular environment of multiple interacting lipids and proteins in HDL may explain why study of PON1 in solution is so hard. Almost all PON1 purification procedures derive from those described at the beginning of the 90′ (Furlong et al., 1991; Gan et al., 1991). These protocols are assumed to provide PON1 pure at ∼ 95%. Using such PON1 preparations, during more than a decade, a wealth of information in the areas of biochemistry, molecular biology and toxicology were collected, allowing partial characterization of the enzyme function (Costa and Furlong, 2002). At the same time, multiple attempts at crystallization of PON1 for solving its 3D structure failed. Finally, the discovery of a co-purified PON1 partner (Fokine et al., 2003), and the determination of the structure of a hybrid recombinant PON1 (Harel et al., 2004) brought new impulse to understand and overcome the difficulty to study PON1 in solution.

2.   HPBP IS NOT A CONTAMINANT OF HUMAN PON1

2.1.   HPBP, an *in vitro* and *in vivo* PON1 Partner

In order to solve the 3D structure of human PON1, we used apparently pure enzyme and obtained crystals (Contreras-Martel et al., 2006). Unexpectedly, the solved structure did not match with the amino-acid sequence of PON1, but corresponded to the structure of a protein having a MW similar to that of PON1. This protein was termed phosphate binding protein (HPBP). This unknown protein co-purified with PON1 (Morales et al., 2006), appeared to be a stabilizing partner (Rochu et al., 2007a). Besides, the presence, in the structure of a hybrid recombinant PON1 expressed in *E. coli*, of a phosphate ion bound to the catalytic calcium in the active site was unexpected too (Harel et al., 2004). Numerous works reported subsequently the presence of various contaminants in "pure" PON1 preparations. Several of these contaminants were found to be responsible of certain catalytic activities previously attributed to PON1 (Connelly et al., 2005; Draganov et al., 2005; Teiber et al., 2004). Other recent works highlighted the importance of the HDL-association for the biological function of PON1 (Gaidukov and Tawfik, 2005; James and Deakin, 2004). Thus, converging data indicate that both activity and stability of PON1 are dramatically dependent on the HDL components, including HPBP. This protein we recently discovered could be involved in phosphatemia-related disorders, including atherosclerosis. The presence of an inorganic phosphate scavenger or receptor in human plasma associated with HDL may be explained by the need to prevent the complexation of this anion with Ca^{2+}. Indeed, hyper-phosphatemia is a cardiovascular-disease risk factor (Amann et al., 2003) because calcium phosphate, along with cholesterol, increases the formation of atherosclerotic plaques in blood vessels (Dorozhkin and Epple, 2002). Consequently, biochemical and physiological characterization of this PON1 partner is mandatory, as well as the environment allowing retaining thermal and storage stability of the enzyme in solution (Berna et al., 2007).

2.2. HPBP, Scavenger of Inorganic Phosphate in Human Plasma

The X-ray structure of HPBP was solved at 1.9 Å resolution (Fig. 1). HPBP shows a fold and a binding site similar to the prokaryotic periplasmic phosphate

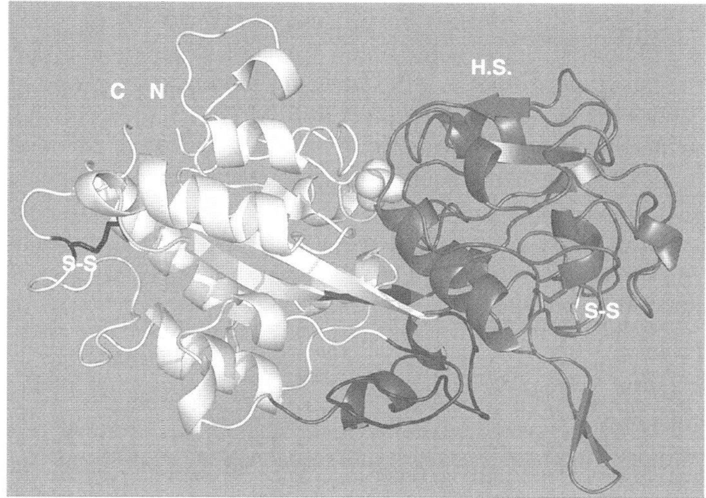

Figure 1. Ribbon representation of HPBP structure showing the two domains (white and black) connected by a hinge, the phosphate molecule (spheres), and the two disulfide bridges (S-S). A motif on the protein surface (H.S.), strongly hydrophobic, is involved in the packing of the crystallographic dimer. N and C represent the protein termini

binding protein (Luecke and Quiocho, 1990) associated with ATP binding cassette transmembrane transporters.

There is limited sequence identity with these proteins, and phosphate-SBPs have never been characterized or predicted from nucleic acid databases in eukaryotes. However, HPBP belongs to the family of ubiquitous eukaryotic proteins named DING (Berna et al., 2002), meaning that phosphate-SBPs are also widespread in eukaryotes. The absence of complete genes for eukaryotic phosphate-SBP from databases is puzzling, but the astonishing 90% sequence conservation of genes between evolutionary distant species suggests that corresponding proteins play an important function. HPBP is the first identified transporter capable of binding phosphate ions in human plasma. The described structure has a site which binds inorganic phosphate (Pi) in the same way as the prokaryotic protein, via the "Venus fly trap" model with the two domains engulfing the Pi as they close the binding cleft via a hinge-bend (and twist) (Brune et al., 1998; Mao et al., 1982).

3.   OPTIMIZED PURIFICATION OF HUMAN PON1

3.1.   The Hidden Necessary Partner

In order to obtain pure PON1 for its crystal structure determination, a first objective was to optimize the purification process and the biochemical characterization of the enzyme. Indeed, the standard purification protocol yielded two proteins with similar MW, PON1 and HPBP. Although these proteins have very different p*I*s, the last DEAE chromatographic step failed to separate them. Besides, harsh conditions (e.g. 9.8 M urea, 4% Triton X100) were needed to separate HPBP and PON1 using 2D SDS-PAGE. This indicates that HPBP is strongly associated with PON1. A 2D SDS-PAGE of purified PON1 (Smolen et al., 1991), showed a major contaminant with p*I* and MW similar to that HPBP, but this contaminant was not considered at the moment. As above-mentioned, our group crystallized HPBP, solved its 3D structure, and showed that the protein binds phosphate tightly, with a submicromolar dissociation constant. HPBP is the first phosphate transporter characterized in human plasma. The PON1-HPBP interaction could impinge directly on the function of both proteins, for example by changing their specificity (Webb, 2006). Characterisation of functions of the freshly discovered protein HPBP has been actively undertaken to provide information on this likelihood.

3.2.   PON1 and HPBP, a Couple with Unfixed Head Making Separation Uneasy

It is worth noting that between individuals, there is a $\sim$ 10- to 40-fold variation in PON1 activity (Davies et al., 1996; Richter and Furlong, 1999). Moreover, within a single phenotype (e.g. PON1Q192), there is a $\sim$ 13-fold variation in PON1 protein level (Costa et al., 2003). In agreement with recent literature data, our results indicate that some PON1 activities are environment-dependent. For example, within plasma

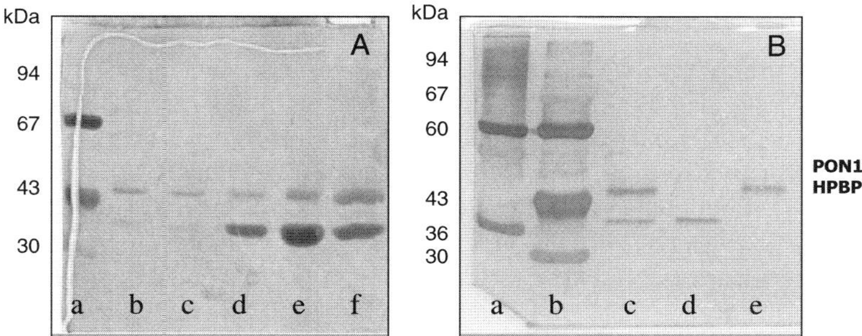

Figure 2. SDS-PAGE (T = 10%). A, Lane a, MW standards; lanes b-f, PON1-containing fractions from plasma bags of different individuals at the terminal step of standard protocol. B, Hydroxyapatite chromatography of PON1-HPBP. Lanes a,b, HMW and LMW standards; c, PON1-HPBP-containing fraction; d, unadsorbed HPBP; e, PON1 eluted by phosphate

bags the PON1/HPBP ratio at the penultimate purification step varies noticeably (Fig. 2A). This is very likely related to the HDL status of the donor. Because human PON1 is a promising catalytic OP-scavenger (Rochu et al., 2007b), it is of importance to determine whether PON1-HPBP association is physiologically and pharmacologically relevant.

Finally, to obtain highly purified PON1, we have consistently modified the purification protocol. Hydroxyapatite chromatography was capable of separating PON1 and HPBP (Fig. 2B) (Renault et al., 2006). In addition, this fundamental step provided evidence about the importance of the molecular environment for the catalytic activity(ies) and stability of the enzyme. PON1 free of HPBP is now available. Highly purified PON1 is intended for crystallization at aim of determining the 3D structure of the human enzyme. In addition, the possible role of HPBP as a helper for stability and/or for OPH activity of PON1 was hypothesized. This prompted us to investigate the balance stability/activity(ies) of PON1.

4. STABILIZATION OF ACTIVE NATURAL HUMAN PON1

4.1. *In vivo* Localization of PON1 and HPBP in Plasma Lipoprotein Fractions

Because the association PON1-HPBP was assumed to be central for activity and stability of PON1, research on these criterions became our first concern. The study was conducted in several complementary directions. It was recently shown that, VLDL and chylomicrons also contain minor but significant amounts of PON1 (Deakin et al., 2005; Fuhrman et al., 2005). We were able to show that HPBP is also present on HDL, LDL, VLDL and chylomicrons (Elias et al., in preparation). These results support the contention that the association PON1-HPBP occurs *in vivo* in human lipoprotein complex particles.

4.2.   Functional State(s) of Natural Human PON1

The crystal structure of the rPON1 enabled a better understanding of PON1 mechanism (Kersonsky and Tawfik, 2006; Rosenblat et al., 2006), and provided a model for PON1 anchoring onto HDL (Harel et al., 2004). PON1 was suggested to be an interfacially activated lipo-lactonase that selectively binds to HDL carrying apolipoprotein A-I (Harel et al., 2004; Rosenblat et al., 2006). Besides, HPBP was alleged to associate with PON1 *in vivo* (Elias et al., in preparation; Rochu et al., 2007a). These facts made essential the characterization of the functional structure of human PON1. This study was achieved using several tools. The oligomeric state(s) of functional PON1, and its molecular association(s) with HPBP were determined.

Capillary electrophoresis (CE) showed that PON1 alone (Fig. 3) or in PON1-HPBP complexes, displays heterogeneous patterns, including a heavy (H) and a light (L) forms. The relative amounts of these forms vary with the amount of HPBP present in the preparation. H is favoured by HPBP (Rochu et al., in preparation). Size exclusion chromatography showed that: 1) PON1-HPBP complex appears to be dimeric in the presence of phosphate, 2) free PON1 and HPBP (in the presence of phosphate) appears to be monomers, 3) PON1 arylesterase activity does not exactly superpose with PON1 monomer (Elias et al., in preparation). Ferguson plots analysis of PAGE under hydrostatic pressure showed that PON1, alone and associated with HPBP displays three active populations with different retardation coefficients (dimer, trimer and tetramer). Dimer and tetramer were major populations (Cléry-Barraud et al., in preparation). Pressures higher than 100 MPa, that generally induce dissociation of protein oligomers (Silva, 1993), failed to change the patterns of PON1. Because pressure-induced dissociation of oligomers results from the breakage of hydrophobic interactions, our findings indicates that hydrophobic interactions are not the major interactions either at the PON1-HPBP

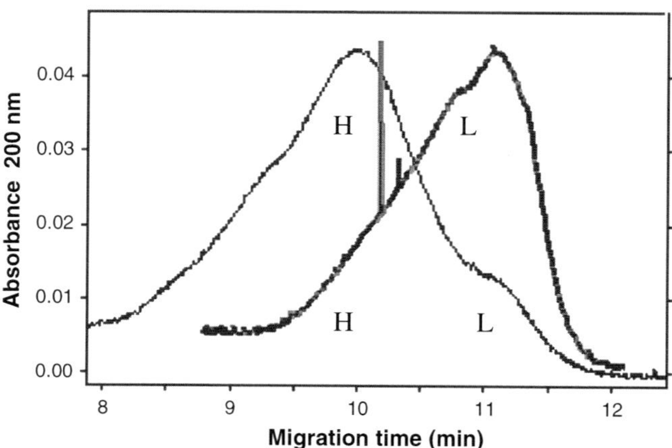

Figure 3. CE migrations at pH 8.2 of free PON1 (left curve) and PON1-HPBP (right curve). Heavy forms (H) were favoured in the absence of HPBP

interface or at PON1-PON1 oligomer interface. In addition, no transient swelling of the protein under moderate pressure (≈ 100–$150\,\text{MPa}$) was observed. This indicates that these pressures did not induce molten globule transition of PON1 and confirms the high stability of this protein.

At length, this first approach at characterizing the dispersity of natural human PON1 oligomerization state, strongly suggests that the enzyme requires at least a dimeric state to be functional, and that this conformation can be obtained by PON1 itself (homodimers) or in association with HPBP (heterodimers). This agrees with the description of special radii of curvature of the PON1-containing HDL particles required for the enzyme to allow expression of its catalytic activity (Cabana et al., 2003). It was hypothesized that a possible dimerization could increase PON1 stability; this in turn may promote more efficient association to HDL (Gaidukov et al., 2006). Dimeric PON1 (homodimer) displays enzyme activity, but failed to display suitable storage stability. On the contrary, PON1-HPBP complex (heterodimer) was shown to be stable (see next section). Further studies will determine whether homo- and heterodimers possess identical catalytic parameters and the same characteristics regarding substrate promiscuity.

4.3. Storage Stability of Natural Human PON1

Storage stability of PON1 was estimated by CE and measurements of the loss of activity of samples incubated at 4 °C. Electropherograms showed that free PON1 (Fig. 4), or PON1 complexed with HPBP, displays a decrease of the amount of its

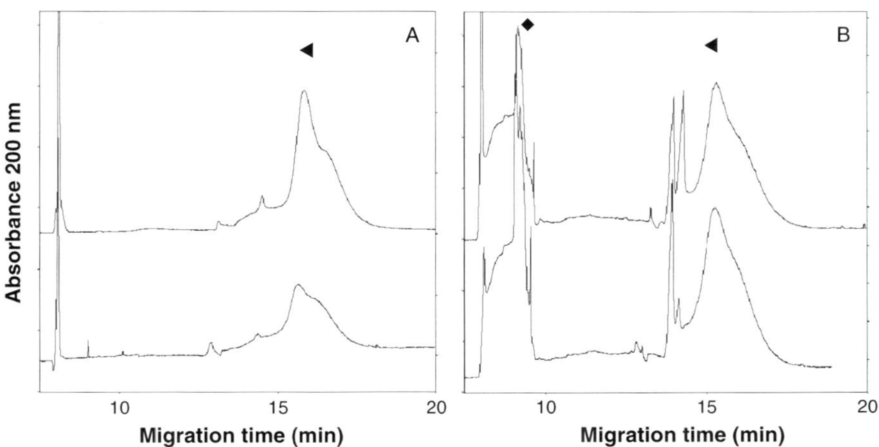

Figure 4. Analysis of storage stability by CE for free PON1 (A), and PON1-HPBP (B), incubated 0 day (upper curves) and 7 days (lower curves) at 4 °C

1st peak, EOF marker; ◄, PON1; ◆, HPBP; minor peaks, buffer components. Noisy baseline, broadening and higher heterogeneity of free PON1 main peak indicate time-dependent conformational alteration of native oligomeric state(s).

main native state and an increase of soluble aggregates. Both phenomena indicate alteration in the molecular integrity of the functional enzyme.

Measurements of residual activity (Fig. 5) demonstrated that both arylesterase and OP-hydrolase activities of the free PON1 were completely abolished after one week, while the PON1-HPBP complex retained its hydrolytic activities for several weeks. It can be noted that a part of the remaining activity may be due to active aggregates (soluble aggregates were observed by CE and active aggregates were characterized by electrophoresis under hydrostatic pressure).

4.4.   Thermal Stability of Natural Human PON1

The thermal stability of PON1 was analyzed using two complementary approaches, i.e., capillary electrophoresis (CE) and differential scanning calorimetry (DSC). CE of PON1 samples performed at different temperatures showed dissociation of a heavy form at $\sim 55\,°C$ and a thermal transition at $\sim 60\,°C$ (Fig. 6A). DSC of PON1 samples (Fig. 6B confirmed that a thermal transition occurs at $61\,°C$. CE and DSC of PON1-HPBP samples provided similar results, but DSC showed a more complex pattern with the main endothermic transition at $61\,°C$ preceded by a clearly individualised endothermic transition at $55\,°C$. This suggests involvement of HPBP in the above-mentioned heavy form. In addition, in phosphate buffer, a decrease in the T_m was observed (data not shown). These results can be related to the presence of HPBP, and to the fact that phosphate is involved in PON1-HPBP association. We recently demonstrated that CE is a convenient tool to analyze stability of free enzymes, whereas DSC allows stability study of enzyme complexes (Rochu et al., 2006).

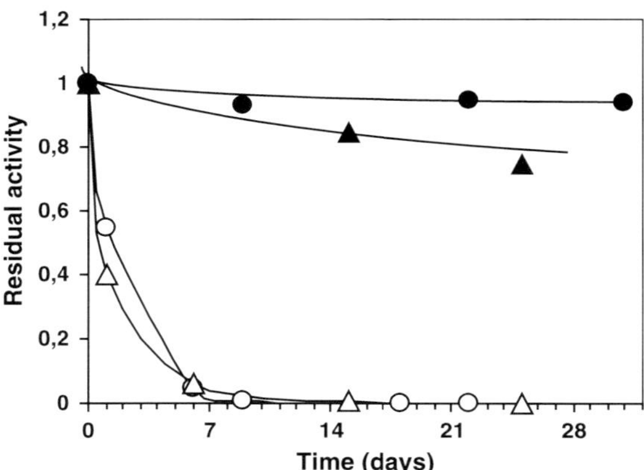

Figure 5. Time-course of residual activity for free PON1 (open symbols) and PON1-HPBP (closed symbols) stored at $4\,°C$. Circles and triangles indicate arylesterase and OPH activity, respectively

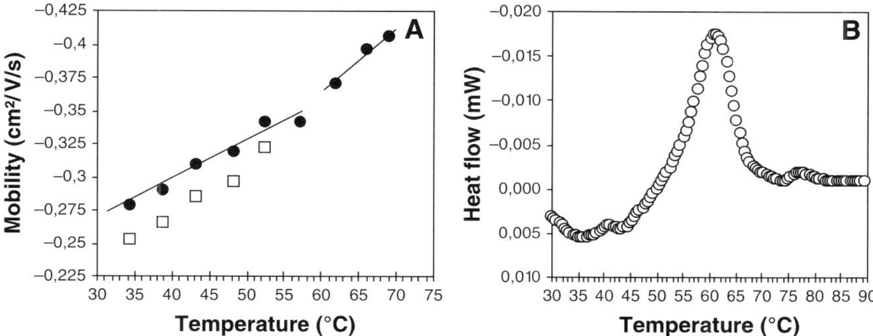

Figure 6. A, Temperature-dependence of PON1 mobility investigated by CE runs performed at varying temperature. Heavy form (□) dissociates completely above 55 °C; a thermal transition occurs near 60 °C. B, Thermogram by DSC for the heat-induced denaturation of PON1 in Tris buffer pH 8, showing mid-transition temperature $(T_m) = 61\,°C$

Thermal stability of PON1 was also investigated by following the thermal inactivation of arylesterase activity. This study confirmed that thermal inactivation of PON1 is complete at 60 °C (Fig. 7).

Finally, these approaches show that with a T_m of $\sim 60\,°C$, PON1 is a remarkable thermally stable OPH. Moreover, in the presence of HPBP, a gain of $\sim 12\,°C$ for the denaturation mid-transition is observed. Clearly, the thermal stability of human PON1 is increased by association with HPBP.

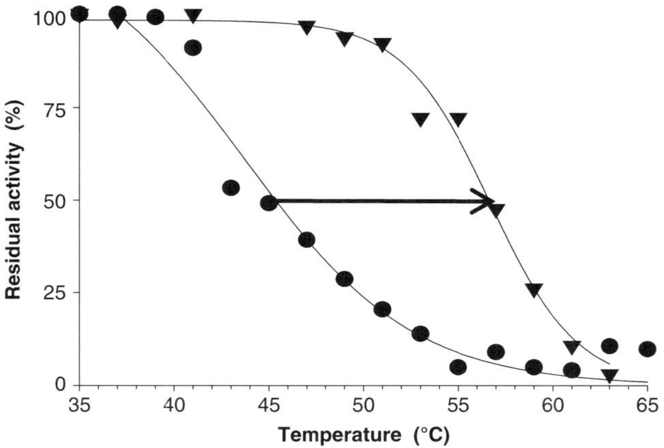

Figure 7. Thermal inactivation of PON1, free (●), or supplemented by HPBP (▼), showing stabilization by HPBP up to 50 °C, and inactivation at $\approx 60\,°C$

5. CONCLUSIONS AND PERSPECTIVES

This review strengthens the critical role of the environment for allowing PON1 to perform adequately and durably its catalytic activity(ies). The prominent role of HPBP as a helper for stability and for OPH activity of PON1 is highlighted. HPBP is a protein having escaped to detection by proteomic and genomic investigations. Although genomics devotees have claimed the human genome completely sequenced, HPBP is also characterized by the absence of its gene in the human genome databanks available at the moment. Independently of the central "omics" questions raised by such a dual gap, these new data are crucial for research at aim of making PON1 a promising catalytic OP-scavenger.

The tandem purification of PON1 and HPBP led to human PON1 free of contaminant, and provided the first protocol for obtaining pure HPBP. Likewise, we established that both proteins are highly unstable in solution after parting of the couple. The analysis of conditions allowing storage of stabilized active PON1 was the subsequent goal. Characterization of the PON1 stability/activity(ies) balance became of the utmost importance, and is a pre-requisite for the development of variants with enhanced hydrolytic activity toward toxic OPs. As suggested by the recurrent association of PON1 with HPBP during all purification steps of current protocols, both proteins were shown to be present in the same lipoprotein fractions of human plasma. Thus, the association PON1-HPBP is thought to be likely functional. Indeed, PON1 exhibits an anti-atherosclerotic activity, and HPBP is assumed to be involved phosphatemia regulation. A synergistic role of both HDL-associated lipoproteins can be reasonably argued. The study of the oligomerization state of PON1, the characterization of its long-term storage stability and the study of its thermal stability showed that HPBP is an essential partner for maintaining a molecular environment matching with preservation of PON1 enzyme activity. HPBP is a newly discovered protein; moreover its crystal structure was solved without previous biochemical characterisation. Study of this protein is still in infancy. Determination of physiological parameters of HPBP and/or research of HPBP deficiency could shed light on the importance of the protein for PON1 function and stability.

A preliminary analysis of the behaviour of the hybrid mammalian rPON1 indicated that functional forms of natural human PON1 and mammalian rPON1 display different properties, regarding their oligomeric structure and stability (Rochu et al., in preparation). Because rPON1 is closer to rabbit PON1 than to human PON1, these results justifies extensive characterization of the natural human enzyme. The strategy is to construct a hybrid gene allowing co-expression of HPBP and PON1 in *E. coli*. At present, knowledge of complete and unambiguous HPBP sequence (Van Dorsselaer et al., in preparation), allowed chemical synthesis of the HPBP gene. Co-expression of both genes is expected to favour crystallization of the human rPON1-rHPBP complex, on purpose of solving the 3D structure of natural human PON1. This step is crucial for engineering and producing new human PON1 mutants of operational catalytic efficiencies to challenge OP exposure.

In conclusion, to be used as a biopharmaceutical against poisoning by chemical warfare agents (CWA) and pesticides, PON1 has to displays several properties (dealing with activity, storage, biological half-life, absence of immunological or behavioural side effects, availability, and cost) (Lenz et al., 2007; Rochu et al., 2007b). At the moment, neither the natural human PON1 nor the mammalian rPON1 fulfil all these pre-requisites. Synergistic efforts on characterization of natural and rPON1 provide information for the future rational design of stable safe and effective PON1-based bioscavenger. Ultimately, maintaining a favourable environment for this administrable "CWA-catalytic scavenger", human rPON1 should preserve or enhance the anti-atherogenic activity of the enzyme. Medical chemical defence and public health researches are thus tightly interconnected.

ACKNOWLEDGEMENTS

The authors thank Dr. Dan S. Tawfik (Weizmann Institute of Science, Israel) for providing the mammalian recombinant PON1-G3C9. This work was supported by DGA contract (PEA 010807) and EMA (LR2006) to P.M. D.R. is under contract with the German Bundesministerium der Verteidigung (M/SAB 1/6/A002).

REFERENCES

Aharoni, A., Gaidukov, L., Khersonsky, O., McQ Gould, S., Roodveldt, C., and Tawfik, D.S., 2005, The 'evolvability' of promiscuous protein functions, *Nat. Genet.* 37, 73–76

Amann, K., Tornig, J., Kugel, B., Gross, M.L., Tyralla, K., El-Shakmak, A., Szabo, A., and Ritz, E., 2003, Hyperphosphatemia aggravates cardiac fibrosis and microvascular disease in experimental uremia, *Kidney Int.* 63, 1296–1301

Berna, A., Bernier, F., Chabrière, E., Perera, T., and Scott, K., 2007, DING Proteins: Novel members of a prokaryotic phosphate-binding protein superfamily which extends into the eukaryotic kingdom, *Int. J. Biochem. Cell Biol.*, in press

Berna, A., Bernier, F., Scott, K., and Stuhlmuller, B., 2002, Ring up the curtain on DING proteins, *FEBS Lett.* 524, 6–10

Blatter, M.C., James, R.W., Messmer, S., Barja, F., and Pometta, D., 1993, Identification of a distinct human high-density lipoprotein subspecies defined by a lipoprotein-associated protein, K-45. Identity with paraoxonase, *Eur. J. Biochem.* 211, 871–879

Brune, M., Hunter, J.L., Howell, S.A., Martin, S.R., Hazlett, T.L., Corrie, J.E.T., and Webb, M.R., 1998, Mechanism of inorganic phosphate interaction with phosphate binding protein from *Escherichia coli*, *Biochemistry* 37, 10370–10380

Cabana, V.G., Reardon, C.A., Feng, N., Neath, S., Lukens, J., and Getz, G.S., 2003, Serum paraoxonase: effect of the apolipoprotein composition of HDL and the acute phase response, *J. Lipid Res.* 44, 780–792

Connelly, P.W., Draganov, D., and Maguire, G.F., 2005, Paraoxonase-1 does not reduce or modify oxidation of phospholipids by peroxinitrite, *Free Rad. Biol. Med.* 38, 164–174

Contreras-Martel, C., Carpentier, P., Morales, R., Renault, F., Chesne-Seck, M.L., Rochu, D., Masson, P., Fontecilla-Camps, J.C., and Chabrière E., 2006, Crystallization of the human phosphate binding protein, *Acta Cryst. F* 62, 67–69

Costa, L.G., Cole, T.B., Jarvik, G.P., and Furlong, C.E., 2003, Functional genomics of the paraoxonase (PON1) polymorphisms: effects on pesticide sensitivity, cardiovascular disease, and drug metabolism, *Ann. Rev. Med.* 54, 371–392

Costa, L.G. and Furlong, C.E., 2002, *Paraoxonase (PON1) in Health and Disease: Basic and Clinical Aspects*, Kluwer Academic Publishers, Dordrecht, The Netherlands

Davies, H.G., Richter, R.J., Keifer, M., Broomfield, C.A., Sowala, J., and Furlong, C.E., 1996, The effect of the human serum paraoxonase polymorphism is reversed with diazoxon, soman and sarin, *Nat. Genet.* 14, 334–336

Deakin, S., Moren, X., and James, R.W., 2005, Very low density lipoproteins provide a vector for secretion of paraoxonase-1 from cells, *Atherosclerosis* 179, 17–25

Dorozhkin, S.V., and Epple, M., 2002, Biological and medical significance of calcium phosphates, *Angew. Chem. Int. Ed. Engl.* 41, 3130–3146

Draganov, D.I., Teiber, J.F., Speelman, A., Osawa, Y., Sunahara, R., and La Du, B.N., 2005, Human paraoxonases (PON1, PON2, and PON3) are lactonases with overlapping and distinct substrate specificities, *J. Lipid Res.* 46, 1239–1247

Fokine, A., Morales, R., Contreras-Martel, C., Carpentier, P., Renault, F., Rochu, D., and Chabrière, E., 2003, Direct phasing at low resolution of a protein co-purified with human paraoxonase (PON1), *Acta Cryst. D* 59, 2083–2087

Fuhrman, B., Volkova, N., and Aviram, M., 2005, Paraoxonase (PON1) is present in postprandial chylomicrons, *Atherosclerosis* 180, 55–61

Furlong, C.E., Richter, R.J., Chapline, C., and Crabb, J.W., 1991, Purification of rabbit and human paraoxonase, *Biochemistry* 30, 10133–10140

Gaidukov, L., Rosenblat, M., Aviram, M., and Tawfik, D.S., 2006, The 192R/Q polymorphs of serum paraoxonase PON1 differ in HDL binding, stimulation of lipo-lactonase, and macrophage cholesterol efflux, *J. Lipid Res.* 47, 2492–2502

Gaidukov, L., and Tawfik, D.S., 2005, High affinity, stability, and lactonase activity of serum paraoxonase PON1 anchored on HDL with apoA-I, *Biochemistry* 44, 11843–11854

Gan, K.N., Smolen, A., Eckerson, H.W., and La Du, B.N., 1991, Purification of human serum paraoxonase/arylesterase, *Drug Metabol. Dispos.* 19, 100–106

Harel, M., Aharoni, A., Gaidukov, L., Brumshtein, B., Khershonsky, O., Yagur, S., Meged, R., Dvir, H., Ravelli, R.G.B., McCarty, A., Toker, L., Silman, I., Sussman, J.L., and Tawfik, D.S., 2004, Structure and evolution of the serum paraoxonase family of detoxifying and anti-atherosclerotic enzymes, *Nature Struct. Mol. Biol.* 11, 412–419

Jakubowski, H., Ambrosius, W.T., and Pratt, J.H., 2001, Genetic determinants of homocysteine thiolactonase activity in humans: implications for atherosclerosis, *FEBS Lett.* 491, 35–39

James, R.W., and Deakin, S.P., 2004, The importance of high-density lipoproteins for paraoxonase-1 secretion, stability, and activity, *Free Rad. Biol. Med.* 37, 1986–1994

Khersonsky, O., and Tawfik, D.S., 2005, Structure-reactivity studies of serum paraoxonase PON1 suggest that its native activity is lactonase, *Biochemistry* 44, 6371–6382

Khersonsky, O., and Tawfik, D.S., 2006, The histidine 115-histidine134 dyad mediates the lactonase activity of mammalian serum paraoxonases, *J. Biol. Chem.* 281, 7649–7656

Lacinski, M., Skorupski, W., Cieslinski, A., Sokolowska, J., Trzeciak, W.H., and Jakubowski, H., 2004, Determinants of homocysteine-thiolactonase activity of the paraoxonase-1 (PON1) protein in humans, *Cell Mol. Biol. (Noisy-le-Grand)* 50, 885–893

Lenz, D.E., Broomfield, C.A., Yeung, D.T., Masson, P., Maxwell, D.M., and Cerasoli, D.M., 2007, Nerve agent bioscavengers: progress in development of a new mode of protection against organophosphorus exposure. In *Chemical Warfare Agents: Toxicity at Low Level* (B. Luckey, S.M. Somani and J.A. Romano), 2nd edition, CRC Press, Boca Raton, Florida, in press

Luecke, H., and Quiocho, F.A., 1990, High specificity of a phosphate transport protein determined by hydrogen bonds, *Nature* 347, 402–406

Mackness, M.I., Arrol, S., Durrington, P.N., 1991, Paraoxonase prevents accumulation of lipoperoxides in low-density lipoprotein, *FEBS Lett.* 286, 152–154

Mao, B., Pear, M.R., McCammon, J.A., and Quiocho, F.A., 1982, Hinge-bending in L-arabinose-binding protein. The "Venus's-flytrap" model, *J. Biol. Chem.* 257, 1131–1133

Morales, R., Berna, A., Carpentier, P., Contreras-Martel, C., Renault, F., Nicodeme, M., Chesne, M.L., Bernier, F., Shaeffer, C., Van-Dorsselaer, A., Fontecilla-Camps, J.C., Masson, P., Rochu, D., and

Chabrière, E., 2006, Serendipitous discovery and X-ray structure of a human phosphate binding apolipoprotein, *Structure* 14, 601–609

Renault, F., Chabrière, E., Andrieu, J.P., Dublet, B., Masson, P., and Rochu, D., 2006, Tandem purification of two HDL-associated partner proteins in human plasma, paraoxonase (PON1) and phosphate binding protein (HPBP) using hydroxyapatite chromatography, *J. Chromatogr.* B 836, 15–21

Richter, R.J., and Furlong, C.E., 1999, Determination of paraoxonase (PON1) status requires more than genotyping, *Pharmacogenetics* 9, 745–753

Righetti, P.G., Castagna, A., Antonucci, F., Piubelli, C., Cecconi, D., Campostrini, N., Rustichelli, C., Antonioli, P., Zanusso, G., Monaco, S., Lomas, L., and Boschetti, E., 2005, Proteome analysis in the clinical chemistry laboratory: Myth or reality ?, *Clin. Chim. Acta* 357, 123–139

Rochu, D., Chabrière, E., and Masson, P., 2007b, Human paraoxonase: a promising approach for pretreatment and therapy of organophosphorus poisoning, *Toxicology*, doi: 10.1016/j.tox. 2006.08.037

Rochu, D., Chabrière, E., Renault, F., Elias, M., Cléry-Barraud, C., and Masson, P., 2007a, Functional states, storage and thermal stability of human paraoxonase: drawbacks, advantages and potentialities, *Toxicology*, doi:10.1016/j.tox.2006.04.010

Rochu, D., Cléry-Barraud, C., Renault, F., Chevalier, A., Bon, C., and Masson, P., 2006, Capillary electrophoresis *versus* differential scanning calorimetry for the analysis of free enzyme *versus* enzyme-ligand complexes: in the search of the ligand-free status of cholinesterases, *Electrophoresis* 27, 442–451

Rosenblat, M., Gaidukov, L., Khersonsky, O., Vaya, J., Oren, R., Tawfik, D.S., Aviram, M., 2006, The catalytic histidine dyad of high density lipoprotein-associated serum paraoxonase-1 (PON1) is essential for PON1-mediated inhibition of low density lipoprotein oxidation and stimulation of macrophage cholesterol efflux, *J. Biol. Chem.* 281, 7657–7665

Shih, D.M., Gu, L., Xia, Y.R., Navab, M., Li, W.F., and Hama, S., 1998, Mice lacking serum paraoxonase are susceptible to organophosphate toxicity and atherosclerosis, *Nature* 394, 284–287

Silva, J.L., 1993, Effects of pressure on large multimeric proteins and viruses. In *High Pressure Chemistry, Biochemistry and Molecular Science* (R. Winter and J. Jonas, eds.), Kluwer Academic Publishers, Dordrecht, the Netherlands, 561–578.

Smolen, A., Eckerson, H.W., Gan, K.N., Hailat, N., and La Du, B.N., 1991, Characteristics of the genetically determined allozymic forms of human serum paraoxonase/arylesterase, *Drug Metab. Dispos.* 19, 107–112

Teiber, J.F., Draganov, D.I., and La Du, B.N., 2003, Lactonase and lactonizing activities of human serum paraoxonase (PON1) and rabbit serum PON3, *Biochem. Pharmacol.* 66, 887–896

Teiber, J.F., Draganov, D.I., and La Du, B.N., 2004, Purified human serum PON1 does not protect LDL against oxidation in the *in vitro* assays initiated with copper or AAPH, *J. Lipid Res.* 45, 2260–2268

Watson, A.D., Berliner, J.A., Hama, S.Y., and La Du, B.N., 1995, Protective effect of high density lipoprotein associated paraoxonase. Inhibition of the biological activity of minimally oxidized low density lipoprotein, *J. Clin. Invest.* 96, 2882–2891

Webb, M.R., 2006, A tale of the unexpected, *Structure* 14, 391–392

PART 4

PONs IN DISEASE OTHER THAN ATHEROSCLEROSIS

CHAPTER 11

PARAOXONASE-1 IN CHRONIC LIVER DISEASES, NEUROLOGICAL DISEASES AND HIV INFECTION

J. MARSILLACH[1], S. PARRA[1], N. FERRÉ[2], B. COLL[1],
C. ALONSO-VILLAVERDE[1], J. JOVEN[1] AND J. CAMPS[1]

[1] Centre de Recerca Biomèdica and Department of Internal Medicine, Hospital Universitari de Sant Joan, Reus, Spain
[2] Department of Clinical Biochemistry and Molecular Genetics, Hospital Clínic Universitari, Barcelona, Spain

Abstract: Over recent years there has been a rapid increase in the number of articles reporting paraoxonase-1 (PON1) alterations in diseases other than arteriosclerosis, and which involve an increased degree of oxidative stress. Chronic liver impairment is associated with decreased serum PON1 activity but with increased serum PON1 concentration; this unusual juxtaposition being explained, perhaps, by the molecular alterations related to collagen synthesis and to the inactivation of the enzyme's active site by lipid peroxides. Polymorphisms of the *PON1* gene have been shown to be associated with an increased development of several neurological diseases and to influence the organism's capacity to protect against neurotoxins. Many infectious diseases, as HIV infection, produce a high oxidative stress, and alterations in PON1 activity and concentration may be related to the course of the disease

Keywords: HIV infection, chronic liver diseases, neurological diseases, oxidative stress, paraoxonase-1

Paraoxonase-1 (PON1) is synthesized mainly by the liver (Pellin et al., 1990). Northern blot analysis have detected PON1 mRNA in human and rabbit liver (Hassett et al., 1991), although more sensitive reverse transcriptase-polymerase chain reaction (RT-PCR) methods have detected PON1 mRNA in mouse liver, kidney, heart, brain, intestine and lung, as well (Primo-Parmo et al., 1996). These results have been confirmed by immunohistochemistry in rat tissue (Rodrigo et al., 2001). Hence it is likely that chronic liver diseases are accompanied by changes in the concentration, in the activity of the enzyme or in the structure of PON1 protein.

Corresponding author: J. Camps, PhD, Centre de Recerca Biomèdica, Institut de Recerca en Ciències de la Salut, Hospital Universitari de Sant Joan, C. Sant Joan s/n, 43201 Reus, Catalunya, Spain.
e-mail.: jcamps@grupsagessa.cat

187

B. Mackness et al. (eds.), The Paraoxonases: Their Role in Disease Development and Xenobiotic Metabolism, 187–198.
© 2008 *Springer.*

1.   EXPERIMENTAL CIRRHOSIS

Oxidative stress plays an important role in the pathogenesis of various liver diseases (Tanikawa and Torimura, 2006). In alcoholic liver disease, in nonalcoholic steato-hepatitis, and in hepatitis C virus infection, oxidative stress is involved in the pathophysiological changes leading to liver cirrhosis and finally to hepatocellular carcinoma. Rat and human liver PON1 are essentially microsomal enzymes associated with vesicles derived from the endoplasmic reticulum (Gil et al., 1993; Gonzalvo et al., 1998). As demonstrated by Mackness and Durrington (1995), PON1 exerts a protective effect against oxidative stress and, hence, it is plausible that there is a relationship between PON1 and oxidative stress in liver diseases. Ferré et al. (2001) observed, in rats with CCl_4-induced cirrhosis, that a decrease in liver microsomal PON1 activity was an early biochemical change related to lipid peroxidation and liver injury. They investigated the relationships between hepatic microsomal PON1 activity, lipid peroxidation and the progress of the disease in this experimental model. Additionally, they monitored the modulation of these processes by the dietary supplementation with zinc, a metal that possesses antioxidant and anti-fibrogenetic properties. They found that PON1 activity decreased while lipid peroxidation increased in CCl_4-treated rats while the addition of zinc was associated with enhanced PON1 activity and a normalisation of lipid peroxidation in these animals. This suggested that PON1 activity may play a role in the defence against free radical production in the hepatic organelles.

2.   HUMAN CHRONIC LIVER IMPAIRMENT

It is accepted that arylesterase and PON1 activities are functions of a single enzyme (Sorenson et al., 1995). In addition, in vitro biochemical studies indicate that hepatic and serum PON1 share many pharmacodynamic properties, as K_M, V_{MAX}, effects of activators and inhibitors, etc. and which support the concept of a common identity for both PON's (Gonzalvo et al., 1998). Early studies had observed a significant decrease of serum arylesterase activity in patients with liver cirrhosis (Burlina and Galzigna, 1974; Burlina et al., 1977; Kawai et al., 1990). This was confirmed by our group (Ferré et al., 2002; Ferré et al., 2005) in patients with various degrees of chronic liver damage. These later studies noted a significant decrease of serum PON1 activity in patients with chronic hepatitis and an even greater decrease in cirrhotic patients compared to a control group. Similar results were obtained in patients with chronic hepatitis by other authors (Kilic et al., 2005).

A genetic association between $PON1_{192}$ polymorphism and chronic hepatitis C virus infection has been identified (Ferré et al., 2005). An explanation for this observation could be found in the polymorphism *per se* being functionally involved in the pathogenesis of chronic hepatitis. PON1 hydrolyses lipid peroxides and there is evidence to suggest that the allele PON1Q is more efficient than the PON1R allele (Aviram et al., 2000). These data suggest that subjects carrying the R allele might be more susceptible to developing diseases deriving from an increased oxidative stress.

In support of these data, the total plasma peroxide concentrations were significantly increased in patients with chronic hepatitis and the concentrations were highest in subjects carrying the RR genotype.

Since standard biochemical tests for liver dysfunction are insufficiently sensitive for a reliable indication of presence or absence of liver disease, histological examination of liver biopsy material has become the best diagnostic test in assessing liver impairment (Sebastiani and Alberti, 2006). However, finding new non-invasive, reliable tests, is an ongoing field of research that possesses clinical relevance. PON1 activity is altered in chronic liver diseases, and results suggest that its measurement may add valuable information in the assessment of liver damage. Serum PON1 activity had a high diagnostic accuracy when distinguishing patients with liver disease from control subjects and, when added to a standard battery of liver function tests, increased the overall sensitivity without impairing the specificity (Ferré et al., 2002; Marsillach et al., 2007).

The value of measuring serum PON1 activity has been studied recently (Xu et al., 2005) in assessing the outcome of liver transplantation in patients with severe liver disease. The serum PON1 activity was low but tended to increase in liver transplanted patients after the hepatic arteries were blocked. Since PON1 activity is closely related to the recovery of liver function, its measurement could provide more accurate information on the success, or otherwise, of the liver transplant.

The measurement of serum PON1 concentration in patients with chronic liver impairment showed intriguing results in that these patients had a high circulating PON1 concentration despite having a low activity (Ferré et al., 2006). These results were confirmed by immunohistochemistry in the same study (Fig. 1) and which showed that PON1 expression in liver tissue was increased in parallel with the degree of liver damage. Also, the patients with higher serum PON1 concentration and of expression had higher lipid peroxidation in serum and in liver biopsy tissues.

An explanation for these findings may be hypothesized in the light of current knowledge of the molecular mechanisms underlying PON1 synthesis and activity. For example, PON1 and collagen share nuclear transcription factors, such as SP1

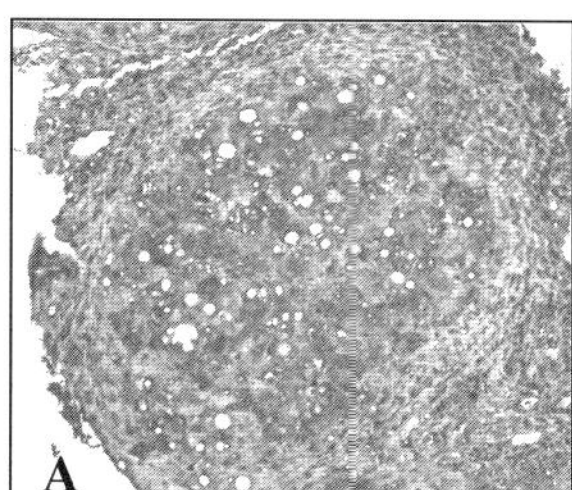
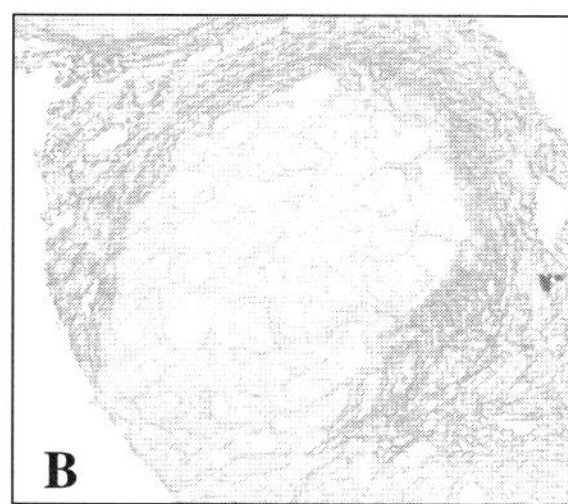
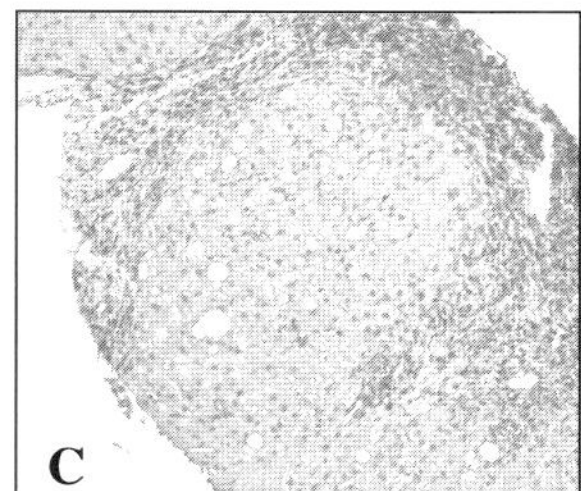

Figure 1. Representative slides of serial sections of liver biopsies from a cirrhotic patient. Tissue samples are stained for PON1 (A), Sirius Red (B) and CD43 (C). Positive staining for Sirius Red and CD43 was restricted to collagen and inflammatory cells, respectively. Positive staining for PON1 appeared basically in parenchymal cells

(Deakin et al., 2003a; García-Ruiz et al., 2002) and sterol regulatory element binding protein (Deakin et al., 2003b; Ferrari et al., 2004), and both genes share a linkage on chromosome 7q (Bowcook et al., 1986; Buchwald et al., 1986). It is likely that chronic stimuli enhancing collagen expression, as occurs in chronic liver disease, would similarly affect PON1. On the other hand, at least two possible mechanisms can be postulated for the decrease observed in PON1 activity. Firstly, the patients presented an increased concentration of total peroxides and, as has been reported, PON1 becomes inactivated after hydrolysing lipid peroxides (Aviram et al., 1999). Secondly, PON1 is bound to the HDL particle and alterations in HDL structure and metabolism would affect PON1 activity (James and Deakin, 2004). Patients with liver disease often exhibit impaired synthesis of several enzymes regulating HDL synthesis (Sabesin et al., 1997) resulting in altered shape and structure of the HDL molecule (Sabesin, 1981; Turner et al., 1979).

HDL particles from patients with liver disease have relative increases in the content of free cholesterol, phospholipids, and a decreased content of esterified cholesterol and apolipoprotein A-I. The resulting abnormal structure and content of the HDL molecule could affect serum PON1 activity. The presence of apolipoprotein A-I in HDL is necessary in maintaining the optimum PON1 activity and for stabilizing the enzyme, and the addition of free cholesterol to HDL *in vitro* lowers PON1 activity (Deakin et al., 2003b) and phospholipids deplete the HDL molecule of its PON1 content (Sorenson et al., 1999). Further, the patient's treatment may also influence PON1 activity since some antibiotics are known to inhibit PON1 activity *in vitro* (Sinan et al., 2006).

Recent evidence indicates that HDL inhibits cell apoptosis and that oxidised HDL loses this capacity (Nofer et al., 2002; Sugano et al., 2000; Watson et al., 1995). Hence, PON1 which is known to protect HDL from oxidation, would encourage the anti-apoptosis potential of HDL (Matsunaga et al., 2001). The hypothesis that PON1 is involved in the regulation of hepatic cell apoptosis has been suggested recently (Ferré et al., 2006). The study observed an increased serum soluble Fas concentration (which is a marker of antiapoptosis), and a decreased Fas positive cell clusters and TUNEL-positive cells (both markers of apoptosis) in the liver biopsies of those patients with liver disease and with a higher PON1 concentration and expression. These data suggest that PON1 influences the antiapoptotic ability of the HDL molecule, perhaps because of its capacity to protect the lipoprotein against oxidation.

3. ALCOHOLIC LIVER DISEASE

The effects of alcohol intake on serum PON1 levels are not well known. Studies have suggested that PON1 genetic polymorphisms are not a major factor in the modulation of serum PON1 levels resulting from alcohol consumption (Costa et al., 2005; Sierksma et al., 2002). Our group conducted a study in alcoholic patients, classified into three sub-groups according to their degree of liver disease (Marsillach et al., 2007). Again, PON1 activity decreased in relation to the degree of liver damage while PON1 concentration increased (Fig. 2).

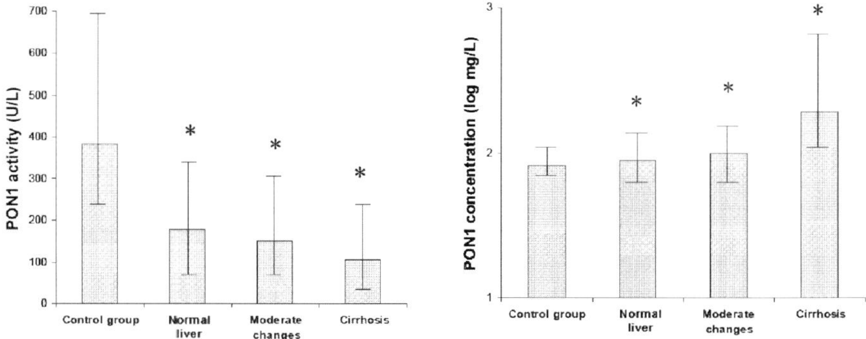

Figure 2. Serum PON1 activity and concentration in the control group and alcoholic patients. Results are presented as medians and 95% CI. * : P < 0.001, with respect to control group

These findings in chronic alcoholic patients are different to those described in normal volunteers reporting moderate alcohol consumption and, in whom, serum PON1 activity and HDL cholesterol were found to be slightly increased (Rao et al., 2003; Sierksma et al., 2002). Our study also found an inverse relationship between PON1 activity and plasma malondialdehyde concentration in alcoholic patients with minimal change disease and which supports the hypothesis of PON1 inactivation following lipid peroxide hydrolysis. We observed a weak but significant direct association between PON1 activity, apolipoprotein A-I, and HDL concentrations. Further, we evaluated the usefulness of adding serum PON1 activity or concentration measurement to the standard panel of liver function tests. The addition of serum PON1 activity increased the diagnostic sensitivity without any impairment of specificity in differentiating between alcoholic patients and control subjects. However the PON1 measurements were unable to discriminate the degrees of liver disease in the different sub-groups of alcoholic patients.

4.   NEUROLOGICAL DISEASES

Alzheimer's disease (AD) is the most frequent form of neurological disease in the elderly. Pathophysiologically, AD is characterised by progressive formation of amyloid plaques and neurofibrillary tangles (Lee et al., 2005; Mohs, 2005). In addition, patients with AD frequently present with vascular complications including cerebrovascular amyloid angiopathy and atherosclerosis (Kalback et al., 2004; Roher et al., 2004; Thal et al., 2003). It seems evident that AD has a significant heritable component (Farrer, 1997), and several studies have investigated associations between *PON1* genetic polymorphisms and AD in different populations. A significant relationship between *PON1$_{192}$* polymorphism and AD has been reported, the frequency of carriers of the R allele being lower than that of the general population (He et al., 2006; Scacchi et al., 2003). The R allele appears protective against AD, and which is contrary to that observed for cardiovascular disease. However, the results are not conclusive since several reports have failed to confirm such findings (Pola et al., 2003; Zuliani et al., 2001). Polymorphisms of *PON2* gene have also been investigated in relation to AD. One study observed a higher frequency of carriers of the C allele of $PON2_{311}$ gene in these patients (Shi et al., 2004). More recently, the association of 29 SNPs in *PON1, PON2* and *PON3* gene cluster has been reported in a study on a large series of 730 Caucasian and 467 African Americans; strong relationships of AD with several of these SNP's have been observed in these populations (Erlich et al., 2006).

Amyloid plaques in patients with AD have a high content of oxidised LDL (Pola et al., 2003). Serum PON1 activity is low in patients with AD, suggesting that a defect in HDL-associated antioxidant capacity plays a role in the pathogenesis of this type of dementia (Paragh et al., 2002). PON1 may be involved, as well, in another mechanism related to the onset of AD. A classical hypothesis on the causes of AD has been that it is the consequence of a loss of cholinergic neurons with decreased acetylcholine levels. Indeed, a successful treatment for AD consists in increasing plasma acetylcholine concentrations using cholinesterase inhibitors (Stanbridge, 2005). However, several investigations have observed that AD contains two subpopulations of patients, defined as "responders" and "non-responders" to cholinesterase inhibitors. Finding tests capable of classifying these patients would be a valuable clinical aid in assigning the appropriate therapy. PON1 possesses multiple biological activities and is a powerful cholinesterase inhibitor (Costa, Vitalone et al., 2005). A recent study indicated that the frequency of R allele carriers of the $PON1_{192}$ gene (QR and RR subjects) is much higher in the responders-to-treatment (70%) than in the non-responder (41%) patients with AD (Pola et al., 2005).

The second most common degenerative disorder is Parkinson disease (PD). It is characterised by bradykinesia, rigidity, and resting tremor. Although the aetiology of PD is unknown, there is clear evidence that both genetic and environmental factors are involved in the development of this disease. Living in a rural area, drinking water from artesian wells or exposure to insecticides increases the risk of PD (Benmoyal-Segal et al., 2005). The $PON1_{192}$ and the $PON1_{55}$ gene polymorphisms are associated with an increased risk for PD. The $PON1_{55}$ M allele is correlated with lower PON1 protein concentration and, therefore, carriers of this allele may have a defect in detoxifying environmental neurotoxins. In addition, individuals carrying the R allele of the $PON1_{192}$ gene have higher enzymatic activities towards paraoxon and chlorpyrilos oxon substrates but lower activity towards diazoxon, soman and sarin substrates. As such, both genes are likely to contribute to susceptibility to PD. As in AD, case-control studies have produced conflicting results. Some studies reported that the M allele of the $PON1_{55}$ gene is a risk factor for PD (Akhmedova et al., 2001; Carmine et al., 2002; Kelada et al., 2003) whilst other studies did not find any significant association (Clarimon et al., 2004; Wang and Liu, 2000). In the $PON1_{192}$ gene studies, most reports agree that there is no relationship between this polymorphism and susceptibility to PD (Zintzaras and Hadjigeorgiou, 2004). A meta-analysis on all previously published studies supports the conclusion of an association of the $PON1_{55}$ gene polymorphism with PD but a lack of association of the $PON1_{192}$ polymorphism with PD (Zintzaras and Hadjigeorgiou, 2004). A recent study suggested that interactions between certain debilitating polymorphisms of the *PON1* and acetylcholine esterase genes are strongly involved with the risk of PD development in rural areas exposed to insecticides (Benmoyal-Segal et al., 2005).

Alterations in PON1 polymorphisms or enzyme protein levels have been investigated in several other neurological diseases. Polymorphisms in the PON gene cluster have been associated with sporadic amyotrophic lateral sclerosis (Saeed et al., 2006; Slowik et al., 2006) and a decreased serum enzymatic activity has been reported in multiple sclerosis. Also, PON1 expression has been found to be up-regulated in foetal livers of Down's syndrome sufferers, which could provide a clue as to why these patients have a lower incidence of arteriosclerosis than the general population (Janel et al., 2006).

5.   HIV INFECTION

HIV-infected patients often develop long-term pro-atherogenic metabolic alterations. These complications may be explained by the infection itself or by secondary effects of antiretroviral therapies. These findings have acquired additional clinical relevance since the introduction of effective therapeutic measures which have changed HIV infection from an acute to a chronic disease. Further, the combination of classical cardiovascular risk factors with some genetic polymorphisms appears to predispose these individuals to a higher risk of premature arteriosclerosis (Alonso-Villaverde et al., 2005; Coll et al., 2005; Coll et al., 2006). There are changes in lipoprotein metabolism in the course of HIV infection, including increased lipid peroxidation, hypocholesterolaemia, hypertriglyceridaemia and low HDL concentration (Rose et al., 2006). Also, patients with higher HDL concentrations appear to have a better disease course than patients with lower HDL concentrations (Alonso-Villaverde et al. 2003).

It is likely that PON1 contributes to these beneficial effects of high HDL concentrations in HIV-infected patients. The patients have lower serum PON1 activity and higher PON1 concentration than the general population, and PON1 activity is inversely correlated with the oxidised LDL concentration (Parra et al., 2007), albeit the allele distributions of the *PON1$_{192}$*, *PON1$_{55}$* and *PON1$_{-107}$* polymorphisms are similar to those of the general population. The finding of a decreased serum enzyme activity together with an increased concentration is in agreement with the findings obtained in patients with chronic liver impairment in whom an increased oxidative environment results in inhibition of PON1's active site while up-regulating its hepatic synthesis. Interestingly, HIV-infected patients co-infected with the hepatitis C virus have significantly lower serum PON1 activity than patients infected with the HIV alone. The possibility that a change in PON1 status can play a role in the course of HIV infection is an area worthy of further investigation. Apolipoprotein A-I inhibits some steps in membrane fusion in HIV infection (Owens et al., 1990) and, as such, higher apolipoprotein A-I concentrations would be associated with lower viral infectivity. Since PON1 is known to increase cholesterol efflux from the cell and the binding of the HDL particle to the ABCA1 receptor (Rosenblat et al.,

2005) the interesting hypothesis follows that PON1 can play an anti-infection role with respect to HIV i.e. membrane metabolism is modulated by the efflux of cholesterol to the HDL particle and this phenomenon would influence HIV replication since the virus needs the cholesterol rafts in the plasma membrane to perform the final assembly and entry into the cell (Ansell et al., 2005; Liao et al., 2003; Nguyen and Hildreth, 2000). An additional issue of considerable importance is whether the replication of the HIV virus itself would be influenced by PON1 status. In the study by Parra et al commented-upon above, a positive association was found between serum PON1 concentration and active viral replication and these data would suggest that PON1 may play a beneficial role in protecting patients from HIV infection, thus opening-up a new and promising area of research into therapies for these patients.

REFERENCES

Akhmedova, S.N., Yakimovsky, A.K., Schwartz, E.I., 2001, Paraoxonase 1 Met-Leu 54 polymorphism is associated with Parkinson's disease. *J. Neurol. Sci.*, 184: 179–82.

Alonso-Villaverde, C., Coll, B., Gomez, F., Parra, S., Camps, J., Joven, J., Masana, L., 2005, The efavirenz-induced increase in HDL-cholesterol is influenced by the multidrug resistance gene 1 C3435T polymorphism. *AIDS*, 19: 341–2.

Alonso-Villaverde, C., Segués, T., Coll-Crespo, B., Pérez-Bernalte, R., Rabassa, A., Gomila, M., Parra, S., González-Esteban, M.A., Jiménez-Expósito, J.M., Masana, L., 2003, High-density lipoprotein concentrations relate to the clinical course of HIV viral load in patients undergoing antiretroviral therapy. *AIDS*, 17: 1173–7.

Ansell, B.J., Watson, K.E., Fogelman, A.M., Navab, M., Fonarow, G.C., 2005, High-density lipoprotein function: recent advances. *J. Am. Coll. Cardiol.*, 46: 1792–8.

Aviram, M., Hardak, E., Vaya, J., Mahmood, S., Milo, S., Hoffman, A., Billicke, S., Draganov, D., Rosenblat, M., 2000, Human serum paraoxonases (PON1) Q and R selectively decrease lipid peroxides in human coronary and carotid atherosclerotic lesions. PON1 esterase and peroxidase-like activities. *Circulation*, 101: 2510–7.

Aviram, M., Rosenblat, M., Billecke, S., Erogul, J., Sorenson, R., Bisgaier, C.L., Newton, R.S., La Du, B., 1999, Human serum paraoxonase (PON1) is inactivated by oxidized low density lipoprotein and preserved by antioxidants. *Free Rad Biol Med*, 26: 892–904.

Benmoyal-Segal, L., Vander, T., Shifman, S., Bryk, B., Ebstein, R., Marcus, E.L., Stessman, J., Darvasi, A., Herishanu,Y., Friedman, A., Soreq, H., 2005, Acetylcholinesterase/paraoxonase interactions increase the risk of insecticide-induced Parkinson's disease. *FASEB J.*, 19: 452–454.

Bowcock, A.M., Crandall, J., Daneshvar, L., Lee, G.M., Young, B., Zunzunegui, V., Craik, C., Cavalli-Sforza, L.L., King, M.C., 1986, Genetic analysis of cystic fibrosis: linkage of DNA and classical markers in multiplex families. *Am J Hum Genet*, 39: 699–706.

Buchwald, M., Zsiga, M., Markiewicz, D., Plavsic, N., Kennedy, D., Zengerling, S., Willard, H.F., Tsipouras, P., Schmiegelow, K., Schwartz, M., et al., 1986, Linkage of cystic fibrosis to the pro alpha 2(I) collagen gene, COL1A2, on chromosome 7. *Cytogenet Cell Genet*, 41: 234–9.

Burlina, A., Galzigna, L., 1974, Serum arylesterase isoenzymes in chronic hepatitis. *Clin Biochem*, 7(3): 202–5.

Burlina, A., Michielin, E., Galzigna, L., 1977, Characteristics and behaviour of arylesterase in human serum and liver. *Eur J Clin Invest*, 7(1): 17–20.

Carmine A., Buervenich, S., Sydow, O., Anvret, M., Olson, L., 2002, Further evidence for association of the paraoxonase 1 (PON1) Met-54 allele with Parkinson's disease. *Mov. Disord.*,17: 764–6.

Clarimon, J., Eerola, J., Hellström, O., Tienari, P.J., Singleton, A., 2004, Paraoxonase 1 *(PON1)* gene polymorphisms and Parkinson's disease in a Finnish population. *Neurosci. Lett.*, 367: 168–70.

Coll, B., Alonso-Villaverde, C., Parra, S., Montero, M., Tous, M., Joven, J., Masana, L., 2005, The stromal derived factor-1 mutated allele (SDF1-3'A) is associated with a lower incidence of atherosclerosis in HIV-infected patients. *AIDS*, 19: 1877–83.

Coll, B., Parra, S., Alonso-Villaverde, C., de Groot, E., Aragonés, G., Montero, M., Tous, M., Camps, J., Joven, J., Masana, L., 2006, HIV-infected patients with lypodystrophy have higher rates of carotid atherosclerosis: the role of monocyte chemoattractant protein-1. *Cytokine*, 34: 51–5.

Costa, L.G., Cole, T.B., Vitalone, C.E., Furlong, C.E., 2005, Measurement of paraoxonase (PON1) status as a potential biomarker of susceptibility to organophosphorate toxicity. *Clin. Chim. Acta*, 352: 37–47.

Costa, L.G., Vitalone, A., Cole, T.B., Furlong, C.E., 2005, Modulation of paraoxonase (PON1) activity. *Biochem Pharmacol*, 69: 541–50.

Deakin, S., Leviev, I., Brulhart-Meynet, M.C., James, R.W., 2003a, Paraoxonase-1 promoter haplotypes and serum paraoxonase: a predominant allele for polymorphic position – 107, implicating the Sp1 transcription factor. *Biochem J*, 372: 643–9.

Deakin, S., Leviev, I., Guernier, S., James, R.W., 2003b, Simvastatin modulates expression of the PON1 gene and increases serum paraoxonase. A role for sterol regulatory elementbinding protein-2. *Arterioscler Thromb Vasc Biol*, 23: 2083–9.

Erlich, P.M., Lunetta, K.L., Cupples, L.A., Huyck, M., Green, R.C., Baldwin, C.T., Farrer, L.A., 2006, Polymprphisms in the PON gene cluster are associated with Alzheimer disease. *Hum. Mol. Genet.*, 15: 77–85.

Farrer, L.A., 1997, Genetics and the dementia patient. *Neurologist*, 3: 13–30.

Ferrari, A., Maretto, S., Girotto, D., Volpin, D., Bressan, G.M., 2004, SREBP contributes to induction of collagen VI transcription by serum starvation. *Biochem Biophys Res Commun*, 313: 600–5.

Ferré, N., Camps, J., Cabre, M., Paul, A., Joven, J., 2001, Hepatic paraoxonase activity alterations and free radical production in rats with experimental cirrhosis. *Metabolism*, 50(9): 997–1000.

Ferré, N., Camps, J., Prats, E., Vilella, E., Paul, A., Figuera, L., Joven, J., 2002, Serum paraoxonase activity: a new additional test for the improved evaluation of chronic liver damage. *Clin Chem*, 48(2): 261–8.

Ferré, N., Marsillach, J., Camps, J., Mackness, B., Mackness, M., Riu, F., Coll, B., Tous, M., Joven, J., 2006, Paraoxonase-1 is associated with oxidative stress, fibrosis and FAS expression in chronic liver diseases. J Hepatol, 45(1): 51–9.

Ferré, N., Marsillach, J., Camps, J., Rull, A., Coll, B., Tous, M., Joven, J., 2005, Genetic association of paraoxonase-1 polymorphisms and chronic hepatitis C virus infection. *Clin Chim Acta*, 361(1–2): 206–10.

García-Ruiz, I., de la Torre, P., Díaz, T., Esteban, E., Fernandez, I., Munoz-Yague, T., Solis-Herruzo, J.A., 2002, Sp1 and Sp3 transcription factors mediate malondialdehyde-induced collagen $\alpha 1(I)$ gene expression in cultured hepatic stellate cells. *J Biol Chem*, 277: 30551–8.

Gil, F., Pla, A., Gonzalvo, M.C., Hernandez, A.F., Villanueva, E., 1993, Rat liver paraoxonase: subcellular distribution and characterization. *Chem Biol Interact.*, 87(1–3): 149–54.

Gonzalvo, M.C., Gil, F., Hernandez, A.F., Rodrigo, L., Villanueva, E., Pla, A., 1998, Human liver paraoxonase (PON1): subcellular distribution and characterization. *J Biochem Mol Toxicol.*, 12(1): 61–9.

Hassett, C., Richter, R.J., Humbert, R., Chapline, C., Crabb, J.W., Omiecinski, C.J., Furlong, C.E., 1991, Characterization of cDNA clones encoding rabbit and human serum paraoxonase: the mature protein retains its signal sequence. *Biochemistry*, 30(42): 10141–9.

He, X.M., Zhang, Z.X., Zhang, J.W., Zhou, Y.T., Tang, M.N., Wu, C.B., Hong, Z., 2006, Gln192Arg polymorphism in paraoxonase 1 gene is associated with Alzheimer disease in a Chinese Han ethnic population. *Chin. Med. J.*, 119: 1204–9.

James, R.W., Deakin, S.P., 2004, The importance of high-density lipoproteins for paraoxonase-1 secretion, stability, and activity. *Free Radic Biol Med*, 37(12): 1986–94.

Janel, N., Christophe, O., Yayha-Graison, E.A., Hamelet, J., Paly, E., Prieur, M., Delezoïde, A.L., Delabar, J.M., 2006, Paraoxonase-1expression is up-regulated in Down Syndromefetal liver. *Biochem. Biophys. Res. Commun.*, 346: 1303–6.

Kalback, W., Esh, C., Castano, E.M., Rahman, E., Kokjohn, T., Luehrs, D.C., Sue, L., Cisneros, R., Gerber, F., Richardson, C., 2004, Atherosclerosis, vascular amyloidosis and brain hypoperfusion in the pathogenesis of sporadic Alzheimer's disease. *Neurol. Res.*, 26: 525–39.

Kawai, H., Sakamoto, F., Inoue, Y., 1990, Improved specific assay for serum arylesterase using a water-soluble substrate. *Clin Chim Acta*, 188(2): 177–82.

Kelada, S.N., Costa-Mallen, P., Checkoway, H., Viernes, H.A., Farin, F.M., Smith-Weller, T., Franklin, G.M., Costa, L.G., 2003, Paraoxonase 1 promoter and coding region polymorphisms in Parkinson's disease. *J. Neurol. Neurosurg. Psychiatry*, 74: 545–8.

Kilic, S.S., Aydin, S., Kilic, N., Erman, F., Aydin, S., Celik, I., 2005, Serum arylesterase and paraoxonase activity in patients with chronic hepatitis. *World J Gastroenterol*, 11(46): 7351–4.

Lee, H.G., Castellani, R.J., Zhu, X., Perry, G., Smith, M.A., 2005, Amyloid-beta in Alzheimer's disease: the horse or the cart? Pathogenic or protective? *Int. J. Exp. Pathol.*, 86: 133–8.

Liao, Z., Graham, D.R., Hildreth, J.E., 2003, Lipid rafts and HIV pathogenesis: virion-associated cholesterol is required for fusion and infection of susceptible cells. *AIDS Res. Hum. Retroviruses*, 19: 675–87.

Mackness, M.I., Durrington, P.N., 1995, HDL, its enzyme and its potential to influence lipid peroxidation. *Atherosclerosis*, 115: 243–253.

Marsillach, J., Ferré, N., Vila, M.C., Lligoña, A., Mackness, B., Mackness, M., Deulofeu, R., Solá, R., Parés, A., Pedro-Botet, J., Joven, J., Camps, J., 2007, Serum paraoxonase-1 in chronic alcoholics: Relationship with liver disease. *Clin Biochem*, 40: 645–50.

Matsunaga, T., Iguchi, K., Nakajima, T., Koyama, I., Miyazaki, T., Inoue, I., Kawai, S., Katayama, S., Hirano, K., Hokari, S., Komoda, T., 2001, Glycated high-density lipoprotein induces apoptosis of endothelial cells via a mitochondrial dysfunction. *Biochem Biophys Res Commun*, 287(3): 714–20.

Mohs, R.C., 2005, The clinical syndrome of Alzheimer's disease: aspects particularly relevant to clinical trials. *Genes Brain Behav.*, 15: 129–33.

Nguyen, D.H., Hildreth, J.E., 2000, Evidence for budding of human immunodeficiency virus type I selectively from glycolipid-enriched membrane lipid rafts. *J. Virol.*, 3264–72.

Nofer, J.R., Kehrel, B., Fobker, M., Levkau, B., Assman, G., von Eckardstein, A., 2002, HDL and arteriosclerosis: beyond reverse cholesterol transport. *Atherosclerosis*, 161: 1–16.

Owens, B., J., Anantharamaiah, G.M., Kahlin, J.B., Srinivas, R.V., Compans, R.W., Segrast, J.P., 1990, Apolipoprotein A-I and its amphipathic helix peptide analogues inhibit human immunodeficiency virus-induced syncytium formation. *J. Clin. Invest.*, 86: 1142–50.

Paragh, G., Balla, P., Katona, E., Seres, I., Egerhazi, A., Degrell, I., 2002, Serum paraoxonase activity changes in patients with Alzheimer's disease and vascular dementia. *Eur. Arch. Psychiatry Clin. Neurosci.*, 252: 63–7.

Parra, S., Alonso-Villaverde, C., Coll, B., Ferré, N., Marsillach, J., Aragonès, G., Mackness, M., Mackness, B., Masana, L., Joven, J., Camps, J., Serum paraoxonase -1 activity and concentration are influenced by HIV infection 2007, *Atherosclerosis*, In press.

Pellin, M.C., Moretto, A., Lotti, M., Vilanova, E., 1990, Distribution and some biochemical properties of rat paraoxonase activity. *Neurotoxicol Teratol*, 12(6): 611–4.

Pola, R., Flex, A., Ciaburri, M., Rovella, E., Valiani, A., Reali, G., Silveri, M.C., Bernabei, R., 2005, Responsiveness to cholinesterase inhibitors in Alzheimer's disease: A possible role for the 192 Q/R polymorphism of the PON-1 gene. *Neurosci. Lett.*, 382: 338–41.

Pola, R., Gaetani, E., Flex, A., Gerardino, L., Aloi, F., Flore, R., Serricchio, M., Pola, P., Bernabei, R., 2003, Lack of association between Alzheimer's disease and Gln-Arg 192 Q/R polymorphism of the PON1 gene in an Italian population. *Dement. Geriatr. Cogn. Disord.*, 15: 88–91.

Primo-Parmo, S.L., Sorenson, R.C., Teiber, J., La Du, B.N, 1996, The human serum paraoxonase/arylesterase gene (PON1) is one member of a multigene family. *Genomics*, 33(3): 498–507.

Rao, M.N., Marmillot, P., Gong, M., Palmer, D.A., Seeff, L.B., Strader, D.B., Lakshman, M.R., 2003, Light, but not heavy alcohol drinking, stimulates paraoxonase by upregulating liver mRNA in rats and humans. *Metabolism*, 52(10): 1287–94.

Rodrigo, L., Hernandez, A.F., Lopez-Caballero, J.J., Gil, F., Pla, A., 2001, Immunohistochemical evidence for the expression and induction of paraoxonase in rat liver, kidney, lung and brain tissue. Implications for its physiological role. *Chem Biol Interact*, 137(2): 123–37.

Roher, A.E., Esh, C., Rahman, A., Kokjohn, T.A., Beach, T.G., 2004, Atherosclerosis of cerebral arteries in Alzheimer disease. *Stroke*, 35: 2623–7.

Rose, H., Woolley, I., Hoy, J., Dart, A., Bryant, B., Mijch, A., Sviridov, D., 2006, HIV infection and high-density lipoprotein: the effect of disease vs. the effect of treatment. *Metabolism*, 55: 90–5.

Rosenblat, M., Vaya, J., Shih, D., Aviram, M., 2005, Paraoxonase 1 (PON1) enhances HDL-mediated macrophage cholesterol efflux via the ABCA1 transporter in association with increased HDL binding to the cells: a possible role for lysophosphatidylcholine. *Atherosclerosis*, 179: 69–77.

Sabesin, S.M., 1981, Lipid and lipoprotein abnormalities in alcoholic liver disease. *Circulation*, 64: 1172–84.

Sabesin, S.M., Hawkins, H.L., Kuiken, L., Ragland, J.B., 1997, Abnormal plasma lipoproteins and lecithin-cholesterol acyltransferase deficiency in alcoholic liver disease. *Gastroenterology*, 72: 510–8.

Saeed, M., Siddique, N., Hung, W.Y., Usacheva, E., Liu, E., Sufit, R.L., Heller, S.L., Haines, J.L., Pericak-Vance, M., Siddique, T., 2006, Paraoxonase cluster polymorphisms are associated with sporadic ALS. *Neurology*, 67: 771–6.

Scacchi, R., Gambina, G., Martini, M.C., Broggio, E., Vilardo, T., Corbo, R.M., 2003, Different pattern of association of paraoxonase Gln192-Arg polymorphism with sporadic late-onset Alzheimer's disease and coronary artery disease. *Neurosci. Lett.*, 339: 17–20.

Sebastiani, G., Alberti, A., 2006, Non invasive fibrosis biomarkers reduce but not substitute the need for liver biopsy. *World J Gastroenterol*, 12(23): 3682–94.

Shi, J., Zhang, S., Tang, M., Liu, X., Li, T., Han, H., Wang, Y., Guo, Y., Zhao, J., Li, H., Ma, C., 2004, Possible association between Cys311Ser polymorphism of paraoxonase 2 gene and late-onset Alzheimer's disease in Chinese. *Brain Res. Mol. Brain Res.*, 120: 201–4.

Sierksma, A., van der Gaga, M.S., van Tol, A., James, R.W., Hendriks, F.J., 2002, Kinetics of HDL cholesterol and paraoxonase activity in moderate alcohol consumers. *Alcohol Clin Exp Res*, 26: 1430–5.

Sinan, S., Kockar, F., Arslan, O., 2006, Novel purification strategy for human PON1 and inhibition of the activity by cephalosporin and aminoglikozide derived antibiotics. *Biochemie*, 88: 565–74.

Slowik, A., Tomik, B., Wolkow, P.P., Partyka, D., Turaj, W., Malecki, M.T., Pera, J., Dziedzic, T., Szczudlik, A., Figlewicz, D.A., 2006, Paraoxonase gene polymorphisms and sporadic ALS. *Neurology*, 67: 766–70.

Sorenson, R.C., Bisgaier, C.L., Aviram, M., Hsu, C., Billecke, S., La Du, B.N., 1999, Human serum paraoxonase/arylesterase's retained hydrophobic N-terminal leader sequence associates with HDL binding phospholipids: apolipoprotein A-I stabilizes activity. *Arterioscler Thromb Vasc Biol*, 19: 2214–25.

Sorenson, R.C., Primo-Parmo, S.L., Kuo, C.L., Adkins, S., Lockridge, O., La Du, B.N., 1995, Reconsideration of the catalytic center and mechanism of mammalian paraoxonase/arylesterase. *Proc Natl Sci USA*, 92: 7187–91.

Stanbridge, J.B., 2004, Pharmacotherapeutic approaches to the treatment of Alzheimer's disease. *Clin. Ther.*, 26: 615–30.

Sugano, M., Tsuchida, K., Makino, N., 2000, High-density lipoprotens protect endothelial cells from tumor necrosis factor-a-induced apoptosis. *Biochem Biophys Res Commun*, 272: 872–876.

Tahl, D.R., Ghebremedhin, E., Orantes, M., Wiestler, O.D., 2003, Vascular pathology in Alzheimer disease: correlation of cerebral amyloid angiopathy and arteriosclerosis/lipohyalinosis with cognitive decline. *J. Neuropathol. Exp. Neurol.*, 62: 1287–301.

Tanikawa, K., Torimura, T., 2006, Studies on oxidative stress in liver diseases: important trends in liver research. *Med Mol Morphol*, 39: 22–7.

Turner, P., Miller, N., Chrystie, I., Coltart, J., Mistry, P., Nicoll, A., Lewis, B., 1979, Splanchnic production of discoidal plasma high-density lipoprotein in man. *Lancet*, 24: 645–6.

Wang, J., Liu, Z., 2000, No association between paraoxonase 1 (PON1) gene polymorphism and susceptibility to Parkinson's disease in a Chinese population. *Mov. Disord.*, 15: 1265–7.

Watson, A.D., Berliner, J.A., Hama, S.Y., La Du, B.N., Faull, K.F., Fogelman, A.M., Navab, M., 1995, Protective effect of high density lipoprotein associated paraoxonase. Inhibition of the biological activity of minimally oxidized low density lipoprotein. *J Clin Invest*, 96: 2882–2891.

Xu, G.Y., Lv, G.C., Chen, Y., Hua, Y.C., Zhu, S.M., Yang, Y.D., 2005, Monitoring the level of serum
paraoxonase 1 activity in liver transplantation patients. *Hepatobiliary Pancreat Dis Int*, 4(2): 178–81.
Zintzaras, E., Hadjigeorgiou, G.M., 2004, Association of paraoxonase 1 gene polymorphisms with risk
of Parkinson's disease: a meta-analysis. *J. Hum. Genet.*, 49: 474–81.
Zuliani, G., Zanca, R., Munari, M.R., Zurlo, A., Vavalle, C., Atti, A.R., Fellin, R., 2001, Genetic polymorphisms in older subjects with vascular or Alzheimer's dementia. *Acta Neurol. Scand.*, 103: 304–8.

CHAPTER 12

AGE-RELATED ALTERATIONS IN PON1

I. SERES, T. FULOP*, G. PARAGH AND A. KHALIL*

First Department of Medicine, Medical and Health Science Center, University of Debrecen, Hungary
** Research Center on Aging, Département de Médecine, Service de Gériatrie Université de*
Sherbroooke, Sherbrooke, Qc, Canada
E-mail:seres@internal.med.unideb.hu

Abstract: Aging is associated with an increased incidence of atherosclerosis. Atherosclerosis is an inflammatory disease initiated by oxidized low density lipoprotein (oxLDL). High density lipoprotein (HDL) is a powerful antioxidant protecting LDL and itself from oxidation as well by detoxifying the hydroperoxides from oxLDL. The enzyme responsible, at least in part, for this antioxidant effect of HDL is its tightly associated paraoxonase (PON1). As PON1 is responsible for the antioxidant effect of HDL and the atherosclerosis related cardiovascular diseases are increasing with aging an alteration either in the PON1 expression and/or activity should occur with aging. Indeed, PON1 serum concentrations do not change with aging while its activity significantly decreases. This chapter will describe the putative causes of this altered PON1 activity at the environmental and genetic levels. The possibility of PON1 activity modulation will be also discussed

Keywords: PON1, aging, longevity, free radicals

1.   INTRODUCTION

Atherosclerosis is a long-term inflammatory disorder affecting the inner walls of arteries with a great number of predisposing factors such as age, sex, diabetes mellitus, physical inactivity and obesity, cigarette smoking or hypertension. The most important being the aging process. Atherosclerosis is characterized by endothelial dysfunction, increased permeability of endothelium to lipoproteins, migration of white blood cells into the walls of arteries, as well as necrosis. LDL particles modified in the arterial wall by free radicals play a critical role in the initiation and progression of the atherosclerotic process. They induce the atherosclerotic inflammatory process by stimulating monocyte infiltration and smooth muscle cell migration and proliferation (Mertens and Holvoet, 2001). The presence of

199

B. Mackness et al. (eds.), The Paraoxonases: Their Role in Disease Development
and Xenobiotic Metabolism, 199–206.

oxidatively modified LDL (oxLDL) has been evidenced in atherosclerotic lesions of animals such as rabbits (Ylä-Herttuala et al., 1989) as well as in humans.

It has been suggested that LDL particles can be protected from free radical-induced oxidation by a HDL linked enzyme, paraoxonase 1 (PON1). PON1 can also inactivate LDL derived oxidized phospholipids after their formation. PON1 is found in several tissues such as liver, kidney, intestine, and also in serum (Aviram et al., 1998). Studies have also shown that PON1 is also able to preserve the anti-atherogenic function of HDL by protecting it directly from oxidation. Thus, PON1 is supposed to possess anti-atherogenic and anti-inflammatory properties, resulting from its ability to destroy modified phospholipids and to prevent accumulation of oxidized lipids in lipoproteins, although the precise mechanism of the PON1 effect is not yet completely elucidated. Considering the aging process as a major risk factor for atherosclerosis it can be supposed that beside the increased formation of the oxLDL there is also a decrease in PON1 activity according to the free radical theory of aging (Harman, 1992). The present chapter will describe our understanding of PON1 activity changes with aging essentially in relation to atherosclerosis development.

2. PON1 AND DEVELOPMENT

Age is a major determinant of PON1 activity. In rodents studies have shown that PON1 activity is very low at birth and increasing thereafter. Similar results were obtained in humans. PON1 was found to be very low at birth and increasing until the age of 6 month and presenting a plateau thereafter (Cole et al., 2003; Mueller et al., 1983). Children, two years or more of age, have the same arylesterase activity as normal adults (Augustinsson and Barr, 1963).

3. PON1 EXPRESSION AND AGING

PON1 prevents the formation of oxidized LDL, inactivates LDL-derived oxidized phospholipids and also protects phospholipids in HDL from oxidation.

The susceptibility of LDL and HDL to lipid peroxidation has been shown to increase with aging. The concentration of the hydroperoxide products is also increasing with aging. This leads to the highly pro-inflammatory and atherogenic effects of oxLDL to the well-known increase in the incidence of atherosclerosis resulting in clinically overt cardio-vascular diseases.

Considering the essential protecting effect of PON1, the first question that has arisen is whether the concentration of PON1 could be one cause of this increased occurrence of atherosclerotic lesions with aging. PON1 is associated to apolipoprotein A1 (apoA-1) in HDL. Frey et al. have demonstrated a decrease in the apoA-1 with aging (Frey et al., 1990), which we confirmed recently (Seres et al., 2004). In spite of this decrease of ApoA-1 others and we could not find altered expression of PON1 in serum or association with HDL during aging (Jarvik et al., 2002; Seres et al., 2004).

4.   PON1 ACTIVITY AND AGING

Most experimental data points to the fact that PON1 expression does not change with aging. Thus, the dramatic increase in cardiovascular diseases with aging can only be explained by other changes altering the anti-atherogenic properties of HDL. The most obvious explanation would be that the PON1 activity changes with aging.

There is a dramatic decrease in PON1 activity with aging (Milochevitch and Khalil, 2001; Seres et al., 2004). This PON1 activity decrease did not seem to be related to the PON1 phenotype. Others (Jarvik et al., 2002) confirmed these findings. Although, the group of Khalil et al. have not observed a change in the PON1 arylesterase activity with aging, another study by Marchegiani et al. have demonstrated a reduction in the PON1s' paraoxonase and arylesterase activities, as well as a significant reduction in the PON1 mass concentration in the elderly compared to young subjects (Marchegiani et al., 2006). This is a very significant finding contributing to our comprehension for the increased cardiovascular disease incidence with aging via increased lipid peroxidation. In accordance with this findings we already knew that the PON1 activity is reduced in atherosclerotic diseases such as acute myocardial infarction or in conditions in which atherosclerosis is common such as diabetes, or familial hypercholesterolemia (Ayub et al., 1999; Mackness et al., 1998).

What can cause the decrease of PON1 activity with aging? Numerous studies have demonstrated an increase in oxidative stress with aging resulting in enhanced free radical production and oxidatively damaged macromolecules (Harman, 1992; Lu et al., 1999). Thus, this inactivation might result from an interaction of oxLDL associated lipids or with other oxidants (e.g. hydrogen peroxide) with the PON1 free sulfhydryl group at cysteine -284 (Cys-284) (Aviram et al., 1999). Indeed, Jaouad et al., 2006 have recently shown that HDL from elderly presents a significant reduction in its antioxidant activity as compared to HDL from young subjects. As the antioxidant activity of HDL is directly related to the PON1 activity it is suggested that PON1 activity alterations account for the decreased HDL antioxidant activity in the elderly. This decreased PON1 activity was related to the decreased Cysteine-284 active site of PON1, which may be due to the age-related oxidative stress conditions. PON1 from the elderly show significantly less sulfhydryl groups than PON1 of young subjects. This could explain the reduction of PON1 antioxidant effect with aging and the consequent decrease of HDL antioxidant activity. However, recently it was shown Cys-284 is not part of the PON1 active site, but is part of a highly conserved stretch that includes active site histidine 285. Thus the Cys-284 is participating to the PON1 stability.

Very recently it was shown by Rozenberg and Aviram (2006) that PON1 can form mixed disulfide between its protein thiol and glutathione called S-glutathionylation. This has been suggested as a mechanism for the protection of the protein from irreversible modifications. In contrast, oxidized glutathione (GSSG) is inhibiting the PON1 activity. It is well established that under oxidative stress as it is occurring with aging reduced/oxidized glutathione (GSH/GSSG) ratio is decreased in the blood and also in tissues. Several studies have shown that the reduced glutathione (GSH) concentration decreases with aging in sera and in white blood cells (Fulop et al., 1985, 1997; Lenton et al., 2000) and concomitantly the GSSG increases. It can be thus suggested that beside the decrease of cystein-284, PON1 can be further inactivated by a mixed disulfide formation between free sulfhydryl group and GSSG as a result of constant oxidative stress. This protective reaction in aging is becoming deleterious because most probably of its irreversibility.

Thus certain antioxidants may increase PON1 activity by preventing its oxidative inactivation. It was shown that its activity was indeed preserved by dietary antioxidants (Aviram et al., 1999, 2005). Khalil and his group, have also shown that 3 week supplementation with olive oil or argan oil, due to their high contents in vitamin E and polyphenols, induces a significant increase in PON1 activities (arylesterase and paraoxonase) (Cherki et al., 2005) The main nutritional antioxidants in the diet were vitamins C and E.

5.   PON1 GENOTYPE AND AGING

As the activity of PON1 was found to be profoundly altered with aging the question arises whether this could be the consequence of genetic changes occurring with the aging process, as no correlation was found with the phenotype. It is well known that aging is a multi-gene process.

PON1 gene polymorphism is due to an amino acid substitution at position 192 in the coding region (Glutamine (Q) to Arginine (R)) as well as to another amino acid substitution at position 55 (Leucine (L) to Methionine (M)) in the coding region as well as several polymorphisms in the promoter region. Alleles at codon 192 (Q and R alleles) and codon 55 (L and M alleles) in the PON1 locus have been associated with the PON1 enzymatic activity and concentration, respectively (Garin et al., 1997; Humbert et al., 1993). In addition, functional PON1 gene promoter polymorphisms at -108C/T and -162G/A are described (Deakin and James, 2004). R alleles containing Arginine and R+ containing genotypes corresponds to the B phenotype. It was shown in many studies that the 192R allele and consequently the B phenotype might increase the risk of cardio- or cerebrovascular diseases (Baum et al., 2006). Thus, as aging is associated essentially to an altered enzyme activity we could expect changes in the locus at position 192. However, it was shown that there is no direct correlation between genotype and activity even inside a group presenting identical genotype (Richter and Furlong, 1999). Heikmans et al. (2000) have shown that PON1 polymorphisms in an elderly population were not associated with mortality of all causes or cardiovascular in their elderly population. This would not exclude the possibility that the polymorphisms are associated with increased risk of fatal cardiovascular disease. It indicates however that this potential increase in risk is limited. Furthermore, other studies relating PON1 activity and genotype are contradictory and depending on the population studied (Mackness et al., 1998; Ombres et al., 1998). This could be due to the increased death of RR or RQ from cardiovascular diseases at an earlier age, mainly if it is related to other risk factors such as hypertension, smoking, hyperlipidemia. Senti et al. (2001) hypothesised that although QQ genotype may be protective against lipid peroxidation and thus

Table 1. Putative causes of PON1 activity decrease with aging

---

Genetic background
Smoking
Sedentary lifestyle
Increased oxidative stress
   Increased free radical production
   Decreased antioxidant defence
Inflammation-aging
   High proinflammatory status (IL-6, IL-10)
   Low efficiency of stress response
Increase of latent chronic diseases
Malnutrition
   under- or over-consumption of macro- or micro-nutrients or calories

---

Table 2. Possible consequences of the decreased PON1 activity with aging

---

Increased lipid peroxidation products
Increased susceptibility of LDL and HDL to oxidation
Decreased antioxidant capacity of HDL
Increased atherosclerosis
Increased incidence of cardiovascular disease
Increased free radical associated diseases (cancer, rheumatoid arthritis,
age-related macular degeneration, osteoarthritis, Alzheimer's disease)

---

against the initiation of the atherosclerosis process, it may be insufficient when an HDL-deficiency state and low PON1 activity reflecting oxidative stress coexist. However these two conditions are frequent in aging.

The results of these studies have lead to the conclusion that not only the genotype but also the phenotype of the PON1 is important (Jarvik et al., 2000). A variety of external factors, such as antioxidant- or cholesterol-rich diets, metals and lifestyle, can positively as well as negatively influence the activity and the concentration of PON1 (Table 1, Table 2).

6. PON1 GENOTYPE AND LONGEVITY AND SUCCESSFUL AGING

A recent study of Marchegiani et al. (2006) tried to relate the role of PON1 variability to survival and successful aging. Previously, the same group (Bonafe et al., 2002; Rea et al., 2004) reported an increase in the frequency of the R allele and R+ carriers both in Irish octo/nonagenarians and Italian centenarians in comparison with their younger counterparts. This led to the conclusion that the R allele had survival advantage through an increased PON1 activity as it decreases mortality in carriers. It is of note that the increase of R allele is rather largely due to the QR heterozygous genotype. For stress responder genes like PON1 the heterozygous genotype may have better adaptive capability to become centenarians (Franceschi et al., 2005; Yashin et al., 2001). It was also noted that most of these persons carried also the M+ genotype. Thus, PON1 genotype could contribute to longevity in a complex manner. Altogether these data suggest that individuals who were at high risk to die at an early age because of cardiovascular diseases due to an R+ PON1 genotype if they survived these R+ individuals have an increased chance to survive into very old age. These findings further support the maintenance in nonagenarians and centenarians of the tight correlation between the genetic variability of the PON1 gene and the enzyme activity. This was also shown for Interleukin-6 (IL-6) in centenarians (Olivieri et al., 2002). This could further support the antagonistic pleiotropic nature of certain genes fitting with the evolutionary theory of aging (Williams and Day, 2003). Nevertheless, the PON1 192 variant is involved in the definition of the longevity phenotype (Bonafe et al., 2002; Franceschi et al., 2005).

7.   CONCLUSION

Aging is the major risk factor for the development of atherosclerosis leading to enhanced incidence of cardiovascular diseases. Oxidised lipids, most importantly ox-LDL plays a major role in the initiation and progression of the atherosclerotic process. HDL plays via its PON1 activity a major antioxidant role. It was shown that this antioxidant activity of HDL is decreased because of the decreased PON1 activity through aging. This decreased activity is depending on the genotype through the lifespan; however this direct association is difficult to demonstrate as post-translational modification also greatly influence the PON1 activity. Nevertheless, the R+ polymorphism is part of the longevity genes. The PON1 posttranslational modifications occurring with aging are mainly related to the increased oxidative stress resulting in increased free radicals production and decreased antioxidant defence. The oxidative stress inactivates either directly or via the S-glutathionylation the active sufhydryl cysteine-284. Other still unknown mechanisms can also contribute to the PON1 activity decrease in HDL with aging. More experimental works are needed to fully elucidate the effect of aging on PON1 changed activities. In the mean time, hopefully, dietary modulation of the oxidative stress by antioxidants could contribute to the restoration of the PON1 activity and reducing the burden of cardiovascular diseases with aging.

ACKNOWLEDGEMENTS

The authors work is supported by the Hungarian Scientific Research Fund (K63025), and Hungarian Health Science Council (ETT 243/2006).

REFERENCES

Augustinsson AB, Barr M.1963, Age variation in plasma arylesterase activity in children. *Clin Chim Acta* 8:568–73.
Aviram M, Kaplan M, Rosenblat M, Fuhrman B. 2005, Dietary antioxidants and paraoxonases against LDL oxidation and atherosclerosis development. *Handb Exp Pharmacol.* 170:263–300.

Aviram M, Rosenblat M, Billecke S, Erogul J, Sorenson R, Bisgaier CL, Newton RS, La Du BN. 1999, Human serum paraoxonase (PON 1) is inactivated by oxidized low density lipoprotein and preserved by antioxidants. *Free Radic. Biol. Med.* 26:892–904.

Aviram M, Rosenblat M, Bisgaier CL, Newton RS, Primo-Parmo SL, La Du BN. 1998, Paraoxonase inhibits high-density lipoprotein oxidation and preserves its functions: a possible peroxidative role for paraoxonase. *J. Clin. Invest.* 101:1581–1590.

Ayub A, Mackness MI, Arrol S, Mackness B, Patel J, Durrington P.N. 1999, Serum paraoxonase after myocardial infarction. Arterioscler. *Thromb. Vasc. Biol.* 19:330–335.

Baum L, Ng HK, Woo KS, Tomlinson B, Rainer TH, Chen X, Cheung WS, Chan DK, Thomas GN, Tong CS, Wong KS. 2006, Paraoxonase 1 gene Q192R polymorphism affects stroke and myocardial infarction risk. *Clin Biochem.* Mar 39(3):191–5.

Bonafe M, Marchegiani F, Cardelli M, Olivieri F, Cavallone L, Giovagnetti S, Pieri C, Marra M, Antonicelli R, Troiano L, Gueresi P, Passeri G, Berardelli M, Paolisso G, Barbieri M, Tesei S, Lisa R, De Benedictis G, Franceschi C. 2002, Genetic analysis of Paraoxonase (PON1) locus reveals an increased frequency of Arg192 allele in centenarians. *Eur J Hum Genet.* 10(5):292–6.

Cherki M, Derouiche A, Drissi A, El Messal M, Bamou Y, Idrissi-Ouadghiri A, Khalil A, Adlouni A. 2005, Consumption of argan oil may have an antiatherogenic effect by improving paraoxonase activities and antioxidant status: Intervention study in healthy men. *Nutr Metab Cardiovasc Dis.* 15(5):352–60.

Cole TB, Jampsa RL, Walter BJ, Arndt TL, Richter RJ, Shih DM, Tward A, Lusis AJ, Jack RM, Costa LG, Furlong CE. 2003, Expression of human paraoxonase (PON1) during development. *Pharmacogenetics* 13:1–8.

Deakin SP, James RW. 2004, Genetic and environmental factors modulating serum concentrations and activities of the antioxidant enzyme paraoxonase-1. *Clinical Science* 107:435–447.

Franceschi C, Olivieri F, Marchegiani F, Cardelli M, Cavallone L, Capri M, Salvioli S, Valensin S, De Benedictis G, Di Iorio A, Caruso C, Paolisso G, Monti D. 2005, Genes involved in immune response/inflammation, IGF1/insulin pathway and response to oxidative stress play a major role in the genetics of human longevity: the lesson of centenarians. *Mech Ageing Dev.* 126(2):351–61.

Frey I, Berg A, Baumstark MW, Collatz KG, Keul J. 1990, Effects of age and physical performance capacity on distribution and composition of high-density lipoprotein subfractions in men. *Eur J Appl Physiol Occup Physiol.* 60(6):441–4.

Fulop T Jr, Foris G, Worum I, Paragh G, Leovey A. 1985, Age related variations of some polymorphonuclear leukocyte functions. *Mech Ageing Dev.* 29(1):1–8.

Fulop T Jr, Fouquet C, Allaire P, Perrin N, Lacombe G, Stankova J, Rola-Pleszczynski M, Gagne D, Wagner JR, Khalil A, Dupuis G. 1997, Changes in apoptosis of human polymorphonuclear granulocytes with aging. *Mech Ageing Dev.* 96(1–3):15–34.

Garin MC, James RW, Dussoix P, Blanche H, Passa P, Froguel P, Ruiz J. 1997, Paraoxonase polymorphism Met-Leu54 is associated with modified serum concentrations of the enzyme. A possible link between the paraoxonase gene and increased risk of cardiovascular disease in diabetes. *J Clin Invest.* 99(1):62–6.

Harman D. 1992, Role of free radicals in aging and disease. *Ann. N Y Acad. Sci.* 673:126–141.

Heijmans BT, Westendorp RG, Lagaay AM, Knook DL, Kluft C, Slagboom PE. 2000, Common paraoxonase gene variants, mortality risk and fatal cardiovascular events in elderly subjects. *Atherosclerosis* 149(1):91–7.

Humbert R, Adler DA, Disteche CM, Hassett C, Omiecinski CJ, Furlong CE. 1993, The molecular basis of the human serum paraoxonase activity polymorphism. *Nat Genet.* 3(1):73–6.

Jaouad L, de Guise C, Berrougui H, Cloutier M, Isabelle M, Fulop T, Payette H, Khalil A. 2006, Age-related decrease in high-density lipoproteins antioxidant activity is due to an alteration in the PON1's free sulfhydryl groups. *Atherosclerosis* 185(1):191–200.

Jarvik GP, Rozek LS, Brophy Hatsukami TS, Richter RJ, Schellenberg GD, Furlong CE. 2000, Paraoxonase (PON1) phenotype is a better predictor of vascular disease than is PON1(192) or PON1(55) genotype. Arterioscler. *Thromb. Vasc. Biol.* 20:2441–2447.

Jarvik GP, Trevanian N, Sai T, McKinstry LA, Wani R, Brophy VH, Richter RJ, Schellenberg GD, Heagerty PJ, Hatsukami TS, Furlong CE. 2002, Vitamins C and E intake is associated with increased paraoxonase activity. *Arterioscler Thromb Vasc Biol.* 22:1329–1333.

Lenton KJ, Therriault H, Cantin AM, Fulop T, Payette H, Wagner JR. 2000, Direct correlation of glutathione and ascorbate and their dependence on age and season in human lymphocytes. Am *J Clin Nutr* 71(5):1194–200.

Lu CY, Lee HC, Fahn HJ, Wei YH. 1999, Oxidative damage elicited by imbalance of free radical scavenging enzymes is associated with large-scale mtDNA deletions in aging human skin. *Mutat Res.* 423(1–2):11–21.

Mackness B, Mackness MI, Arrol S, Turkie W, Julier K, Abuasha B, Miller JE, Boulton AJ, Durrington PN. 1998, Serum paraoxonase (PON1) 55 and 192 polymorphism and paracxonase activity and concentration in non-insulin dependent diabetes mellitus. *Atherosclerosis* 139(2):341–9.

Mackness MI, Mackness B, Durrington PN, Fogelman AM, Berliner J, Lusis AJ, Navab M, Shih D, Fonarow GC. 1998, Paraoxonase and coronary heart disease. *Curr Opin Lipidol.* 9(4):319–24.

Marchegiani F, Marra M, Spazzafumo L, James RW, Boemi M, Olivieri F, Cardelli M, Cavallone L, Bonfigli AR, Franceschi C. 2006, Jun;Paraoxonase activity and genotype predispose to successful aging. *J Gerontol A Biol Sci Med Sci.* 61(6):541–6.

Mertens A, Holvoet P. 2001, Oxidized LDL and HDL: antagonists in atherothrombosis. *FASEB J.* 15(12):2073–84.

Milochevitch C, Khalil A. 2001, Study of the paraoxonase and platelet activating factor acetyl hydrolase activities with aging. Prostaglandins Leukot. *Essent. Fatty Acids* 65:241–246.

Mueller RF, Hornung S, Furlong CE, Anderson J, Giblett ER, Motulsky AG. 1983, Plasma paraoxonase polymorphism: a new enzyme assay, population, family, biochemical and linkage studies. *Am J Hum Genet* 35:393–408.

Olivieri F, Bonafe M, Cavallone L, Giovagnetti S, Marchegiani F, Cardelli M, Mugianesi E, Giampieri C, Moresi R, Stecconi R, Lisa R, Franceschi C. 2002, The -174 C/G locus affects in vitro/in vivo IL-6 production during aging. *Exp Gerontol.* 37(2–3):309–14.

Ombres D, Pannitteri G, Montali A, Candeloro A, Seccareccia F, Campagna F, Cantini R, Campa PP, Ricci G, Arca M. 1998, The gln-Arg192 polymorphism of human paraoxonase gene is not associated with coronary artery disease in italian patients. Arterioscler *Thromb Vasc Biol.* 18(10):1611–6.

Rea IM, McKeown PP, McMaster D, Young IS, Patterson C, Savage MJ, Belton C, Marchegiani F, Olivieri F, Bonafe M, Franceschi C. 2004, Paraoxonase polymorphisms PON1 192 and 55 and longevity in Italian centenarians and Irish nonagenarians. A pooled analysis. *Exp Gerontol.* 39(4): 629–35.

Richter RJ, Furlong CE. 1999, Determination of paraoxonase (PON1) status requires more than genotyping. *Pharmacogenetics.* 9(6):745–53.

Rozenberg O, Aviram M. 2006, S-Glutathionylation regulates HDL-associated paraoxonase 1 (PON1) activity. *Biochem Biophys Res Commun.* 351(2):492–8.

Senti M, Tomas M, Vila J, Marrugat J, Elosua R, Sala J, Masia R. 2001, Relationship of age-related myocardial infarction risk and Gen/Arg 192 variants of the human paraoxonase 1 gene: the REGICOR study, *Atherosclerosis* 156:443–449.

Seres I, Paragh G, Deschene E, Fulop T Jr, Khalil A. 2004, Study of factors influencing the decreased HDL associated PON1 activity with aging. *Exp Gerontol.* 39(1):59–66.

Williams PD, Day T. 2003, Antagonistic pleiotropy, mortality source interactions, and the evolutionary theory of senescence. *Evolution Int J Org Evolution.* 57(7):1478–88.

Yashin AI, Ukraintseva SV, De Benedictis G, Anisimov VN, Butov AA, Arbeev K, Jdanov DA, Boiko SI, Begun AS, Bonafe M, Franceschi C. 2001, Have the oldest old adults ever been frail in the past? A hypothesis that explains modern trends in survival. *J Gerontol A Biol Sci Med Sci.* 56(10):B432–42.

Yla-Herttuala S, Palinski W, Rosenfeld ME, Parthasarathy S, Carew TE, Butler S, Witztum JL, Steinberg D. 1989, Evidence for the presence of oxidatively modified low density lipoprotein in atherosclerotic lesions of rabbit and man. *J Clin Invest.* 84(4):1086–95.

PART 5

PON AND TOXICOLOGY

CHAPTER 13

PARAOXONASE (PON1) AND ORGANOPHOSPHATE TOXICITY

L.G. COSTA[1,2], T.B. COLE[1,3], K.L. JANSEN[1] AND C.E. FURLONG[3]

[1] *Dept. of Environmental and Occupational Health Sciences, University of Washington, Seattle, WA, USA*

[2] *Dept. of Human Anatomy, Pharmacology and Forensic Medicine, University of Parma, Italy*

[3] *Depts of Genome Sciences and Medicine (Medical Genetics), University of Washington, Seattle, WA, USA*

Abstract: Paraoxonase (PON1) is a high density lipoprotein-associated enzyme capable of hydrolyzing multiple substrates, including several organophosphorus (OP) insecticides and nerve agents, oxidized lipids and a number of drugs or pro-drugs. Several polymorphisms in the PON1 gene have been described, which have been shown to affect either the catalytic efficiency of hydrolysis or the expression level of the enzyme. Animal studies have shown that PON1 is an important determinant of the toxicity of certain OPs. Evidence for this was provided by cross-species comparisons, by administration of exogenous PON1 and by experiments in PON1 knockout and transgenic mice. Low PON1 plays also a role in the higher susceptibility of the young to OP toxicity. Recent findings also suggest that PON1 may modulate the toxicity resulting from exposure to mixtures of OP compounds

Keywords: Paraoxonase, PON1 status, organophosphate, paraoxon, chlorpyrifos oxon, diazoxon

1.   INTRODUCTION

The first investigator credited to report that organophosphorus (OP) compounds could be hydrolyzed enzymatically was Mazur (1946). This finding led to a series of studies by Aldridge in the 1950s on the hydrolysis of paraoxon (E600). He proposed the definition of "A-esterases" for the enzymes hydrolyzing OPs and arylesters, such as p-nitrophenyl acetate, and described the general characteristics of the serum paraoxonase (PON1) (Aldridge, 1953). Earlier studies on different human populations showed that the hydrolytic activity of serum PON1 was polymorphically distributed (Eckerson et al., 1983; Geldmacher-von Malllinckrodt and Diepgen, 1988; Mueller et al., 1983; Playfer et al., 1976). Further research demonstrated that the molecular basis of the polymorphism was a Q192R substitution, with the

B. Mackness et al. (eds.), The Paraoxonases: Their Role in Disease Development and Xenobiotic Metabolism, 209–220.

$PON1_{R192}$ alloform having high paraoxon hydrolase activity (Adkins et al., 1993; Humbert et al., 1993). Additional polymorphisms in the PON1 promoter region can influence the level of expression of this enzyme (Brophy et al., 2001; 2002; Leviev and James, 2000; Suehiro et al., 2000). It was assumed that individuals with high paraoxonase activity would be less sensitive to the toxicity of paraoxon and of other OPs identified *in vitro* as substrates for this enzyme. In this chapter, we examine the approaches used to elucidate the role of PON1 in the detoxication of OPs and in the modulation of their toxicity. In particular, the role of PON1 in the acute toxicity of OPs, their developmental toxicity and neurotoxicity, and the toxicity of mixtures of OPs will be discussed.

2.   OPs: METABOLISM AND TOXICITY

OPs are triesters of phosphoric acid with a general structure shown in Fig. 1. OPs were developed in the 1940s and most are currently used as insecticides, though some have been utilized as drugs (e.g. trichlorphon for schistosomiasis or ecothiophate for glaucoma), or as nerve agents in chemical warfare (e.g. sarin or soman) (Lotti, 2000). Exposure to OPs is associated with three distinct syndromes: a cholinergic syndrome, as a consequence of inhibition of acetylcholinesterase (AChE) and accumulation of acetylcholine at muscarinic and nicotinic cholinergic receptors throughout the body; a still poorly defined intermediate syndrome, characterized by weakness of respiratory, neck and proximal limb muscles, which occurs hours to days after the onset of severe cholinergic over-stimulation; and a delayed polyneuropathy, which is caused only by certain OPs, and can be described as a dying back axonopathy, whose primary event appears to be inhibition of another enzyme, neuropathy target esterase (NTE) (Lotti, 2000). Only OPs with a P = O moiety can interact with Ache or NTE. Since most OP insecticides are organophosphorothioates (i.e. they have a P = S moiety; see Fig. 1), they require metabolic activation to their corresponding oxygen analogs. Such activation occurs mostly in the liver and is mediated by various isozymes of cytochrome P450, some of which can

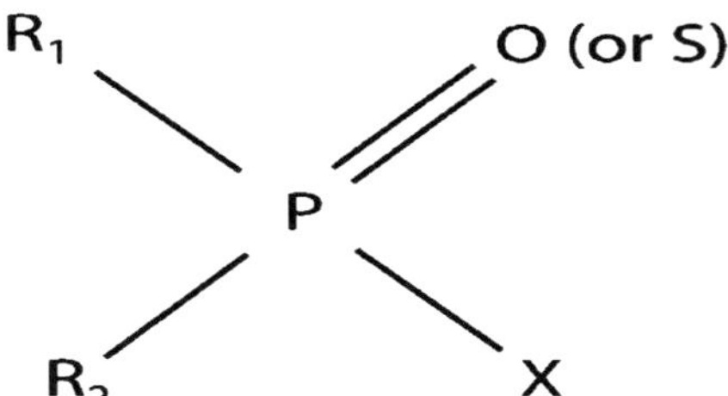

Figure 1. General structure of OPs, where X is the so called "leaving group" that is displaced when the OP phosphorylates AChE, and is the most sensitive to hydrolysis; R_1 and R_2 are most commonly alcoxy groups (OCH_3 or OC_2H_5), though other substitutes are also possible; either an oxygen or a sulfur (in this case the compound should be defined as a phosphorothioate) are also attached to the phosphorus with a double bond

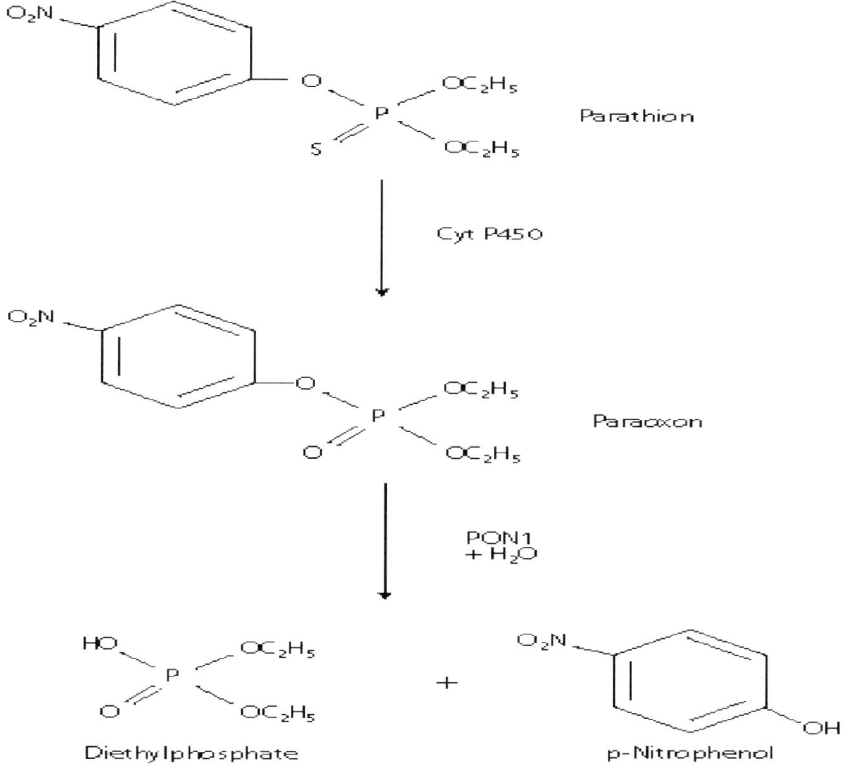

Figure 2. Activation of parathion by cytochromes P450 and hydrolysis of paraoxon by PON1

also detoxify OPs (Jokanovic, 2001). In addition, certain OPs can be detoxified hydrolytically by the action of A-esterases. One of these A-esterases is PON1, which can hydrolyze the oxygen analogs of various commonly used OP insecticides (e.g. paraoxon, diazoxon, chlorpyrifs oxon), as well as nerve agents, such as soman or sarin (Geldmacher-von Mallinckrodt and Diepgen, 1988). Figure 2 shows the activation of parathion to paraoxon by cytochromes P450, and hydrolysis of paraoxon by PON1. The products of such hydrolysis, p-nitrophenol and diethylphosphate, are then excreted in the urine. PON1 is synthesized in the liver and is secreted into the circulation, where it is strongly bound to high density lipoproteins (HDL) (Costa and Furlong, 2002).

3. ROLE OF PON1 IN MODULATING THE ACUTE TOXICITY OF OPs

While the ability of PON1 in hydrolyzing a number of OP substrates *in vitro* is well established, evidence that the enzyme plays a role in modulating the toxicity of these same OPs *in vivo* has emerged more slowly. Evidence for the role of

PON1 in OP toxicity was derived from cross-species comparisons, from animal experiments utilizing exogenous PON1, and from more recent studies with PON1 knockout and transgenic mice. Initial indirect evidence was provided by comparison across animal species which differ in the levels of their plasma PON1 activity. Birds, that have very low PON1 activity (Brealey et al., 1980; Furlong et al., 2000; Machin et al., 1976), display a much higher sensitivity, compared to rats, to the acute toxicity of some OPs (e.g. pirimiphos-ethyl, diazinon) (Brealey et al., 1980). Among mammals, rabbits have a 5–20-fold higher plasma PON1 activity than rats (Aldridge, 1953; Costa et al., 1987; Zech and Zurcher, 1974), and were four-fold less sensitive to the toxicity of paraoxon (Costa et al., 1987). Though several other factors, such as rates of absorption and distribution of OPs, rates of activation and of detoxication by other metabolic pathways, and intrinsic susceptibility of target enzymes, can affect the overall toxicity outcome, these initial observations provided some supporting, albeit indirect, evidence, to the hypothesis that low serum PON1 activity would lead to an increased sensitivity to the acute toxic effects of OPs (Costa et al., 2002).

A more direct approach consisted in the direct administration of exogenous PON1 to rats or mice. Background and impetus for this approach was provided by an early study by Main (1956), who showed that intravenous administration of partially purified PON1 from rabbit serum to rats, would protect them from the toxicity of paraoxon. Subsequent experiments in rats and mice utilizing pure rabbit PON1 (Furlong et al., 1991), confirmed and expanded this early observation. Administration (via the tail vein) of the enzyme to rats increased serum PON1 activity toward paraoxon by 9-fold, and that toward chlorpyrifos-oxon by 50-fold (Costa et al., 1990). Thirty minutes after PON1 injection, rats were challenged with an acute dose of paraoxon or chlorpyrifos oxon given by the iv, dermal, ip or oral route, at doses causing similar degrees of AChE inhibition in plasma, red blood cells, brain and diaphragm. Four hours later, at sacrifice, AChE activity measurements indicated a much lower degree of inhibition in animals that had been pre-treated with PON1. Protection was more evident in case of chlorpyrifos oxon, and was more prominent in two target tissues for OP toxicity, brain and diaphragm (Costa et al., 1990). Of practical relevance is that a substantial protective effect was present when OP exposure occurred by the dermal route, which represents an important route of exposure for occupationally exposed workers.

An additional series of experiments was carried out in mice, as much less purified enzyme is required for injection, and since this species was deemed to be ideal for studies involving genetically modified animals with altered PON1 levels (see below). An initial experiment followed the same protocol previously utilized in rats, and provided evidence that iv administration of pure rabbit PON1 increased serum chlorpyrifos oxonase activity by 30–40 fold, and protected animals toward AChE inhibition by dermally applied chlorpyrifos oxon (Li et al., 1993). As with iv injection of rabbit PON1, the increased serum PON1 activity was short lasting $(t_{1/2} = \sim 6h)$, further experiments were aimed at investigating other routes of administration for PON1. Administration of PON1 by the iv + ip routes increased plasma enzyme activity toward chlorpyrifos oxon by 35-fold and extended the half-life to 30 hours. An even longer half-life, albeit with lower peak activity levels, was found when PON1 was given by the iv + im route (Li et al., 1993).

Rabbit PON1 also provided protection against the toxicity of the parent compound, chlorpyrifos, when the OP exposure occurred 30 min. after iv injection of PON1 (Li et al., 1993), or 24 h after an iv + ip administration of PON1 (Li et al., 1995). Further experiments also showed that PON1, when given 30 min. after dermal administration of chlorpyrifos, prevented the reduction of AChE activity in all tissues; when PON1 was injected three hours after chlorpyrifos, a protective effect was still seen in brain and diaphragm (Li et al., 1995). Altogether, this series of studies indicated that by artificially increasing serum levels of PON1 (by injection of purified rabbit enzyme), it was possible to decrease the acute toxicity of certain OPs. Of relevance is also the finding that PON1 exerted a protective effect when given after OP exposure, suggesting a potential use in OP poisoning, possibly in combination with other conventional treatments. This series of experiments provided convincing evidence that high serum PON1 levels protect against the toxicity of specific OP compounds.

More recently, PON1 knockout and transgenic animals have provided important new tools to investigate the role of PON1 in modulating OP toxicity. PON1 knockout ($PON1^{-/-}$) mice were produced by targeted disruption of exon 1 of the PON1 gene, and have normal appearance and body weights (Shih et al., 1998). Plasma from $PON1^{-/-}$ mice has no detectable hydrolytic activity toward paraoxon and diazoxon, and very limited chlorpyrifos-oxonase activity. A similar pattern of activity is also found in liver, indicating that both plasma and liver PON1 are encoded by the same gene (Li et al., 2000). PON1 hemozygous mice ($PON1^{+/-}$) have approximately 40% of plasma and liver PON1 activity compared with wild-type mice ($PON1^{+/+}$). As expected, PON1 knockout mice did not differ from wild type animals in their sensitivity to demeton-S-methyl, an OP insecticide with a structure similar to malathion, which is not a substrate for PON1 (Li et al., 2000). As also predicted, $PON^{-/-}$ mice showed a dramatically increased sensitivity to chlorpyrifos oxon and diazoxon (Li et al., 2000; Shih et al., 1998). $PON1^{+/-}$ mice showed an intermediate sensitivity to diazoxon toxicity (Li et al., 2000). PON1 null mice showed only a slight increase in sensitivity to the toxicity of chlorpyrifos and diazinon (Li et al., 2000; Shih et al., 1988). The most surprising observation was that PON1 null mice did not show an increased sensitivity to paraoxon, the substrate after which the enzyme was named, in spite of having no paraoxonase activity in plasma and liver (Li et al., 2000).

Additional experiments were designed to determine whether administration of exogenous PON1 to $PON1^{-/-}$ mice, to restore serum PON1, would also restore resistance to OP toxicity. For this purpose, either human pure $PON1_{Q192}$ or $PON1_{R192}$ was injected, by the iv route, into PON1 knockout mice; the effects of various OPs on brain and diaphragm AChE activity was then determined. $PON1_{R192}$ provided significantly better protection than $PON1_{Q192}$ toward chlorpyrifos oxon (Table 1), a finding confirmed by a subsequent study by Cowan et al. (2001), who administered recombinant adenoviruses containing PON1-LQ or PON1-LR genes to BALB/c

Table 1. Inhibition of brain AChE activity by chlorpyrifos oxon in mice

Chlorpyrifos oxon (1.5–2.0 mg/kg), dissolved in acetone, was applied dermally to shaved skin on the back of the animal. Brain AChE activity was measured four hours after chlorpyrifos oxon. PON1 (916 U of arylesterase activity) was injected i.p. four hours prior to chlorpyrifos oxon.

Mouse	Brain AChE activity (% inhibition)
PON1 knockout	85–92
PON1 knockout + hPON1 R192	17
PON1 knockout + hPON1 Q192	70
hPON1 R192-TG	15
hPON1 Q192-TG	65

Data from Li et al. (2000) and Cole et al. (2005).

mice before challenge with chlorpyrifos oxon. Both alloforms were equally effective in protecting against the toxicity of diazoxon (Li et al., 2000), while neither $PON1_{R192}$ nor $PON1_{Q192}$ afforded protection against paraoxon toxicity (Li et al., 2000).

The results of these experiments in PON1 knockout mice prompted a re-examination of the *in vitro* catalytic efficiencies of the two human PON1 alloforms under more physiological salt concentrations (lower NaCl concentration). Results from kinetic analysis of substrate hydrolysis by purified human alloforms provided indeed an explanation for the *in vivo* finding. In the case of chlorpyrifos oxon, the catalytic efficiency of the $PON1_{R192}$ alloform was significantly higher than that of the $PON1_{Q192}$ alloform and, as said, provided better protection against chlorpyrifos oxon toxicity when injected into PON $-/-$ mice (Table 1). Catalytic efficiency was still high in case of diazoxon, albeit lower than with chlorpyrifos oxon, but no alloform-specific difference was evident. With paraoxon, the $PON1_{R192}$ alloform was much more efficient than the $PON1_{Q192}$ alloform, however its overall catalytic efficiency was too low to protect against exposure (Li et al., 2000). This confirms the hypothesis (Chambers et al., 1994; Pond et al., 1995) that PON1 is not efficient at hydrolyzing paraoxon at low concentrations, suggesting that PON1 may not degrade paraoxon efficiently *in vivo*, and that other pathways (e.g. cytochromes P450, carboxylesterase, and plasma cholinesterase) are primarily responsible for detoxifying paraoxon *in vivo*.

Additional experiments carried out in PON1 transgenic mice (mice expressing either human PON_{Q192} or human $PON1_{R192}$ on a knockout background, and mice carrying the human $PON1_{R192}$ on top of mPON1) provided further evidence for such conclusions. A transgenic mouse line that carries the human $PON1_{R192}$ allele over its mouse PON1 was tested for its sensitivity to paraoxon. These mice, whose serum paraoxonase activity was 3.5-fold higher than wild type mice, showed similar sensitivity to paraoxon as wild type mice (Li et al., 2000). On the other hand, $hPON1_{R192}$-TG mice (expressing human $PON1_{R192}$ on a knockout background) were significantly less sensitive to the toxicity of chlorpyrifos oxon than $hPON1_{Q192}$-TG mice, despite having the same level of PON1 protein in liver and plasma (Table 1) (Cole et al., 2005). Of relevance is also the finding that $hPON1_{R192}$-TG mice were less sensitive than $hPON1_{Q192}$-TG animals to the toxicity of the parent compound chlorpyrifos (Cole et al., 2005).

Altogether, these animal experiments allow the following conclusions with regard to the role of PON1 in modulating the toxicity of the three OP compounds tested: 1) In the case of chlorpyrifos oxon, both the level of expression and the Q192R genotype are important determinants of susceptibility, highlighting the importance of assessing PON1 status in potentially exposed individuals; 2) with regard to diazoxon, protection or susceptibility is dictated primarily by the level of expression of PON1, independently of the Q192R genotype stressing the importance of knowing PON1 levels; 3) PON1 status does not appear to play a role in modulating sensitivity to paraoxon toxicity.

4. ROLE OF PON1 IN MODULATING THE DEVELOPMENTAL TOXICITY AND NEUROTOXICITY OF OPs

Although genetic determinants play a primary role in determining an individual's PON1 status, the contribution of other factors in modulating PON1 activity may also be important. Environmental, pharmacological, dietary and life-style factors, as well as certain disease conditions, have been found to influence PON1 activity (Costa et al., 2005). Age is also a major determinant of PON1 activity. Studies in rodents have shown that serum and liver PON1 activity is very low at birth, and increases up to postnatal day 21, with a parallel increase in liver mRNA (Karanth and Pope, 2000; Li et al., 1997; Mortensen et al., 1996; Moser et al., 1998). A similar increase was also seen in transgenic mice expressing either the human $PON1_{R192}$ or the $PON1_{Q192}$ alloforms under the control of the human PON1 regulatory sequences, indicating conservation of the developmental regulatory elements between human and mouse PON1 (Cole et al., 2003). Studies in humans have also shown that serum PON1 activity is very low at birth and increases over time, reaching a plateau between 6 and 15 months of age (Augustinsson and Barr, 1963; Chen et al., 2003; Cole et al., 2003; Ecobichon and Stephens, 1973; Mueller et al., 1983).

Low PON1 activity during development could represent a relevant risk factor for increased susceptibility to the toxicity of certain OP insecticides. There is ample evidence that OP toxicity is influenced by age, with young animals being more sensitive than adults to the effects of acute exposure (Harbison, 1975; Moser et al., 1998; Pope and Liu, 1997). While intrinsic differences in brain AChE do not account for the age-related differences in sensitivity, as indicated by *in vitro* studies (Benke and Murphy, 1975; Pope and Chakraborti, 1992), lower metabolic detoxication abilities of young animals appear to be a major determinant for their increased susceptibility (Benke and Murphy, 1975; Murphy, 1982). In particular, studies with chlorpyrifos have indicated that a lower hydrolytic detoxication by PON1 accounts for the differential age-related sensitivity in acute toxicity (Mortensen et al., 1996; Moser et al., 1998; Padilla et al., 2000). Studies in PON1 knockout mice have shown that they are significantly more sensitive to the effects of chlorpyrifos oxon on brain AChE than wild-type mice of the same age (Cole et al., unpublished).

The finding of low PON1 activity in neonates (Cole et al., 2003) suggested that PON1 levels may be even lower before birth, as indeed indicated by data showing a 24% lower activity in premature babies using phenylactate as a substrate (33–36 weeks of gestation) compared to term babies (Ecobichon and Stephens, 1973). In addition, an expectant mother with low PON1 status would be predicted not to be able to provide protection for her fetus against exposure to some OPs (Cole et al., 2003). In a recent study, PON1 status was established for 130 Latina women and their newborns (Furlong et al., 2006). Among newborns, levels of PON1 (measured as arylesterase activity) varied by 26-fold (4.3–110.7 U/ml), and among mothers by 14-fold (19.8–281.4 U/ml). On average, newborns' PON1 levels were four-fold lower than the mothers' PON1 levels. Average PON1 levels in newborns were comparable with hPON1 levels in transgenic mice expressing $PON1_{Q192}$ or $PON1_{R192}$, allowing for prediction of relative sensitivity to chlorpyrifos oxon and diazoxon. For diazoxon, the predicted variability in sensitivity between newborns and mothers was 65-fold, while it was 131–164 fold for chlorpyrifos oxon (Furlong et al., 2006). These findings indicate that most newborns and many of the mothers in this cohort would be more susceptible to the adverse effects of certain OPs due to their PON1 status. One study provides support for this hypothesis. Offspring of mothers with low PON1 activity exposed *in utero* to chlorpyrifos had significantly smaller head circumference compared to those born to mothers with high PON1 activity or those not exposed to chlorpyrifos (Berkowitz et al., 2004). Since small head size has been found to be predictive of subsequent cognitive ability, these findings suggest that prenatal exposure to chlorpyrifos may have even more detrimental long-lasting effects in offspring of mothers with low PON1 activity.

5. ROLE OF PON1 IN MODULATING THE TOXICITY OF MIXTURES OF OPs

As discussed earlier, PON1 status significantly modulates the toxicity of certain OPs. However, in real life, individuals are often exposed to more than one OP at a given time. Though all OPs are generally thought to act through a common general mechanism of toxicity, i.e. inhibition of AChE, for an appropriate risk assessment one should take into account possible interactive effects. Recent work has started to evaluate the possibility that PON1 status may modulate the toxicity of mixtures of OPs. The enzyme carboxylesterase (CarE) catalyzes the hydrolysis of the OP insecticide malathion, and of its active metabolite malaoxon, that is not a substrate for PON1 (Jokanovic, 2001). When CarE is inhibited, e.g. by triothotolyl phosphate, the toxicity of malathion is greatly enhanced (Cohen and Murphy, 1971). In addition to this hydrolytic role, CarE serves also as an irreversible binding site for certain OPs, such as paraoxon, diazoxon or chlorpyrifos oxon. This binding does not have toxicological consequences per se, but contributes to the scavenging of the OP, which is thus not available for inhibiting its primary target AChE. More importantly in this context, binding of an OP to CarE inhibits its ability to detoxify insecticide substrates as well as any normal physiological function of this enzyme.

The role of PON1 in modulating the interaction of chlorpyrifos oxon and malaoxon has been recently investigated (Jansen et al., unpublished). Administration of a relatively low dose of chlorpyrifos oxon (0.75 mg/kg) to mice causes minimal inhibition of brain AChE activity, but significantly inhibits plasma CarE activity. Inhibition of both enzymes is greater in PON1 knockout mice, and greater in $hPON1_{Q192}$-TG mice than in $hPON1_{R192}$-TG animals. These differences can be ascribed to the reported role of PON1 in modulating the detoxication, and hence the biological effects of chlorpyrifos oxon *in vivo*. When wild-type mice exposed to chlorpyrifos oxon are challenged with various doses of malaoxon, the toxicity of the latter is increased. This potentiation is significantly greater in PON1 knockout mice, and greater in transgenic mice expressing the $hPON1_{Q192}$ allozyme than those expressing $hPON1_{R192}$ (Jansen et al., unpublished). These initial studies indicate that PON1 status can influence the outcome of interactions between OPs when at least one OP is affected by PON1 status. Given the often observed exposures to multiple OPs, this area of investigation will be of relevance for further studies.

6.   CONCLUSIONS

PON1 owes its name to one of its most studied substrates, paraoxon, and most of the earlier studies had focused entirely on its role in OP metabolism and toxicity. In the past fifteen years, PON1 has been shown to play a role in lipid metabolism and hence in cardiovascular disease, and in biotransformation of drugs (Costa and Furlong, 2002; Costa et al., 2003; Draganov and La Du, 2004), and research in these areas has been very active. Yet, given the wide use of OPs and their high toxicity, investigations on the role of PON1 in modulating their toxic effects remains of utmost importance. Indeed, PON1 polymorphisms in relationship to adverse effects of OPs remain a notable example of gene-environment interaction (Costa and Eaton, 2006). This chapter has highlighted current knowledge on the role of PON1 in modulating the acute and developmental toxicities of OPs, and in the toxicity of OP mixtures. Research should further probe the role of PON1 particularly with regard to developmental effects of OPs, given the continuous concerns on the potential role of chemical exposures in neurodevelopmental disorders. The possibility that PON1 status affects the outcome of exposure to mixtures of OPs is also of relevance, given that such mixed exposures are often encountered in real life. In addition to animal and mechanistic studies, there is the need for controlled human studies, in which PON1 status is correlated with the degree of exposure, biomarkers of effects, and with signs and symptoms of toxicity.

ACKNOWLEDGEMENTS

Research by the authors was supported by grants from the National Institutes of Health (ES04696, ES07033, ES11387, ES09883 and ES09601/EPA-R826886).

REFERENCES

Adkins, S., Gan, K. N., Mody, M., LaDu, B. N. (1993). Molecular basis for the polymorphic forms of human serum paraoxonase/arylesterase: glutamine or arginine at position 191, for the respective A or B allozymes. *Am. J. Hum. Genet.* **53**, 598–608.

Aldridge, W. N. (1953). Serum esterases I. Two types of esterase (A and B) hydrolyzing p-nitrophenyl acetate, proprionate and butyrate and a method for their determination. *Biochem. J.* **53**, 110–117.

Augustinsson, K. B., Barr, M. (1963). Age variation in plasma arylesterase activity in children. *Clin. Chim. Acta.* **8**, 568–573.

Benke, G. M., and Murphy, S. D. (1975). The influence of age in the toxicity and metabolism of methylparathion and parathion in male and female rats. *Toxicol. Appl. Pharmacol.* **31**, 254–269.

Berkowitz, G. S., Wetmur, J. G., Birman-Deych, E., Obel, J., Lapinski, R. H., Godbold, J. H., Holzman, I. R., Wolff, M. S. (2004). In utero pesticide exposure, maternal paraoxonase activity, and head circumference. *Environ. Health Perspect.* **112**, 388–391.

Brealey, C. B., Walker, C. M., Baldwin, B. C. (1980). A-esterase activities in relation to the differential toxicity of pirimiphos-methyl to birds and mammals. *Pestic. Sci.* **11**, 546–554.

Brophy, V. H., Jampsa, R. L., Clendenning, J. B., McKinstry, L. A., Furlong, C. E. (2001). Effects of 5′ regulatory – region polymorphisms on paraoxonase gene (PON1) expression. *Am. J. Hum. Genet.* **68**, 1428–1436.

Brophy, V. H., Jarvik, G. P., Furlong, C. E. (2002). PON1 polymorphisms. In *Paraoxonase (PON1) in Health and Disease: Basic and Clinical Aspects*, (L. G. Costa, C. E. Furlong, Eds.), Kluwer Academic Publishers, Norwell, MA, pp. 53–77.

Chambers, J. E., Ma, R., Boone, J. S., Chambers, H. W. (1994). Role of detoxication pathways in acute toxicity of phosphorothioate insecticides in the rat. *Life Sci.* **54**, 1357–1364.

Chen, J., Kumar, M., Chan, W., Berkowitz, G. S., Wetmur, J. G. (2003). Increased influence of genetic variation on PON1 activity in neonates. *Env. Health Perspect.* **111**, 1403–1410.

Cohen, S. D., Murphy, S. D. (1971). Malathion potentiation and inhibition of various carboxylic esters by triorthotolyl phosphate (TOTP) in mice. *Biochem Pharmacol.* **20**, 575–587.

Cole, T. B., Jampsa, R. L., Walter, B. J., Arndt, T. L., Richter, R. J., Shih, D. M., Tward, A., Lusis, A. J., Jack, R. M., Costa, L. G., Furlong, C. E. (2003). Expression of human paraoxonase during development. *Pharmacogenetics* **13**, 1–8.

Cole, T. B., Walter, B. J., Shih, D. M., Tward, A. D., Lusis, A. J., Timchalk, C., Richter, R. J., Costa, L. G., Furlong, C. E. (2005). Toxicity of chlorpyrifos and chlorpyrifos oxon in a transgenic mouse model of the human paraoxonase (PON1) Q192R polymorphism. *Pharmacogenet. Genom.* **15**, 589–598.

Costa, L. G., Cole, T. B., Jarvik, G. P., Furlong, C. E. (2003). Functional genomics of the paraoxonase (PON1) polymorphisms: effect on pesticide sensitivity, cardiovascular disease and drug metabolism. *Annu. Rev. Med.* **54**, 371–392.

Costa, L. G., Eaton, D. L. (2006). *Gene-Environment Interactions. Fundamentals of Ecogenetics.* John Wiley & Sons, Hoboken, NJ, p. 557.

Costa, L. G., Furlong, C. E., Eds. (2002). *Paraoxonase (PON1) in Health and Disease: Basic and Clinical Aspects.* Kluwer Academic Publishers, Norwell, MA, p. 210.

Costa, L. G., Li, W. F., Richter, R. J., Shih, D. M., Lusis, A. J., Furlong, C. E. (2002). PON1 and organophosphate toxicity. In *Paraoxonase (PON1) in Health and Disease: Basic and Clinical Aspects.* (L. G. Costa, C. E. Furlong, Eds.), Kluwer Academic Press, Norwell, MA, pp. 165–183.

Costa, L. G., McDonald, B. E., Murphy, S. D., Omenn, G. S., Richter, R. J., Motulsky, A. G., Furlong, C. E. (1990). Serum paraoxonase and its influence on paraoxon and chlorpyrifos-oxon toxicity in rats. *Toxicol. Appl. Pharmacol.* **103**, 66–76.

Costa, L. G., Richter, R. J., Murphy, S. D., Omenn, G. S., Motulsky, A. G., Furlong, C. E. (1987). Species differences in serum paraoxonase correlate with sensitivity to paraoxon toxicity. In *Toxicology of Pesticides: Experimental, Clinical and Regulatory Per-spectives.* (L. G. Costa, C. L. Galli, S. D. Murphy, Eds.), Springer-Verlag, Heidelberg, pp. 263–266.

Costa, L. G., Vitalone, A., Cole, T. B., Furlong, C. E. (2005). Modulation of paraoxonase activity. *Biochem Pharmacol.* **69**, 541–550.

Cowan, J., Sinton, C. M., Varley, A. W., Wians, F. H., Haley, R. W., Munford, R. S. (2001). Gene therapy to prevent organophosphate intoxication. *Toxicol. Appl. Pharmacol.* **173**, 1–6.

Draganov, D. I., La Du, B. N. (2004). Pharmacogenetics of paraoxonases: a brief review. *Naunyn-Schmiedeberg's Arch. Pharmacol.* **369**, 78–88.

Eckerson, H. W., Wyte, C. M., LaDu, B. N. (1983). The human serum paraoxonase/arylesterase polymorphism. *Am. J. Hum. Genet.* **35**, 1126–1138.

Ecobichon, D. J., Stephens, D. S. (1973). Perinatal development of human blood esterases. *Clin. Pharmacol. Ther.* **14**, 41–47.

Furlong, C. E., Holland, N., Richter R. J., Bradman, A., Ho, A., Eskenazi, B. (2006). PON1 status of farmworker mothers and children as a predictor of organophosphate sensitivity. *Pharmacogent. Genom.* **16**, 183–190.

Furlong, C. E., Li, W. F., Richter, R. J., Shih, D. M., Lusis, A. J., Alleva, E., Costa, L. G. (2000). Genetic and temporal determinants of pesticide sensitivity: role of paraoxonase (PON1). *Neurotoxicology* **21**, 91–100.

Furlong, C. E., Richter, R. J., Pline, C., Crabb, J. W. (1991). Purification of rabbit and human serum paraoxonase. *Biochemistry* **30**, 10133–10140.

Geldmacher-von Mallinckrodt, M., Diepgen, T. L. (1988). The human serum paraoxonase: polymoprphism and specificity. *Toxicol. Environ. Chem.* **18**, 79–186.

Harbison, R. D. (1975). Perinatal development of human blood esterases. *Clin. Pharmacol. Ther.* **14**, 41–47.

Humbert, R., Adler, D. A., Disteche, C. M., Omiecinski, C. J., Furlong C. E. (1993). The molecular basis of the human serum paraoxonase polymorphisms. *Nature Genet.* **3**, 73–76.

Jokanovic, M. (2001). Biotransformation of organophosphorus compounds. *Toxicology* **166**, 139–160.

Karanth, S., Pope, C. (2000). Carboxylesterase and A-esterase activities during maturation and aging: relationship to the toxicity of chlorpyrifos and parathion in rats. *Toxicol. Sci.* **58**, 282–289.

Leviev, I., James R. W. (2000). Promoter polymorphisms of human paraoxonase PON1 gene and serum paraoxonase activities and concentrations. *Arterioscler. Thromb. Vasc. Biol.* **20**, 516–521.

Li, W. F., Costa, L. G., Furlong, C. E. (1993). Serum paraoxonase status: a major factor in determining resistance to organophosphates. *J. Toxicol. Environ. Hlth.* **40**, 337–346.

Li, W. F., Costa, L. G., Richter, R. J., Hagen, T., Shih, D. M., Tward, A., Lusis, A. J., Furlong, C. E. (2000). Catalytic efficiency determines the in vivo efficacy of PON1 for detoxifying organophosphates. *Pharmacogenetics* **10**, 767–779.

Li, W. F., Furlong, C. E., Costa, L. G. (1995). Paraoxonase protects against chlorpyrifos toxicity in mice. *Toxicol. Lett.* **76**, 219–226.

Li, W. F., Matthews, C., Disteche, C. M., Costa, L. G., Furlong, C. E. (1997). Paraoxonase (PON1) gene in mice: sequencing, chromosomal localization and developmental expression. *Pharmacogenetics* **7**, 137–144.

Lotti, M. (2000). Organophosphorus compounds. In *Experimental and Clinical Neurotoxicology.* (P. S. Spencer, H. Schaumburg, A. C. Ludolph, Eds.), Oxford, Oxford University Press, pp. 898–925.

Machin, A. F., Anderson, P. H., Quick, M. P., Woddel, D. F., Skibniewska, K. A., Howells, L. C. (1976). The metabolism of diazinon in the liver and blood of species of varying susceptibility to diazinon poisoning. *Xenobiotica* **6**, 104.

Main, A. R. (1956). The role of A-esterase in the acute toxicity of paraoxon, TEEP and parathion. *Can. J. Biochem. Physiol.* **34**, 197–216.

Mazur, A. (1946). An enzyme in animal tissue capable of hydrolyzing the phosphorus-fluorine bond of alkyl fluorophosphates. *J. Biol. Chem.* **164**, 271–289.

Mortensen, S. R., Chanda, S. M., Hooper, M. J., Padilla, S. (1996). Maturational differences in chlorpyrifos-oxonase activity may contribute to age-related sensitivity to chlorpyrifos. *J. Biochem. Toxicol.* **11**, 279–287.

Moser, V. C., Chanda, S. M., Mortensen, S. R., Padilla, S. (1998). Age- and gender-related differences in sensitivity to chlorpyrifos in the rat reflect developmental profiles of esterase activities. *Toxicol. Sci.* **46**, 211–222.

Mueller, R. F., Hornung, S., Furlong, C. E., Anderson, J., Giblett, E. R., Motulsky, A.G. (1983). Plasma paraoxonase polymorphism: a new enzyme assay, population, family biochemical and linkage studies. *Am. J. Hum. Genet.* **35**, 393–408.

Murphy, S. D. (1982). Toxicity and hepatic metabolism of organophosphate insecticides in developing rats. *Banbury Report* **11**, 125–136.

Padilla, S., Buzzard, J., Moser, V. C. (2000). Comparison of the role of esterases in the differential age-related sensibility to chlorpyrifos and metamidophos. *Neurotoxicology* **21**, 49–56.

Playfer, J. R., Eze, L. C., Bullen, M. F., Evans, D. A. (1976). Genetic polymorphism and interethnic variability of plasma paraoxonase activity. *J. Med. Genet.* **13**, 337–342.

Pond, A. L., Chambers, H. W., Chambers, J. E. (1995). Organophosphate detoxication potential of various rat tissues via A-esterase and aliesterase activities. *Toxicol. Lett.* **70**, 245–252.

Pope, C. N., Chakraborti, T. K., (1992). Dose-related inhibition of brain and plasma cholinesterase in neonatal and adult rats following sublethal organophosphate exposure. *Toxicology* **73**, 35–43.

Pope, C. N., Liu, J. (1997). Age-related differences in sensitivity to organophosphorus pesticides. *Environ. Toxicol. Pharmacol.* **4**, 309–314.

Shih, D. M., Gu, L., Xia, Y. R., Navab, M., Li, W. F., Hama, S., Castellani, L. W., Furlong, C. E., Costa, L. G., Fogelman, A. M., Lusis, A. J. (1998). Mice lacking serum paraoxonase are susceptible to organophosphate toxicity and atherosclerosis. *Nature* **394**, 284–287.

Suehiro, T., Nakamura, T., Inoue, M., Shiinoki, T., Ikeda, Y., Kumoin, Y., Shindo, M., Tanaka, H., Hashimoto, K. (2000). A polymorphism upstream from the human paraoxonase (PON1) gene and its association with PON1 expression. *Atherosclerosis* **150**, 295–298.

Zech, R., Zurcher, K. (1974). Organophosphate splitting serum enzymes in different mammals. *Comp. Biochem. Physiol. B.* **48**, 427–433.

CHAPTER 14

IMPLICATIONS OF PARAOXONASE-1 (PON1) ACTIVITY AND POLYMORPHISMS ON BIOCHEMICAL AND CLINICAL OUTCOMES IN WORKERS EXPOSED TO PESTICIDES

A.F. HERNÁNDEZ[1], O. LÓPEZ[1], G. PENA[2], J.L. SERRANO[3], T. PARRÓN[3], L. RODRIGO[1]. F. GIL[1] AND A. PLA[1]

[1] Department of Legal Medicine and Toxicology. University of Granada Medical School. Avda. Madrid, 11. 18071 Granada. Spain
[2] Centro de Salud Motril Este, SAS, Granada
[3] Delegación Provincial de Salud, Almería

Abstract: Paraoxonase-1 (PON1) is known to play an important role in the individual susceptibility to environmental chemicals, particularly pesticides. The major results of our studies on biochemical and clinical end-points of workers long-term exposed to pesticides in a large intensive agriculture area from Southeast Spain are presented herein and compared with several other epidemiologic studies performed in different scenarios. In addition of being an individual marker of susceptibility, PON1 can be also considered a biological indicator of exposure to pesticides, since workers spraying these agents (chiefly OPs) showed decreased enzyme levels. Besides, long-term exposure to pesticides appears to indirectly elicit higher levels of PON1, which might be regarded as enzyme induction. On the other hand, carriers of the PON1 192R allele showed lower levels of erythrocyte cholinesterase and catalase, but a higher glutathione reductase activity. Regarding clinical outcomes, workers with the PON1 R allele had less risk of reporting a previous episode of pesticide poisoning as well as a lower risk of pesticide-related symptomatology. Exposure to low doses of pesticides which are metabolically activated in the liver seems to elicit subtle and early biochemical changes of hepatotoxicity. It is concluded that epidemiological studies addressing health or biochemical outcomes of workers occupationally exposed to pesticides should determine PON1 genotypes and phenotypes (activities), as these biomarkers may help in identifying those individuals at increased risk of developing pesticide toxicity or who are showing early effects after pesticide exposure

Keywords: paraoxonase, pesticides, cholinesterases, biomarkers, occupational diseases

B. Mackness et al. (eds.), The Paraoxonases: Their Role in Disease Development
and Xenobiotic Metabolism, 221–237.
© 2008 Springer.

1.   INTRODUCTION

PON1 is a glycoprotein synthesised primarily in the liver and a portion is secreted into serum where it is associated with high-density lipoprotein (HDL) particles. It is a member of a family of proteins that also includes PON2 and PON3, the genes of which are located on the long arm of human chromosome 7. The PONs family are hydrolases showing wide substrate specificity, although only PON1 is an efficient esterase towards many organophosphorus (OP) compounds (Costa et al., 2005a). Neither PON2 nor PON3 have catalytic activity toward OPs, but PON2 may have general antioxidant properties (Ng et al., 2001), while PON 3 has lactonase activity (Draganov et al., 2000). In contrast to PON1 and PON3, which are expressed primarily in the liver and associated with HDL in the circulation, PON2 is widely expressed in a number of tissues and remains intracellular (Ng et al., 2005).

PON1 is a polymorphic enzyme. Earlier studies reported that serum PON1 activity towards paraoxon (POase) exhibited a polymorphic distribution (bimodal or trimodal) in humans of Caucasian origin, so that individuals with high, interme-diate or low paraoxonase activity were identified (Eckerson et al., 1983; Mueller et al., 1983). The hypothesis that low metabolizers may be more sensitive to the toxicity of specific OP compounds was confirmed in the past, supporting the role of PON1 polymorphisms in determining the susceptibility to OP toxicity (Costa et al., 2003a).

The most important coding region polymorphism, *PON1* 192 Q/R confers different enzyme levels and catalytic activity in a substrate-dependent manner, which has been related to the differential sensitivity of individuals to the toxic effects of OPs. Various population studies have reported large variations in the allele frequencies for the *PON1* 192 Q/R polymorphism showing great interethnic variability (Cataño et al., 2006). Gene frequencies of the PON1 192 Q allele range from 0.75 for Caucasians of Northern European origin to 0.31 for some Asian populations (Brophy et al., 2002). In Asian and African-American popula-tions, the *PON1* 192 R allele predominates over the *PON1* 192 Q, while in Western populations the reverse is observed (Draganov and La Du, 2004). This different allele distribution may have important implications upon environmental or occupational exposure to pesticides. The higher the hydrolytic activity of the allele, the greater is its protective effect, supporting the metabolic impor-tance of high hydrolytic capacity. Of all the polymorphisms characterised in the 5′-regulatory region, the *PON1* −108 C/T has the most significant effect on PON1 levels in serum. Thus, the *PON1* −108 C allele produces levels of PON1 about twice as high as those seen with the −108 T allele (Costa et al., 2003a; Furlong et al., 2005).

The catalytic efficiency with which PON1 degrades toxic OPs is what determines the degree of protection afforded by PON1 against physiological or xenobiotic insults (Costa et al., 2005b). Thus, higher concentrations of PON1 provide better protection. Therefore, for adequate risk assessment it is important to know PON1 levels and activity (Costa et al., 2005b).

2. TOXICOLOGICAL IMPLICATIONS OF PESTICIDE METABOLISM FOR BIOMONITORING PURPOSES

Pesticides represent a large and important class of environmental chemicals; among them, insecticides are of most concern because of their high acute toxicity (Costa, 1997). One of the main classes of insecticides are OPs, which are in use since the mid 1940s (Gallo and Lawryk, 1991), although nowadays their use is restricted by several international agencies.

Pesticides are metabolised chiefly in the liver, but also in the gut and other organs, to derivatives that are more easily eliminated from the body. Animal studies have revealed that cytochrome P450 (CYPs) 3A4/5, 2C8, 1A2, 2C19 and 2D6, paraoxonases (PON1 and PON2), or glutathione S-transferases (GSTM1, GSTT1 and GSTP1) are the enzymes primarily involved in the metabolism of pesticides (Liu et al., 2006; Mutch and Williams, 2006), although the profile of participating isoforms is different for each agent.

A number of organophosphorothioate insecticides are detoxified in part via a two-step pathway involving bioactivation of the parent compound by the CYP systems, followed by hydrolysis of the resulting oxygenated metabolite (oxon) by serum and liver PON1 (Furlong et al., 2000). The marked inter-individual variation in expression of the various CYP isozymes may result in sub-populations of individuals that produce higher systemic oxon levels with increased susceptibility to OP toxicity (Mutch and Williams, 2006). Therefore, genetic polymorphisms in the enzymes involved in the bioactivation and detoxication of pesticides can largely influence their toxicity and would increase or decrease the sensitivity to certain pesticides. For most polymorphisms that alter responses to chemical hazards, the genetic difference does not produce a qualitatively different response, but rather induces a shift in the dose-response relationship (Kelada et al., 2003). Stratification of a studied health outcome or biomarker by relevant genotype (or phenotype) may allow for detection of different levels of risk among subgroups of exposed persons (Rothman et al., 2001).

Application of pesticides can result in exposure by either dermal or respiratory routes, and can produce illness even with low-grade depressions in blood cholinesterases (Gordon and Richter, 1991). As different OPs may inhibit erythrocyte cholinesterase (AChE) or plasma cholinesterase (PChE) to a different extent, the determination of both enzymes remains a very valid way to assess exposure to, and early biological effects of OP exposure (Costa et al., 2005a). AChE is considered better than PChE for the assessment of chronic exposure to OPs since the exposure-related decrease in AChE activity most likely reflects continued exposure to irreversibly inhibiting OPs in the exposed population because of its lower recovery rate compared to that of PChE (Kamel and Hoppin, 2004). OPs also inhibit plasma PChE, but the effect lasts at most a few weeks and is therefore less useful for assessing chronic exposure (Kamel and Hoppin, 2004) with the exception of intensive agriculture workers as they are continuously exposed to pesticides on a regular basis, thus precluding PChE from being fully recovered (Hernández et al., 2006). On the other hand, serum PChE activity has been reported to be a slightly more sensitive indicator of mixed exposure than red blood cell AChE activity (Richter et al., 1992).

In a previous study we have found that the number of years working as pesticides applicators was the only variable significantly associated in the multivariate analysis with a decrease of AChE greater than 15% of basal levels, used as a surrogate of exposure (OR: 1.14, 95% CI: 1.04–1.25), indicating a pesticide-induced cumulative inhibition of AChE in the exposed population (Hernández et al., 2005). AChE, in addition of being inhibited by OP or carbamates, may be also indirectly affected by the oxidative stress induced by other pesticide groups (i.e., organochlorine, pyrethroids) leading to lipoperoxidation of biological membranes. Thus, membrane-bound enzymes such as AChE may be eventually affected by this process (Banerjee et al., 1999; Hernández et al., 2005). OPs are also capable to produce oxidative stress, although it is unclear whether this may be a consequence of AChE inhibition or a separate effect (Giodano et al., 2006).

3.   CHANGES IN PON1 AFTER EXPOSURE TO PESTICIDES

Certain physiological and pathological conditions, as well as some environmental factors can increase or decrease PON1 activity (Costa et al., 2005b). Occupational exposure to OPs may be another environmental factor which can modulate PON1 as well, as farm-workers applying this type of pesticides have significantly lower PON1 activity towards paraoxon (POase) (Hernández et al., 2003). This obser-vation was further extended to overall pesticides, not only OPs, as farm-workers spraying different pesticides during the course of a grown season showed signif-icantly lower levels of POase activity (Hernández et al., 2004). These results are consistent with those reported by Kuang et al. (2006), who found decreased activ-ities of carboxylesterases, PChE, AChE and PON1 in different level in Chinese subjects long-term exposed to OPs. Therefore, the possibility arises of significant PON1 changes in humans chronically exposed to different pesticides, not only OPs, indicating that its serum levels could be modulated by long-term pesticide exposure. The metabolic activation of pesticides (including OPs) to highly reactive interme-diates might account for the decreased POase activity, as the enzyme can be inacti-vated by these compounds after oxidative stress challenge (Aviram et al., 1999).

Changes in POase activity have been also reported after acute poisoning by OP pesticides. Twenty-eight patients acutely poisoned by OPs were found to have lower PON1 activity than controls, but 6 months after poisoning the levels of eight patients had returned to normal, suggesting temporary inhibition of PON1 (Sozmen et al., 2002).

In a previous study, exposure to pesticides resulted in higher levels of POase and PChE in farm-workers as compared to controls at the post-exposure period, considered as baseline. This finding suggests an adaptive biological response indicating certain degree of enzyme induction in workers long-term exposed to pesticides (Hernández et al., 2004). This observation is consistent with the reported up-regulation of PON1 following chronic OP exposure (Browne et al., 2006), although this effect was only seen in carriers of the R allele, an isoform which

Table 1. Serum PON1 activity measured by using rates of hydrolysis against phenylacetate (AREase), paraoxon (POase), paraoxon in the presence of high salt concentration (ssPOase) and diazoxon (DZOase). Enzyme activities from controls and pesticide applicators (sprayers) were measured at two periods with different exposure to pesticides

Enzyme	Overall sample		PON1 192 R allele		PON1 192 QQ genotype	
	Control	Pesticide applicator	Control	Pesticide applicator	Control	Pesticide applicator
	Mean ± S.D.	Mean ± S.D.	Mean ± S.D.	Mean ± S.D.	Mean ± S.D.	Mean ± S.D.
AREase						
Exposure	108.4 ± 19.6	107.0 ± 23.3	100.1 ± 17.5	99.7 ± 20.1	118.0 ± 17.7	114.4 ± 24.1
Post-exposure	110.7 ± 20.8	104.2 ± 24.6	105.6 ± 19.5	98.4 ± 24.3	117.2 ± 17.4	104.6 ± 20.8*
POase						
Exposure	238.2 ± 114.0	217.6 ± 107.2	320.2 ± 95.4	289.9 ± 101.8	142.5 ± 18.7	145.3 ± 46.7
Post-exposure	262.9 ± 125.6	260.3 ± 122.7	356.4 ± 105.6	303.0 ± 107.6	167.6 ± 63.7	172.0 ± 66.2
ssPOase						
Exposure	452.8 ± 274.0	384.8 ± 226.6	669.1 ± 187.8	557.2 ± 179.1*	200.4 ± 32.6	215.9 ± 111.4
Postexposure	435.4 ± 259.5	393.7 ± 252.1	636.2 ± 175.6	556.0 ± 199.0	215.8 ± 136.0	199.9 ± 124.2
DZOase						
Exposure	7.4 ± 1.9	6.6 ± 2.2*	6.3 ± 1.5	5.5 ± 1.7	8.7 ± 1.4	7.7 ± 2.1
Postexposure	7.2 ± 2.4	7.2 ± 2.6	5.6 ± 1.4	6.2 ± 2.5	8.8 ± 2.0	8.1 ± 2.2

Units: AREase (kU/l), remaining enzymes (U/l).

Comparisons were made by the Student's t-test or Mann-Whitney test.

* Significance between pesticide applicators and control (p < 0.05).

Table 2. Serum PON1 activity measured by using rates of hydrolysis against phenylacetate (AREase), paraoxon (POase), paraoxon in the presence of high salt concentration (ssPOase) and diazoxon (DZOase). Enzyme activities are expressed controlling for AChE decrease above or below 15% from postexposure to the higher exposure periods of a growing season

Enzyme	Overall sample		PON1 192 R allele		PON1 192 QQ genotype	
	AChE < 15% Mean ± S.D.	AChE > 15% Mean ± S.D.	AChE < 15% Mean ± S.D.	AChE > 15% Mean ± S.D.	AChE < 15% Mean ± S.D.	AChE > 15% Mean ± S.D.
AREase						
Exposure	106.0 ± 23.7	110.4 ± 19.9	98.5 ± 20.9	102.8 ± 15.1	113.6 ± 24.3	118.5 ± 21.5
Post-exposure	106.8 ± 24.7	102.2 ± 19.2	102.8 ± 25.2	99.2 ± 21.1	111.0 ± 23.9	105.4 ± 16.7
POase						
Exposure	224.7 ± 107.9	215.7 ± 110.0	301.5 ± 93.0	283.5 ± 113.6	147.9 ± 53.1	143.2 ± 33.5
Post-exposure	260.2 ± 121.4	255.5 ± 125.9	343.1 ± 98.8	337.8 ± 117.4	177.2 ± 78.4	164.0 ± 47.4
ssPOase						
Exposure	425.3 ± 259.0	381.6 ± 225.5	636.9 ± 180.3	542.2 ± 189.8*	213.6 ± 106.4	209.9 ± 97.4
Postexposure	422.6 ± 257.2	373.2 ± 243.7	617.9 ± 170.2	545.5 ± 212.6	220.3 ± 154.0	189.0 ± 93.2
DZOase						
Exposure	7.0 ± 2.2	6.8 ± 2.0	5.9 ± 1.7	5.8 ± 1.5	8.1 ± 2.0	7.8 ± 1.9
Postexposure	7.1 ± 2.5	7.2 ± 2.5	5.8 ± 2.3	6.2 ± 2.2	8.3 ± 2.2	8.4 ± 2.2

Units: AREase (kU/l), remaining enzymes (U/l).

Comparisons were made by the Student's t-test or Mann-Whitney test.

* Significance between decrease in AChE > 15% and < 15% (p < 0.05)

may have a potential biosensor effect that leads to subsequent increases in enzyme activity after pesticide exposure (Browne et al., 2006).

With the aim of evaluating whether or not PON1 activity towards different substrates undergoes changes (either enzyme induction or inhibition) as a function of the PON1 192 polymorphism, we reanalysed previous data and results are shown in Tables 1 and 2. From all PON1 activities measured, only diazoxonase (DZOase) was significantly decreased in the exposed population at the period of higher exposure to pesticides. However, when controlling for PON1 192 polymorphism, individuals with the PON1 192 R allele showed significantly lower salt-stimulated paraoxonase (ssPOase) at the period of higher exposure to pesticides. In contrast, farm-workers with the PON1 192 QQ genotype presented significantly lower AREase activity but at the postexposure period, which is more difficult to interpret. The lower enzyme activities suggest that pesticides may affect PON1, this effect being chiefly shown by carriers of the PON1 192R allele, whom might be considered more susceptible. However, and in contrast to previous studies, direct evidence of higher PON1 activity was not observed, thus failing to support the notion of an induction effect as previously reported for carriers of the PON1 192 R allele (Browne et al., 2006).

On the other hand, no significant differences were found in the different PON1 activities studied when controlled for a decrease in AChE greater than 15%, used as a surrogate for pesticide exposure (Table 2). However, after stratifying by the PON1 192 polymorphism, it was found that individuals with the R allele showed significantly lower levels of ssPOase activity at the period of high exposure to pesticides. This finding suggests that exposure to anticholinesterase pesticides decreases PON1 activity, particularly when the enzyme activity was measured by using paraoxon as a substrate, as reported elsewhere for OPs in either acute poisoning (Sozmen et al., 2002) or chronic exposure (Hernandez et al., 2003). Taken into consideration that the lipid peroxidation induced by pesticides other than OPs may affect membrane-bound enzymes, this mechanism can also contribute to the lower AChE (Banerjee et al., 1999; Hernández et al., 2005). Thus, the observed effect should not be exclusively attributed to either OPs or carbamate (anticholinesterase) pesticides, but can also be extended to those pesticides capable to generate oxidative stress.

4. IMPLICATIONS OF THE PON1 R ALLELE IN PESTICIDE TOXICITY

Several epidemiological studies have highlighted the involvement of PON1 in xenobiotic susceptibility in different scenarios: Gulf War veterans, sheep dippers, farm-workers occupationally exposed to pesticides, and acute poisoning by nerve gases in terrorist attacks.

a) Clinical studies of paraoxonase have linked OP exposure with the **Gulf War Syndrome (GWS)**. Gulf War Veterans were found to be exposed to poisonous anticholinesterases in combination with other chemical agents. Compared to controls, ill veterans with neurological impairment were more likely to have the R allele than to be homozygous Q for the PON1 gene (Haley et al., 1999). US veterans with GWS had significantly lower serum levels of the Q allozyme than healthy veterans, whereas British veterans with GWS showed lower levels of both the Q and R allozymes compared with healthy controls (Haley, 2003). Also, somanase and sarinase activities of US veterans with GWS ran parallel to the PON1 type Q isozyme levels, suggesting than they may have been exposed to some environmental or chemical toxin with a similar preference for hydrolysis by the PON1 type Q isozyme (La Du et al., 2001).

b) Some **"sheep dipping" studies** have found a higher proportion of the PON1 192R allele and lower rates of diazoxon hydrolysis in farmers with chronic ill health attributed to OP exposure as compared to controls. These findings suggest that their ill health may be explained by a lower ability to detoxify diazoxon, the active metabolite of diazinon which is the OP currently used in the "sheep dipping" process (Mackness et al., 2003). Therefore, carriers of the PON1 R allele seem to have a higher susceptibility to certain anticholinesterase pesticides, such as diazinon (Cherry et al., 2002).

c) In our studies on **farm-workers long-term exposed to pesticides**, most of individuals showing a PChE depression greater than 25% of basal levels were carriers of the PON1 R allele (Hernández et al., 2004). This finding was further evaluated in a logistic regression analysis in which the PON1 R allele was found to be the unique significant predictor of PChE depression greater than 25% (OR 4.97; 95% CI: 1.58–15.58) independently of certain individual and occupational covariates as well as plasma phenotypes of PChE (Hernández et al., 2004).

We have also reported that carriers of the PON1 R allele exhibited lower AChE activity as compared to individuals with the QQ genotype (Hernández et al., 2005). This finding not necessarily means that carriers of the R allele may have biological disadvantages upon long-term low-dose exposure to pesticides. In contrast, it may result from genotype-dependent inverse relationships between AChE and PON1 activities involving exposure to low doses of OPs. In individuals with low PON1 levels the *ACHE-PON1* gene interaction would up-regulate AChE due to inhibition of circulating AChE, whereas those with high PON1 levels will be relatively protected, avoiding up-regulation and maintaining low blood AChE levels (Bryk et al., 2005).

Recently, we have found that certain erythrocyte antioxidant enzymes, namely catalase and glutathione reductase were dependent on PON1 polymorphisms, as farm-workers with the R allele showed lower levels of catalase and higher levels of glutathione reductase (López-Guarnido, 2005). This preliminary observation confirms that farm-workers are exposed to more oxidative stress as a result of pesticide exposure, as evidenced by changes in their antioxidant status (i.e., decreased levels of erythrocyte superoxide dismutase, glutathione reductase and catalase). The decreased activity of superoxide dismutase and glutathione reductase may result from electrophillic attack of the generated free radicals (Kono and Fridovich, 1982), indicating the failure of the total antioxidant defence mechanism to protect target tissues. It appears possible that the oxidative stress observed in farm-workers long-term exposed to pesticides could contribute to lifetime accumulation of oxidative damage and, thus, may be involved in the pathogenesis of various pathologies associated with the chronic exposition to pesticides, including cancer and neurodegenerative disorders, which are clearly related to oxidative damage.

d) In the 1995 **terrorist attack** in the Tokyo subway, several thousand people were exposed to the nerve agent sarin, an OP which is metabolized preferentially by the PON1 Q alloform (Davies et al., 1996). However, among 10 of the 12 victims, 7 expressed the PON1 Q allele -six heterozygotes and one homozygote- (Yamada et al., 2001). Thus, the genotype that confers high hydrolyzing activity toward sarin did not appear to provide protection from acute poisoning. In addition, the catalytic efficiency of sarin hydrolysis by PON1 is low, and the exposure to sarin was massive, most likely overcoming any potential protection afforded by the PON1 192Q allele (Costa et al., 2003b).

From a clinical standpoint, we have found that farm-workers with the PON1 QQ genotype more likely have had a previous intoxication by pesticides (OR = 2.64, CI 95%: 1.09–5.45). The remaining PON1 polymorphisms at positions 55, −108 and −909 failed to be associated. This finding is consistent with previous studies reporting that the PON1 Q allele (genotypes QQ or QR) was an independent predictor of chronic toxicity by pesticides in farm-workers (Lee et al., 2003) and also this allele was more frequent in patients with acute OP poisoning than in controls (Sozmen et al., 2002). In addition, we found that carriers of the PON1 R allele were at roughly 4-fold less risk of having a previous episode of acute pesticide poisoning than homozygous for the PON1 Q allele (OR: 0.28, 95% CI: 0.08–1.02) independently of several individual or occupational covariates (Hernández et al., 2005). In other study, we found that two variables, namely farm-workers with the PON1 R allele and the utilisation of personal protective equipment during mixing/loading of pesticides were the only predictors associated with a lower risk of pesticide-related symptomatology in a multivariate analysis after controlling for working conditions and other potential confounders (OR 0.33, 95% CI 0.11–0.98 and OR 0.17, 95% CI 0.03–0.84; respectively).

These observations point that individuals carrying the PON1 R allele may have some kind of pathophysiological advantages upon long-term low-dose exposure to pesticides, so that they might tolerate relatively higher concentrations of these chemicals as compared with carriers of the QQ genotype, indicating a sort of "occupational selection pressure".

Controversial results have been reported by Padungtod et al. (1999) who examined the effects of PON1 genotypes on male reproductive outcomes in Chinese pesticide-factory workers. In that study, exposed workers with the PON1 R allele had significantly lower sperm count and lower percentage of sperm with normal morphology than the non-exposed carriers of the PON1 R allele. Also, exposed PON1 R carriers had significantly higher serum LH levels. Discrepancies might reflect different distributions of PON1 genotypes or enzyme activities in populations from diverse geographical areas, or qualitative/quantitative differences in the profile of pesticide use.

5. PROMOTER POLYMORPHISM (INEFFICIENT PON1 -108TT GENOTYPE) AND RISK OF ADVERSE OUTCOMES

In a previous study (Hernández et al., 2003) we found that the PON1 polymorphism at position 55 was in linkage disequilibrium with that at position 192 as described elsewhere (Blatter-Garin et al., 1997; Brophy et al., 2001; Mackness et al., 1998), and with those of the regulatory region at positions -108 and -909 (Brophy et al., 2001). The two promoter region polymorphisms (-108 and -909) also were in linkage disequilibrium to each other. By contrast, the *PON1* 192 polymorphism failed to be in linkage disequilibrium with those from the promoter region (-108 and -909). It can be concluded that from these four PON1 polymorphisms only PON1 192 and -108 have sufficient relevance to be included in epidemiological studies addressing the contribution of PON1 to adverse health outcomes, as they have the greatest effect on PON1 activity.

Because of the catalytic efficiency toward phenylacetate (arylesterase, AREase) or diazoxon (DZOase) is not affected by the PON1 192 polymorphism, they are the preferred substrates to provide a better indication of overall PON1 activity levels (Costa et al., 2005b). Thus, AREase may provide an estimate of the relative sensitivity to OP compounds (Furlong et al., 2006). Since the *PON1* -108 TT genotype is associated with lower rates of AREase independently of the *PON1* 192 polymorphism (Cataño et al., 2006; Hernández et al., 2003), subjects homozygous for this particular genotype may be at an increased risk to any OP compound. If so, it would be a new biomarker of susceptibility against these agents. To further evaluate the hypothesis that the PON1 -108 TT genotype may be a risk factor for certain biochemical and clinical outcomes, we have monitored a cohort of 81 pesticide-applicators and 37 controls at two different periods of a growing season (exposure and postexposure).

Multiple linear regression analysis of PON1 activities toward different substrates is shown in Tables 3 and 4. It is noteworthy that pesticide applicators had significantly lower levels of DZOase than controls at the high exposure period. The decrease of AChE greater than 15% was a determinant of higher levels of AREase and ssPOase at the same period, thus refining the association found in Table 2. Interestingly, those individuals with the PON1 192 R allele that also showed a decrease in AChE higher than 15% (interaction term) exhibited significantly lower levels of ssPOase at the period of high exposure to pesticides. It was also found the utilization of personal protective equipment during mixing/loading of pesticides was a determinant of higher levels of POase and ssPOase at either exposure period, indicating that the use of individual protection measures prevents pesticides from being absorbed and precludes PON1 inactivation as reported previously (Hernández et al., 2003). By contrast, the antecedent of a previous episode of poisoning by pesticides and pesticide-related symptomatology were predictors of higher POase levels at the post-exposure period. Remarkably, individuals with unusual PChE phenotypes exhibited higher levels of ssPOase.

Overall, these results suggest that long-term exposure to pesticides elicits a decrease in PON1 activity while workers are exposed to these compounds (exposure

Table 3. Multiple linear regression of serum PON1 activities at two periods with different pesticide exposure (I). Models include controls and pesticide applicators

Enzyme	Exposure					Post-exposure				
	Adj R^2	Predictors	B	P	Correlation	Adj R^2	Predictors	B	P	Correlation
AREase	0.335	Age	−0.46	0.068	−0.203	0.186	Age	−0.63	0.024	−0.250
		Alcohol	8.56	0.081	−0.194		Alcohol	12.93	0.019	0.259
		PON1 −108 TT	−23.38	< 0.001	0.537		PON1 −108 TT	−17.90	< 0.001	−0.403
		PON1 192 R	−12.88	0.058	−0.210					
		AChE > 15%	12.41	0.057	0.211					
POase	0.499	Tobacco	−31.54	0.078	−0.196	0.553	Tobacco	−34.02	0.085	−0.197
		PON1 192 R	157.78	< 0.001	0.542		PON1 192 R	186.45	< 0.001	0.589
		PON1 −108 TT	−65.55	< 0.001	−0.401		PON1 −108 TT	−72.39	< 0.001	−0.416
		Age	−1.85	0.071	−0.201					
ssPOase	0.664	Tobacco	−64.34	0.039	−0.228	0.646	Tobacco	−69.68	0.040	−0.229
		PON1 192 R	434.61	< 0.001	0.715		PON1 192 R	424.35	< 0.001	0.677
		PON1 −108 TT	−104.92	0.001	−0.374		PON1 −108 TT	−105.18	0.001	−0.354
		Age	−3.47	0.052	−0.216		Alcohol	73.70	0.050	0.219
		PChE phenotype	80.29	0.069	0.202					
		AChE > 15%	75.45	0.100	0.183					
		Interaction term	−129.85	0.040	−0.228					
DZOase	0.409	Exposure	−1.53	0.021	−0.256	0.341	Age	−0.07	0.024	−0.253
		PON1 192 R	−2.41	< 0.001	−0.444		PON1 192 R	−2.19	0.006	−0.305
		PON1 −108TT	−1.28	< 0.001	−0.394		PON1 −108 TT	−1.54	0.002	−0.344
		Alcohol	0.69	0.090	0.189		Alcohol	1.28	0.026	0.249
							BMI	0.11	0.053	0.217

The model was adjusted for the following variables: age, gender (0: male; 1: female), BMI, exposure to pesticides (1: sprayer; 0: non-sprayer), smoking habit (0: no; 1: yes), alcohol consumption (0: no; 1: yes), PON1 192 polymorphism (1: allele R; 0: genotype QQ), PON1 −108 TT polymorphism (1: genotype TT; 0: allele C); PChE phenotype (1: unusual variants; 0: usual variant), GST-M1 (1: null allele; 0: functional allele), GST-T1 (1: null allele; 0: functional allele) decrease of AChE > 15% at the high exposure period with respect to the low exposure period (1: yes; 0: decrease < 15%), interaction term (between PON1 192 R allele and AChE decrease > 15%).

Adj. R^2: Adjusted Regression coefficient. Correlation means partial correlation.

Table 4. Multiple linear regression of serum PON1 activities at two periods with different pesticide exposure (II). Models include only pesticide applicators

Enzyme	Exposure					Post-exposure				
	Adj R^2	Predictors	B	P	Correlation	Adj R^2	Predictors	B	P	Correlation
AREase	0.378	Alcohol	10.86	0.078	0.257	0.186	Alcohol	14.18	0.037	0.32
		PON1 −108 TT	−22.49	< 0.001	−0.548		PON1 −108 TT	−16.10	0.006	−0.394
		PON1 192 R	−26.80	0.022	−0.329		PPE mixing/loading	16.02	0.012	0.362
POase	0.586	PON1 192 R	162.37	< 0.001	0.489	0.713	PON1 192 R	219.03	< 0.001	0.624
		PON1 −108 TT	−65.22	0.001	−0.450		PON1 −108 TT	−70.04	0.001	−0.494
		Symptomatology	38.51	0.083	0.253		Symptomatology	46.36	0.033	0.319
		Previous poisoning	49.05	0.020	0.335		Previous poisoning	59.80	0.007	0.394
		PPE mixing/loading	45.97	0.033	0.308		PPE mixing/loading	65.85	0.003	0.437
ssPOase	0.765	PON1 192 R	444.84	< 0.001	0.715	0.740	PON1 192 R	470.23	< 0.001	0.673
		PON1 −108 TT	−107.20	0.001	−0.483		PON1 −108 TT	−117.50	0.001	−0.454
		AChE > 15%	76.67	0.108	0.235		Previous poisoning	73.58	0.048	0.286
		PPE mixing/loading	69.20	0.046	−0.290		PPE mixing/loading	81.19	0.035	0.305
DZOase	0.412	PON1 192 R	−3.85	< 0.001	0.494	0.391	PON1 192 R	−3.22	0.015	−0.353
		PON1 −108TT	−1.18	0.011	−0.363		PON1 −108 TT	−1.52	0.010	−0.370
		GST-T1	0.86	0.054	0.280		Alcohol	1.88	0.009	0.376
		Previous poisoning	−0.88	0.071	0.263					

The model was adjusted for the following variables: age, gender (0: male; 1: female), BMI, exposure to pesticides (1: sprayer; 0: non-sprayer), smoking habit (0: no; 1: yes), alcohol consumption (0: no; 1: yes), PON1 192 polymorphism (1: allele R; 0: genotype QQ), PON1 −108 TT polymorphism (1: genotype TT; 0: allele C); PChE phenotype (1: unusual variants; 0: usual variant), GST-M1 (1: null allele; 0: functional allele), GST-T1 (1: null allele; 0: functional allele) decrease of AChE > 15% at the high exposure period with respect to the low exposure period (1: yes; 0: decrease < 15%), interaction term (between PON1 192 R allele and AChE decrease > 15%), years exposed to pesticides as applicators, utilisation of OP (1: yes, 0: no), utilisation of personal protective equipment (PPE) when mixing/loading pesticides (1: yes, 0: no); utilisation of PPE during the spraying of pesticides (1: yes, 0: no).

Adj. R^2: Adjusted Regression coefficient. Correlation means partial correlation.

period). In contrast, those individuals with pesticide-induced adverse clinical outcomes or with an unusual PChE phenotype undergo a sort of enzyme induction in the post-exposure period, which seem to be consistent with the study of Browne et al. (2006). Thus, increased PON1 activity can be considered a protective factor that would benefit the carrier from pesticide use. Some pesticides (i.e., carbaryl and thiabendazole) have been reported to induce CYP1A1, an effect that appears to be mediated via the xenobiotic responsive element (XRE) because both pesticides specifically activate various fusion constructs containing XRE sequences, such as CYP 1A and GST (Delescluse et al., 2001). Some environmental factors, such as dietary polyphenols and 3-methycholanthrene, have shown to increase PON1 activity via activation of a XRE-like sequence within the PON1 promoter region (Gouedard et al., 2004). Therefore, it can be hypothesised that PON1 is up-regulated via the XRE as previously suggested (Rodrigo et al., 2001), so that the higher PON1 levels observed in subjects reporting certain adverse clinical outcomes may be attributed to this mechanism.

Given that many pesticides are metabolically activated by xenobiotic-metabolising enzymes mainly in the liver microsomes, highly reactive molecules may very likely arise in liver cells. These molecules, if not properly neutralised soon after their generation, can interact with relevant hepatocellular targets and elicit early toxicity at a biochemical level. This eventuality can be monitored by using conventional biomarkers of liver damage in serum, such as transaminases (ALT and AST), gamma-glutamyl transferase (GGT) and alkaline phosphatase. The two latter enzymes are biochemical markers of cholestasis. In addition to monitoring these enzymes, we also assessed whether polymorphisms in certain pesticide-metabolising enzymes (such as PON1, PChE, GSTM1 and GSTT1) put individuals at an increased risk of affecting liver biochemistry parameters (data not shown). It was found that pesticide applicators had significantly higher levels of ALT and that the decrease in AChE greater than 15% was a determinant of higher levels of GGT. In addition, workers using OPs had higher levels of ALT and GGT, and the cumulative lifetime exposure to pesticides (years exposed) was directly associated with higher levels of alkaline phosphatase. However, the interaction term between the PON1 R allele and decrease in AChE greater than 15% failed to be significantly associated with any of the liver biochemistry parameters studied. These results provide support for a very subtle liver dysfunction at a biochemical level. The association of OP exposure with an increase in GGT is in accordance with previous studies (Banerjee et al., 1999; Parrón et al., 1996) and may be consistent with an enzyme induction effect (Park and Breckenridge, 1981) rather than with a cholestatic change, as a parallel increase in alkaline phosphatase was not observed. In addition, carriers of the PON1 192 R allele seem to be at an increased risk of hepatotoxicity as they showed higher ALT levels. In turn, subjects with the PON1 −108 TT genotype presented lower levels of ALT (near the significance level of 0.05), which calls for further investigation to confirm this effect and elucidate the underlying mechanism.

These results extend the information gathered from a recent study examining cytolysis markers in serum from greenhouse workers long-term exposed to pesticides (Hernandez et al., 2006) which has found that pesticide exposure was associated with increased serum AST levels, and decreased serum LDH and amino-oxidase activities. These findings also supported a subtle pesticide-induced liver dysfunction at a biochemical level, but overall indicated absence of clinically significant hepatotoxicity.

As regard to polymorphisms in other pesticide-metabolizing enzymes, carriers of unusual PChE phenotypes had higher levels of ALT, AST and GGT and lower levels of alkaline phosphatase. Subjects with null GSTT1 had significantly higher levels of ALT and GGT than those with functional GSTT1. Therefore, carriers of unusual PChE phenotypes and null GSTT1 are at increased risk of hepatotoxicity upon exposure to pesticides or any other environmental chemical.

6.   CONCLUSION

From this review it can be concluded that carriers of PON1 192 R allele seem to be at a high risk of adverse biochemical outcomes but at a less risk of adverse clinical outcomes. In addition, exposure to low doses of pesticides which are metabolically activated in the liver may lead to early biochemical changes of hepatotoxicity, but the toxicological relevance of this observation is uncertain and should be evaluated in more depth.

The toxicological role of PON1 continues to attract attention, reflecting growing public concern with environmental impacts on health. Although further research is required to clarify and confirm the role of PON1 activity (and PON1 functional genotype) in long-term exposure to pesticides, the current evidence on PON1 is high enough to be considered a useful biochemical parameter. Therefore, we propose that PON1 should be systematically measured together with cholinesterases in the Health Surveillance Programme of pesticide-exposed workers.

REFERENCES

Aviram, M., Rosenblat, M., Billecke, S., Erogul, J., Sorenson, R., Bisgaier, C.L., Newton, R.S., La Du, B., 1999, Human serum paraoxonase (PON1) is inactivated by oxidized low density lipoprotein and preserved by antioxidants. *Free Radical Biol Med* 26: 892–904.

Banerjee, B.D., Seth, V., Bhattacharya, A., Pasha, S.T., Chakraborty, A.K., 1999, Biochemical effects of some pesticides on lipid peroxidation and free-radical scavengers. *Toxicol Lett* 107: 33–47.

Blatter-Garin, M.C., James, R.W., Dussoix, P., Blanche, H., Passa, P., Froguel, P., Ruiz, J., 1997, Paraoxonase polymorphism Met-Leu54 is associated with modified serum concentrations of the enzyme. A possible link between the paraoxonase gene and increased risk of cardiovascular disease in diabetes. *J Clin Invest* 99: 62–66.

Brophy, V.H., Jampsa, R.L., Clendenning, J.B., McKinstry, L.A., Jarvik, G.P., Furlong, C.E., 2001, Effects of 5′ Regulatory-region polymorphisms on paraoxonase-gene (*PON1*) expression. *Am J Hum Genet* 68: 1428–1436.

Brophy, V.H., Jarvik, G.P., Furlong, C.E., 2002, PON1 polymorphisms. In *Paraoxonase (PON1) in Health and Disease: Basic and Clinical Aspects* (L.G. Costa, C.E. Furlong, eds.), Kluwer Academic Publishers, Norwell, USA, pp. 53–77.

Browne, R.O., Moyal-Segal, L.B., Zumsteg, D., David, Y., Kofman, O., Berger, A., Soreq, H., Friedman, A., 2006, Coding region paraoxonase polymorphisms dictate accentuated neuronal reactions in chronic, sub-threshold pesticide exposure. *FASEB J* 20: 1733–1735.

Cataño, H.C., Cueva, J.L., Cárdenas, A.M., Izaguirre, V., Zavaleta, A.I., Carranza, E., Hernández, A.F., 2006, Distribution of paraoxonase-1 gene polymorphisms and enzyme activity in a Peruvian population. *Environ Mol Mutag* 47: 699–706.

Cherry, N., Mackness, M., Durrington, P., Povey, A., Dippnall, M., Smith, T., Mackness, B., 2002, Paraoxonase (PON1) polymorphisms in farmers attributing ill health to sheep dip. *Lancet* 359: 763–764.

Costa, L.G., 1997, Basic toxicology of pesticides. *Occup Med* 12: 251–268.

Costa, L.G., Cole, T.B., Furlong, C.E., 2003b, Polymorphisms of paraoxonase (PON1) and their significance in clinical toxicology of organophosphates. *J Toxicol Clin Toxicol* 41: 37–45.

Costa, L.G., Cole, T.B., Vitalone, A., Furlong, C.E., 2005a, Measurement of paraoxonase (PON1) status as a potential biomarker of susceptibility to organophosphate toxicity. *Clin Chim Acta* 352: 37–47.

Costa, L.G., Richter, R.J., Li, W.F., Cole, T., Guizzetti, M., Furlong, C.E., 2003a, Paraoxonase (PON1) as a biomarker of susceptibility for organophosphate toxicity. *Biomarkers* 8: 1–12.

Costa, L.G., Vitalone, A., Cole, T.B., Furlong, C.E., 2005b, Modulation of paraoxonase (PON1) activity. *Biochem Pharmacol* 69: 541–550.

Dantoine, T., Debord, J., Merle, L., Charmes, J.P., 2003, From organophosphate compound toxicity to atherosclerosis: role of paraoxonase 1. *Rev Med Interne* 24: 436–442.

Davies, H., Richter, R.J., Kiefer, M., Broomfield, C., Sowalla, J., Furlong, C.E., 1996, The human serum paraoxonase polymorphism is reversed with diazoxon, soman and sarin. *Nat Genet* 14: 334–336.

Deakin, S.P., James, R.W., 2004, Genetic and environmental factors modulating serum concentrations and activities of the antioxidant enzyme paraoxonase-1. *Clin Sci* 107: 435–447.

Delescluse, C., Ledirac, N., Li, R., Piechocki, M.P., Hines, R.N., Gidrol, X., Rahmani, R., 2001, Induction of cythochrome P450 1A1 gene expression, oxidative stress, and genotoxicity by carbaryl and thiabendazole in transfected human HepG2 and lymphoblastoid cells. *Biochem Pharmacol* 61: 399–407.

Draganov, D.I., La Du, B.N., 2004, Pharmacogenetics of paraoxonases: a brief review. *Naunyn Schmiedeberg's Arch. Pharmacol* 369: 78–88.

Draganov, D.I., Stetson, P.L., Watson, C.E., Billecke, S.S., La Du, B.N., 2000, Rabbit serum paraoxonase 3 (PON3) is a high density lipoprotein-associated lactonase and protects low density lipoprotein against oxidation. *J Biol Chem* 275: 33435–33442.

Eckerson, H.W., Wyte, C.M., La Du, B.N., 1983, The human serum paraoxonase/arylesterase polymorphism. *Am J Human Genet* 35: 1126–1138.

Furlong, C.E., Cole, T.B., Jarvik, G.P., Pettan-Brewer, C., Geiss, G.K., Richter, R.J., Shih, D.M., Tward, A.D., Lusis, A.J., Costa, L.G., 2005, Role of paraoxonase (PON1) status in pesticide sensitivity: genetic and temporal determinants. *Neurotoxicology* 26: 651–659.

Furlong, C.E., Holland, N., Richter, R.J., Bradman, A., Ho, A., Eskenazi, B., 2006, PON1 status of farmworker mothers and children as a predictor of organophosphate sensitivity. *Pharmacogenet Genomics* 16: 183–190.

Furlong, C.E., Li, W.F., Richter, R.J., Shih, D.M., Lusis, A.J., Alleva, E., Costa, L.G., 2000, Genetic and temporal determinants of pesticide sensitivity: role of paraoxonase (PON1). *Neurotoxicology* 21: 91–100.

Gallo, M.A., Lawryk, N.J., 1991. Organic phosphorus pesticides. In *Handbook of Pesticide Toxicology. Classes of Pesticides* (E.R. Hayes Jr., E.R. Laws, eds.), vol. 2, Academic Press, San Diego, USA, pp. 917–1123.

Giordano, G., Afsharinejad, Z., Guizzetti, M., Vitalone, A., Kavanagh, T.J., Costa, L.G., 2006, Organophosphorus insecticides chlorpyrifos and diazinon and oxidative stress in neuronal cells in a genetic model of glutathione deficiency, *Toxicol Appl Pharmacol* doi:10.1016/j.taap.2006.09.016.

Gordon, M., Richter, E.D., 1991, Hazards associated with aerial spraying of organophosphate insecticides in Israel. *Rev Environ Health* 9: 229–238.

Gouedard, C., Barouki, R., Mosel, Y., 2004, Dietary polyphenols increase Paraoxonase 1 gene expression by an aryl hydrocarbon receptor dependent mechanism. *Mol Cell Biol* 24: 5209–5222.

Haley, R.W., 2003, Gulf war syndrome: narrowing the possibilities. *Lancet Neurol* 2: 272–273.

Haley, R.W., Billecke, S., La Du, B.N., 1999, Association of low PON1 type Q (type A) arylesterase activity with neurologic symptom complexes in Gulf War veterans. *Toxicol Appl Pharmacol* 157: 227–233.

Hernández, A.F., Gómez, M.A., Pena, G., Gil, F., Rodrigo, L., Villanueva, E., Pla, A., 2004, Effect of long-term exposure to pesticides on plasma esterases from plastic greenhouse workers. *J Toxicol Environ Health A* 67: 1095–1108.

Hernández, A.F., Gomez, M.A., Pérez, V., García-Lario, J.V., Pena, G., Gil, F., López, O., Rodrigo, L., Pino, G., Pla, A., 2006, Influence of exposure to pesticides on serum components and enzyme activities of cytotoxicity among intensive agriculture farmers. *Environ Res* 102: 70–76.

Hernández, A.F., López, O., Rodrigo, L., Gil, F., Pena, G., Serrano, J.L., Parrón, T., Álvarez, J.C., Lorente, J.A., Pla, A., 2005, Changes in erythrocyte enzymes in humans long-term exposed to pesticides Influence of several markers of individual susceptibility. *Toxicol Lett* 159: 13–21.

Hernández, A.F., Mackness, B., Rodrigo, L., López, O., Pla, A., Gil, F., Durrington, P.N., Pena, G., Parrón, T., Serrano, J.L., Mackness, M.I., 2003, Paraoxonase activity and genetic polymorphisms in greenhouse workers with long term pesticide exposure. *Hum Exp Toxicol* 22: 565–574.

Kamel, F., Hoppin, J.A., 2004, Association of Pesticide Exposure with Neurologic Dysfunction and Disease. *Environ Health Perspect* 112: 950–958.

Kelada, S.N., Eaton, D.L., Wang, S.S., Rothman, N.R., Khoury, M.J., 2003, The role of genetic polymorphisms in environmental health. *Environ Health Perspect* 111: 1055–1064.

Kono, Y., Fridovich, I., 1982, Superoxide radical inhibits catalase. *J Biol Chem* 257: 5751–5754.

Kuang, X.Y., Zhou, Z.J., Ma, X.X., Yao, F., Wu, Q.E., Chen, B., 2006, Activity of esterases and effect of genetic polymorphism in workers exposed to organophosphorus pesticides. *Zhonghua Lao Dong Wei Sheng Zhi Ye Bing Za Zhi* 24: 333–336.

La Du, B.N., Billecke, S., Hsu, C., Haley, R.W., Broomfield, C.A., 2001, Serum paraoxonase (PON1) isozymes: the quantitative analysis of isozymes affecting individual sensitivity to environmental chemicals. *Drug Metab Dispos* 29: 566–569.

Lee, B.W., London, L., Paulauskis, J., Myers, J., Christiani, D.C., 2003, Association between human paraoxonase gene polymorphism and chronic symptoms in pesticide-exposed workers. *J Occup Environ Med* 45 : 118–122.

Liu, Y.J., Huang, P.L., Chang, Y.F., Chen, Y.H., Chiou, Y.H., Xu, Z.L., Wong, R.H., 2006, GSTP1 genetic polymorphism is associated with a higher risk of DNA damage in pesticide-exposed fruit growers. *Cancer Epidemiol Biomarkers Prev* 15: 659–666.

López-Guarnido, O., 2005, Ph.D. Thesis *Influencia de la exposición crónica a plaguicidas sobre diversos marcadores bioquímicos (esterasas y enzimas antioxidantes) en trabajadores de invernadero de la costa oriental de Andalucía.* University of Granada, Spain.

Mackness, B., Durrington, P., Povey, A., Thomson, S., Dippnall, M., Mackness, M., Smith, T., Cherry, N., 2003, Paraoxonase and susceptibility to organo-phosphorus poisoning in farmers dipping sheep. *Pharmacogenetics* 13: 81–88.

Mackness, B., Mackness, M.I., Arrol, S., Turkie, W., Julier, K., Abuasha, B., Millar, J.E., Boulton, A.J., Durrington, P.N., 1998, Serum paraoxonase (PON1) 55 and 192 polymorphism and paraoxonase activity and concentration in non-insulin dependent diabetes mellitus. *Atherosclerosis* 139: 341–349.

Mueller, R.F., Hornung, S., Furlong, C.E., Anderson, J., Giblett, E.R., Motulsky, A.G., 1983, Plasma paraoxonase polymorphism: a new enzyme assay, population, family, biochemical, and linkage studies. *Am J Hum Genet* 35: 393–408.

Mutch, E., Williams, F.M., 2006, Diazinon, chlorpyrifos and parathion are metabolised by multiple cytochromes P450 in human liver. *Toxicology* 224: 22–32.

Ng, C.J., Shih, D.M., Hama, S.Y., Villa, N., Navab, M., Reddy, S.T., 2005, The paraoxonase gene family and atherosclerosis. *Free Radical Biol Med* 38: 153–163.

Ng, C.J., Wadleigh, D.J., Gangopadhyay, A., Hama, S., Grijalva, V.R., Navab, M., Fogelman, A.M., Reddy, S.T., 2001, Paraoxonase-2 is a ubiquitously expressed protein with antioxidant properties and is capable of preventing cell-mediated oxidative modification of low density lipoprotein. *J Biol Chem* 276: 44444–44449.

Padungtod, C., Niu, T., Wang, Z., Savitz, D.A., Christiani, D.C., Ryan, L.M., Xu, X., 1999, Paraoxonase polymorphism and its effect on male reproductive outcomes among Chinese pesticide factory workers. *Am J Ind Med* 36: 379–387.

Park, B.K., Breckenridge, A.M., 1981, Clinical implications of enzyme induction and enzyme inhibition. *Clin Pharmacokinet* 6: 1–24.

Parrón, T., Hernández, A.F., Pla, A., Villanueva, E., 1996, Clinical and biochemical changes in greenhouse sprayers chronically exposed to pesticides. *Hum Exp Toxicol* 15: 957–963.

Richter, E.D., Chuwers, P., Levy, Y., Gordon, M., Grauer, F., Marzouk, J., Levy, S., Barron, S., Gruener, N., 1992, Health effects from exposure to organophosphate pesticides in workers and residents in Israel. *Isr J Med Sci* 28: 584–598.

Rodrigo, L., Hernández, A.F., López-Caballero, J.J., Gil, F., Pla, A., 2001, Immunohistochemical evidence for the expression and induction of paraoxonase in rat liver, kidney, lung and brain tissue. Implications for its physiological role. *Chem Biol Interact* 137: 123–137.

Rothman, N., Wacholder, S., Caporaso, N.E., Garcia-Closas, M., Buetow, K., Fraumeni, J.F. Jr., 2001, The use of common genetic polymorphisms to enhance the epidemiologic study of environmental carcinogens. *Biochim Biophys Acta* 1471: C1–10.

Sözmen, E.Y., Mackness, B., Sözmen, B., Durrington, P., Girgin, F.K., Aslan, L., Mackness, M., 2002, Effect of organophosphate intoxications on human serum paraoxonase. *Hum Exp Toxicol* 21: 247–252.

Yamada, Y., Takatori, T., Nagao, M., Iwase, H., Kuroda, N., Yanagida, J., Shinozuka, T., 2001, Expression of paraoxonase isoform did not confer protection from acute sarin poisoning in the Tokyo subway terrorist attack. *Int J Leg Med* 115: 82–84.

PART 6

PON-GENETICS AND REGULATION

CHAPTER 15

TRANSCRIPTIONAL REGULATION OF THE PARAOXONASE GENES

S.P. DEAKIN AND R.W. JAMES

Division of Endocrinology, Diabetes and Nutrition, University Hospital, Geneva, Switzerland

Abstract: The paraoxonase (PON) gene family includes 3 members: PON1, PON2 and PON3, aligned in tandem on chromosome 7 in humans and chromosome 6 in mice. They are highly similar to each other and are thought to result from gene duplication. PON1 and PON3 are mainly expressed in the liver but have also been detected in kidney and intestine. PON2, thought to be the oldest in evolutionary terms, is more widely expressed and has been detected in many tissues including brain, liver, kidney, testis, intestine and heart. It is also strongly expressed in macrophages. Clearly transcriptional control of the PON genes is important in determining their function in the various tissues in which they are expressed. The majority of available data concentrates on the control of PON1 gene expression while control of the PON2 and PON3 genes is less well studied. This chapter will discuss transcriptional regulation of the paraoxonase genes with particular reference to recent advances at the molecular level

Keywords: promoter polymorphism, transcriptional regulation, statin, fibrate, polyphenol, bile acids, inflammation, cytokines, sterol regulatory element binding protein (SREBP) -2, Sp1, aryl hydrocarbon receptor (Arh), peroxisome proliferator activated receptor alpha (PPARα), farnesoid X receptor (FXR), fibroblast growth factor (FGF), nuclear factor-κB (NF-κB), nicotinamide adenine dinucleotide phosphate (NADPH) oxidase

1.   PON1 PROMOTER POLYMORPHISMS

Much of the research to date on Paraoxonase 1 gene expression has focused on single nucleotide polymorphisms (SNP) found within its promoter. Numerous SNPs have been identified within the PON1 promoter that have varying degrees of influence over gene expression (Brophy et al., 2001a; Carlson et al., 2006; Leviev and James, 2000; Suehiro et al., 2000). The best studied SNPs are located at positions -909 (C/G), -823 (A/G), -162 (A/G), -126 (C/G) and -108 (C/T) (Fig. 1a).

Luciferase reporter gene experiments in HepG2 cells have shown that promoters containing SNPs GAAC as opposed to CGGT at positions -909, -832, -162 and

B. Mackness et al. (eds.), The Paraoxonases: Their Role in Disease Development and Xenobiotic Metabolism, 241–250.

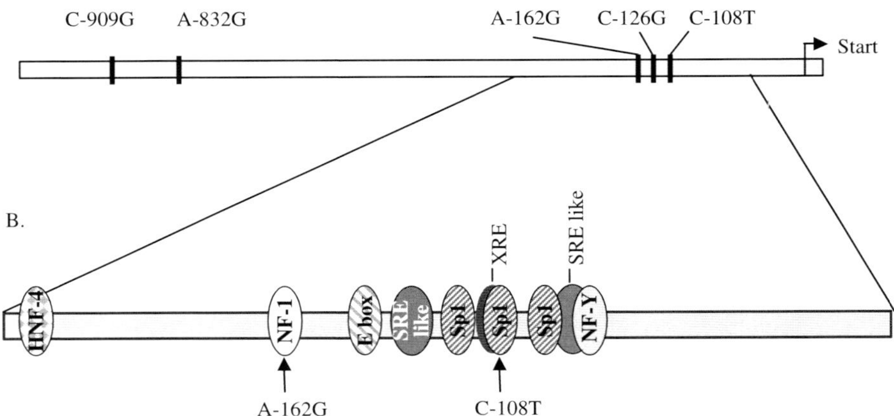

Figure 1. The PON1 promoter A) Locations of polymorphisms in the promoter of the PON1 gene. B) Enlargement of the approximately 200 base pair region of the promoter that is sufficient and necessary for transcription of the PON1 gene. The −162 and −108 polymorphisms and potential transcription factor binding sites are marked. NF-Y=nuclear factor-Y

−108 respectively, are up to two times more active. Genotyping of individuals has confirmed that these variations in promoter activity correlate with significant differences in serum PON1 concentration and activity (Brophy et al., 2001a; Leviev and James, 2000).

There is significant linkage disequilibrium between the PON1 promoter polymorphisms hampering the identification of clinically significant SNPs. However haplotype analysis of two populations indicates that the C(-108)T SNP is the main contributor to serum PON1 variation accounting for 23–24% of total variation. There is a slight contribution from the A(-162)G site (1.1% total variation) but the SNPs at −909 and −823 made little or no difference to serum PON1 levels (Brophy et al., 2001b; Deakin et al., 2003a).

The C(-108)T SNP is located in the centre of a consensus binding site for the ubiquitous transcription factor Sp1 (Fig. 1b). The binding site is abolished in the presence of the T variant suggesting that an alteration in Sp1 binding may be responsible for the effect of the SNP on promoter activity. Indeed *in vitro* DNA binding studies have clearly shown that interaction of Sp1 with the promoter is weaker in the presence of C than of T (Deakin et al., 2003a; Deakin et al., 2007). Co-transfection of the C variant of the promoter with a plasmid expressing Sp1 resulted in strong increase in expression. This up-regulation was mediated at least in part by protein kinase C that is known to activate Sp1 (Osaki et al., 2004). These data support the hypothesis that Sp1 binding is important in the transcriptional regulation of the PON1 gene.

The SNP A(-162)G also lies over a potential transcription factor binding site. In this case the nuclear factor-1 (NF-1) site is disrupted by the low activity G variant and maintained by the high activity A variant possibly explaining the effect on gene expression (Brophy et al., 2001a).

2. PHARMACOLOGICAL REGULATION OF PON1 GENE EXPRESSION

2.1. Statins

Due to its potential protective role in atherosclerosis the effect of lipid lowering drugs on PON1 expression is of considerable interest.

3-hydroxy-3-methylglutaryl-CoA reductase inhibitors (statins) are the most widely used lipid lowering drugs and can also display a number of pleiotropic properties including an antioxidant effect. Several small clinical trials have investigated the effect of statin treatment on human serum PON1. The majority of which have observed an increase in concentration and/or activity (Deakin and James, 2004).

In *in vitro* luciferase reporter gene assays simvastatin up-regulated PON1 promoter activity in a concentration dependent manner in HepG2 cells. This activation was attenuated by mevalonate and farensyl pyrophosphate, intermediates downstream of HMG Co-A reductase in the cholesterol synthesis pathway. Subsequent studies in HEK293 cells confirmed the simvastatin effect and showed that pitvastatin and atorvastatin can also up-regulate PON1 promoter activity suggesting that this is a class specific property (Deakin et al., 2003b; Ota et al., 2005). In contrast, Gouédard et al. found that treatment of HuH-7 cells with simvastatin, pravastatin or fluvastatin resulted in an inhibition of promoter activity of up to 40% (Gouédard et al., 2003). These conflicting results may be due to differences in the cell lines used or in the version of the promoter studied (see below and Deakin et al., 2007).

Using various deletions of the promoter in reporter assays, the statin responsive element was localised to a 127bp region at the proximal end of the promoter which contains both the C(-108)T and A(-162)G SNPs. In electrophoretic mobility shift assays (EMSAs) simvastatin treatment of cells increased binding of nuclear extracts to this region of the promoter suggesting that statins can directly influence binding of transcription factors to the PON1 promoter (Deakin et al., 2003b). Within the statin responsive element, there are two sequences with homology to the sterol regulatory element (SRE) to which the sterol regulatory element binding proteins (SREBPs) can bind (Fig. 1b). SREBP2 is particularly implicated in control of cholesterol metabolism and is known to be up-regulated in HepG2 cells treated with statin. Co-transfection of SREBP2 with the PON1 promoter strongly increased promoter activity by up to 18-fold in HepG2 cells. The region required for SREBP2 induction of the promoter was the same as that required for response to simvastatin, suggesting that the statin effect on PON1 may be mediated by increased interaction of SREBP2 with the promoter. SREBP2 is known to interact and co-operate with Sp1 during transcriptional activation and further studies revealed that it is able to strongly enhance the binding of Sp1 to the PON1 promoter (Deakin et al., 2003b).

As mentioned above the C(-108)T SNP lies within an Sp1 binding site and disrupts Sp1 binding to the promoter. This suggests that any interaction of SREBP2 with Sp1 would also be influenced by the polymorphism. DNase1 protection experiments with the PON1 promoter show that although Sp1 binding to the C allele of the promoter is enhanced by SREBP2, there is no effect of SREBP2 on the T allele. Co-transfections of PON1 promoter constructs with SREBP2 confirmed that only the C allele is up-regulated by SREBP2. No increase in promoter activity was detected when the T allele was used. If the response of the PON1 promoter to simvastatin treatment is mediated by SREBP2 one would expect that the C(-108)T polymorphism would also effect activation by statins. Reporter gene assays support this hypothesis, finding that the -108C allele of the PON1 promoter is more strongly up-regulated by simvastatin than the T allele (Deakin et al., 2007). *In vivo* a small patient study found that patients homozygous for the -108C allele showed an increase in PON1 activity and concentration after treatment with simvastatin, while those homozygous for the T allele did not (Deakin et al., 2007). Taken together these data suggest a pharmacogenetic interaction between statin and the PON1 gene. Patients with the C allele of the C(-108)T SNP are likely to derive greater benefit from statin therapy with respect to an increase in PON1 activity.

2.2.   Fibrates

Fibrates are a second class of drug widely used for serum lipid management. They act as agonists of the peroxisome proliferator activated receptor alpha (PPARα) a transcription factor involved in lipid metabolism. Treatment of HuH7 cells with fenofibric acid activated the PON1 promoter by up to 70% in luciferase reporter gene assays. This was confirmed by real time PCR analysis. Several other fibrates were tested with variable results. Bezafibrate slightly activated the promoter while clofibric acid and gemfibrozil did not, suggesting that up-regulation of the PON1 promoter is not a class effect (Gouédard et al., 2003).

These results are to some extent supported by epidemiological data in which treatment with fenofibrate but not gemfibrozil or bezafibrate increased PON1 enzyme activity. There is also, however, a study which found a positive influence of gemfibrozil treatment on PON1 activity (Deakin and James, 2004).

There are several potential PPARα binding sites on the PON1 promoter, however over-expression of PPARα in HuH7 cells had no effect on basal PON1 promoter activity and actually inhibited promoter activation by fenofibric acid. In further preliminary experiments over-expression of the retinoid X receptor (RXR), which forms a dimer with PPARα during transactivation, also failed to influence the PON1 promoter. Overall these data suggest that fenofibric acid does not increase PON1 gene expression via a direct interaction with PPARα/RXR. The actual mechanism of promoter activation is still to be elucidated (Gouédard et al., 2003).

3. DIETARY REGULATION OF PON1 GENE EXPRESSION

3.1. Polyphenols

It is generally accepted that consumption of fruit, vegetables and possibly red wine is beneficial for protection against heart disease. These foods contain polyphenolic compounds that are thought to be one of the factors responsible for their athero-protective effect. Recent studies have indicated that wine consumption and certain polyphenols present in wine or fruit juice can increase serum PON1 activity in both mice and humans (Deakin and James, 2004). PON1 is susceptible to inactivation by oxidation and it is possible that the antioxidant polyphenols help to maintain enzyme activity. In addition to being antioxidants, polyphenol compounds are able to modulate gene expression via interactions with the nuclear receptor AhR (aryl hydrocarbon receptor). AhR is a ligand activated transcription factor. Classically its ligands include synthetic xenobiotics and polycyclic aromatic hydrocarbons but more recently dietary polyphenols such as naringenin, flavone and quercetin have been identified as potential receptor agonists.

In HuH7 cells treated with naringenin, flavone and quercetin, northern blot analysis and reporter gene assays showed that PON1 gene expression was increased up to 2-fold. Truncated constructs of the promoter fused to a reporter gene identified a 20bp sequence between nucleotides −126 and −108 to be the region predominantly responsible for activation by polyphenols. Mutation of this region lead to a significant reduction in promoter up-regulation by quercetin, further confirming its importance (Gouédard et al., 2004a).

AhR gene regulation is mediated by binding of the receptor to xenobiotic responsive elements (XREs) that have a consensus sequence of GCGTG. Although there are no exact matches to the XRE in the PON1 promoter there are several motifs similar to the consensus sequence, one of which is located between positions −112 and −108 (Fig. 1b). Over-expression of AhR in transient transfection experiments resulted in an increase in activation of the PON1 promoter by naringenin, flavone and quercetin. Conversely down-regulation of AhR expression with siRNA abolished the pholyphenol mediated response. A direct interaction between AhR and the polyphenol responsive element of the PON1 promoter was demonstrated using EMSAs and binding of AhR to the promoter was increased in the presence of quercetin (Gouédard et al., 2004a).

Resveratrol a polyphenolic phytoalxin found in grapes and wine has also been shown to activate PON1 gene expression via AhR (Gouédard et al., 2004b).

Overall these data show that, at least *in vitro*, polyphenols are able to up-regulate the PON1 promoter by acting as ligands for AhR, which interacts directly with the promoter. This suggests that consumption of wine and fruit juices can positively influence PON1 gene expression.

As mentioned above, ligands for AhR also include synthetic xenobiotics and polycyclic aromatic hydrocarbons. Some of these, e.g. 3-methylcholanthrene and to a lesser extent TCDD (2, 3, 7, 8-tetrachlorobenzo(p)dioxin) are also able to up-regulate the PON1 promoter (Gouédard et al., 2004a). This is intriguing given the role of the PON1 enzyme in detoxifying certain xenobiotic toxins.

4.   ATHEROGENIC DIET AND BILE ACIDS

In both rabbits and transgenic mouse models exposure to a high fat/high cholesterol diet resulted in a significant drop in PON1 activity and/or mass. Although the effect of a pro-atherogenic diet on human PON1 is less clear several studies report a reduction in PON1 activity in response to diets with a high fat content (Deakin and James, 2004).

In a study of atherosclerosis susceptible C57BL/6J mice it was noted that the cholic acid content of the atherogenic diet was the main factor responsible for the fall in PON1 activity. This observation led to the hypothesis that bile acids may have a negative effect on PON1 gene expression (Shih et al., 1996).

Bile acids are endogenous agonists for the nuclear hormone receptor, FXR (farnesoid X receptor) that, in co-operation with RXR, regulates the expression of genes involved in bile acid and lipid homeostasis. Activation of FXR/RXR by bile acids induces the expression of fibroblast growth factor FGF-15 or its human homologue FGF-19 which signals through the cell surface receptor tyrosine kinase, fibroblast growth factor receptor-4 (FGFR-4).

In wild type mice a diet containing 1% cholic acid resulted in a 37% reduction in PON1 mRNA, while in FXR knockout mice PON1 expression was unaffected. *In vitro* treatment of the human cell line HepG2 with FXR agonists induced a 65% decrease in PON1 mRNA. FGFR-4 knockout mice were also unable to repress PON1 mRNA in response to bile acids. Treatment of HepG2 cells with recombinant FGF-19 repressed PON1 mRNA levels in a dose dependant manner and this repression was attenuated by a c-Jun N-terminal kinase (c-JNK) inhibitor. Together these data suggest that bile acids repress PON1 gene expression via FXR induction of FGF15/19. Binding of FGF15/19 to the FGFR-4 activates the c-JNK pathway which may then play a role in inhibiting PON1 expression (Gutierrez et al., 2006; Shih et al., 2006).

The exact mechanism by which FGFR-4 and cJNK activation regulates PON1 promoter activity is still unclear. Promoter reporter gene assays in HepG2 cells localized FXR-dependent repression to a region between -230bp and -96bp however there is no FXR responsive element within this sequence. There is a putative binding site for the hepatocyte nuclear factor-4 (HNF-4) between −206 and −194 (Fig. 1b). In other studies HNF-4 was inactivated via the FXR pathway. It is therefore possible that inactivation of HNF-4 is involved in the inhibition of PON1 promoter activity (Shih et al., 2006).

In conclusion an increase in the bile acid pool, in response to a diet high in saturated fats and cholesterol may be responsible for the observed reduction in serum PON1 activity and concentration via the mechanism described above.

5. INFLAMMATORY REGULATION OF PON1 GENE EXPRESSION

Increasing evidence links inflammation to an elevated risk of atherosclerosis. During inflammation and particularly during the acute phase response there is a reduction in several HDL proteins important in reverse cholesterol transport and inhibiting oxidation. These include cholesterol ester transfer protein, lecithin cholesterol acyltransferase, hepatic lipase, ApoAI and PON1. It is thought that reduction in these proteins accompanied by an increase in proteins such as ApoJ and serum amyloid A (SAA) changes the HDL from an anti-inflammatory to a pro-inflammatory particle (Van Lenten et al., 2006).

Several studies provide evidence that hepatic PON1 mRNA is down-regulated in response to inflammatory cytokines. Treatment with lipopolysaccharide and oxidized LDL, known to induce an immune response, reduced PON1 mRNA in both animal models and human hepatocyte cells lines. A similar effect was see with the cytokines tumor necrosis factor alpha (TNFα), interleukin-1 (IL-1) and interleukin-6 (IL-6) (Feingold et al., 1998; Kumon et al., 2002; Kumon et al., 2003; Van Lenten et al., 2001). Injection of oxidized phospholipids into wild type mice repressed PON1 gene expression. This effect was abolished in IL-6 knockout mice suggesting a central role for IL-6 in the PON1 response (Van Lenten et al., 2001). Interestingly, the sequence surrounding the -162 PON1 promoter polymorphism displays some homology with an IL-6 responsive element, however it is not yet clear if this region is involved in cytokine mediated down-regulation of PON1 mRNA (Brophy et al., 2001b).

A recent study by Han and colleagues examined the regulation of ApoAI, PON1 and SAA. Stimulation of several murine hepatic cell lines with a mixture of cytokines (TNFα, IL-1β and IL-6) resulted in an increase in expression of SAA in parallel with a decrease in ApoAI and PON1 mRNA. There are nuclear factor-κB (NF-κB) binding sites on the promoter of SAA but not on the ApoAI or PON1 promoters. Treatment of murine hepatic cells with cytokines in the presence of NF-κB inhibitors or repressors inhibited the increase of SAA mRNA. Surprisingly the down-regulation of PON1 and ApoAI was also attenuated. Similar results were obtained in a mouse transgenic model (Han et al., 2006).

PPARα is known to limit inflammation by interfering with NF-κB activation. In contrast to NF-κB, there are PPARα response elements in the PON1 and ApoAI promoters but not in the SAA promoter. Incubation of cells with the cytokine mix, in combination with PPARα agonists reversed the cytokine effects, inhibiting both the increase in SAA mRNA and decrease in PON1 and ApoAI mRNA. Interestingly there was only a minor effect of the PPARα agonists on PON1 and ApoAI expression in the absence of cytokine (Gouédard et al., 2003; Han et al.,

2006). The results suggest that cytokines both up-regulate SAA and down-regulate PON1 and ApoAI via NF-κB. Activation of PPARα antagonizes NF-κB leading to an attenuation of the inflammatory response. It is not clear if PPARα interacts directly with the PON1 promoter or if it acts via an as yet unidentified factor.

6.   TRANSCRIPTIONAL REGULATION OF PON2 AND PON3

Paraoxonases 2 and 3 are not as extensively studied as PON1 and consequently there is less available data on their transcriptional regulation.

Paraoxonase-3 was the last of the paraoxonases to be characterized. To date almost nothing is known about its transcriptional regulation, however it appears that it not influenced by oxidative stress or by an atherogenic diet (Reddy et al., 2001).

PON2 is almost ubiquitously expressed *in vivo*, however its expression in macrophages has been the most thoroughly investigated. A study of human monocyte macrophages taken from hypocholesterolemic patients revealed that PON2 mRNA was lower in patient monocytes compared to those of controls. Atorvastatin therapy increased PON2 mRNA content both *in vivo* and *in vitro* suggesting that like PON1, PON2 can be regulated by statins. The mechanism for the statin effect on PON2 has not yet been elucidated but it may be linked with the ability to lower the cholesterol content of the cell. Incubation of monocytes from healthy controls with acetylated LDL increased cellular cholesterol content by up to 77% and was accompanied by a corresponding decrease in PON2 mRNA (Rosenblat et al., 2004).

In contrast to cellular cholesterol, cellular oxidative stress up regulated PON2 expression. ApoE deficient mice are exposed to considerably greater oxidative stress than their wild type littermates. Peritoneal macrophages from these mice have increased levels of PON2 mRNA. Similarly PON2 expression was increased in peritoneal macrophages from wild type mice treated with oxidized phospholipids and other pro-oxidant agents (Rosenblat et al., 2003).

Differentiation of human monocytes to macrophages is associated with an increase in the production of superoxide anions via the reduced nicotinamide adenine dinucleotide phosphate (NADPH) oxidase. As PON2 mRNA is increased by oxidative stress it was hypothesized that PON2 expression may be induced during macrophage differentiation. PON2 mRNA was measured during differentiation of the human pre-monocytic cell line THP-1 and was found to increase gradually as differentiation progressed. Treatment of the THP-1 cells with vitamin E reduced superoxide anion production by 21%. In parallel PON2 expression decreased by 15% suggesting that the increase in PON2 mRNA during macrophage differenti-ation is indeed linked to superoxide anion formation. Similar results were obtained *in vivo* using mouse models (Shiner et al., 2004).

The role of NADPH oxidase was confirmed using mice lacking the enzyme activity. In these mice PON2 mRNA extracted from peritoneal macrophages was significantly lower than in controls and its induction was weaker during maturation (Shiner et al., 2004).

The PON2 promoter contains 3 potential response elements for the transcription factor AP-1. The AP-1 subunit c-Jun is activated by c-JNK. Inhibition of c-JNK resulted in a dose-dependent decrease in macrophage PON2 expression suggesting that AP-1 may be involved in the control of PON2 expression (Shiner et al., 2004).

Monocyte differentiation into macrophages is an inflammatory event and can play a role in the initiation of atherosclerosis. It is possible that the increase in PON2 is a mechanism to compensate against the rise in oxidative stress that occurs during differentiation.

Transcriptional regulation of the PON genes is clearly complex, involving several different pathways which may or may not interact depending upon the *in vivo* situation. More work is required to fully elucidate the regulatory mechanisms controlling PON transcription. This should improve our understanding of the interaction of external factors such as drugs and diet with the PON genes. It may also give us some insight into the elusive physiological roles of these intriguing enzymes.

REFERENCES

Brophy, V. H., Hastings, M. D., Clendenning, J. B., Richter, R. J., Jarvik, G. P. and Furlong, C. E. (2001a). Polymorphisms in the human paraoxonase (PON1) promoter. *Pharmacogenetics* 11, 77–84.

Brophy, V. H., Jampsa, R. L., Clendenning, J. B., McKinstry, L. A., Jarvik, G. P. and Furlong, C. E. (2001b). Effects of 5' regulatory-region polymorphisms on paraoxonase-gene (PON1) expression. *Am J Hum Genet* 68, 1428–36.

Carlson, C. S., Heagerty, J. H., Hatsukami, T. S., Richter, R. J., Ranchalis, J., Lewis, J., Bacus, T. J., McKinstry, L. A., Schellenberg, G. D., Rieder, M., Nickerson, D., Furlong, C. E., Chait, A. and Jarvik, J. P. (2006). TagSNP analyses of the PON gene cluster: effects on PON1 activity, LDL oxidative susceptibility, and vascular disease. *J. Lipid Res.* 47, 1014–1024.

Deakin, S., Leviev, I., Brulhart Meynet, M. C. and James, R. W. (2003a). Paraoxonase-1 promoter haplotypes and serum paraoxonase: a predominant role in vivo for polymorphic position −107 implicating the transcription factor Sp1. *Biochem J* 372, 643–649.

Deakin, S., Leviev, I., Guernier, S. and James, R. W. (2003b). Simvastatin modulates expression of the PON1 gene and increases serum paraoxonase: a role for sterol regulatory element-binding protein-2. *Arterioscler Thromb Vasc Biol* 23, 2083–9.

Deakin, S. P., Guernier, S. and James, R. W. (2007). Parmacogenetic interaction between paraoxonase-1 gene promoter polymorphism C-107T and statin. *Pharmacogenetics and Genomics* 17: 451–7.

Deakin, S. P. and James, R. W. (2004). Genetic and environmental factors modulating serum concentrations and activities of the antioxidant enzyme paraoxonase-1. *Clin. Sci. (Lond.)* 107, 435–447.

Feingold, K. R., Memon, R. A., Moser, A. H. and Grunfeld, C. (1998). Paraoxonase activity in the serum and hepatic mRNA levels decrease during the acute phase response. *Atherosclerosis* 139, 307–15.

Gouédard, C., Barouki, R. and Morel, Y. (2004a). Dietary polyphenols increase paraoxonase 1 gene expression by an aryl hydrocarbon receptor-dependant mechanism. *Mol Cell biol.* 24, 5209–5222.

Gouédard, C., Barouki, R. and Morel, Y. (2004b). Induction of the Paraoxonase-1 Gene Expression by Resveratrol. *Arterioscler. Thromb. Vasc. Biol.* 24, 2378–2383.

Gouédard, C., Kcum-Besson, N., Barouki, R. and Morel, Y. (2003). Opposite regulation of the human paraoxonase-1 gene PON-1 by fenofibrate and statins. *Mol Pharmacol* 63, 945–56.

Gutierrez, A., Ratliff, E. P., Andres, A. M., Huang, X., McKeehan, W. L. and Davies, R. A. (2006). Bile acids decrease hepatic paraoxonase 1 expression and plasma high-density lipoprotein lebels via FXR-Mediated Signaling of FGFR4. *Arterioscler. Thromb. Vasc. Biol.* 26, 301–306.

Han, C. Y., Chiba, T., Campbell, J. S., Fausto, N., Chaisson, M., Orasanu, G., Plutzky, J. and Chait, A. (2006). Reciprocal and Coordinate Regulation of Serum Amyloid A Versus Apolipoptorein A-I and Paraoxonase-I by Inflammation in Murine Hepatocytes. *Arterioscler. Thromb. Vasc. Biol.* 26, 1806–1813.

Kumon, Y., Nakauchi, Y., Suehiro, T., Shiinoki, T., Tanimoto, N., Nakamura, T., Hashimoto, K. and Sipe, J. D. (2002). Proinflammatory cytokines but not acute phase serum amyloid A or c-reactive protein, downregulate paraoxonase (PON1) expression by HepG2 cells. *Amyloid* 9, 160–164.

Kumon, Y., Suehiro, T., Ikeda, Y. and Hashimoto, K. (2003). Human paraoxonase-1 gene expression by HepG2 cells is downregulated by interleukin-1beta and tumor necrosis factor-alpha, but is upregulated by interleukin-6. *Life Sci.* 73, 2807–2815.

Leviev, I. and James, R. W. (2000). Promoter polymorphisms of the human paraoxonase PON1 gene and serum paraoxonase activities and concentrations. *Arterioscler Thromb Vasc Biol* 20, 516–21.

Osaki, F., Ikeda, Y., Suehiro, T., Ota, K., Tsuzura, S., Arii, K., Kumon, Y. and Hashimoto, K. (2004). Roles of Sp1 and protein kinase C in regulation of human serum paraoxonase 1 (PON1) gene transcription in HepG2 cells. *Atherosclerosis* 176, 279–287.

Ota, K., Suehiro, T., Arii, K., Ikea, Y., Kumon, Y., Osaki, F. and Hashimoto, K. (2005). Effect of pitavastatin on transactivation of human serum paraoxonase 1 gene. *Metabolism* 54, 142–150.

Reddy, S. T., Wadleigh, D. J., Grijalva, V., Ng, C., Hama, S., Gangopadhyay, A , Shih, D. M., Lusis, A. J., Navab, M. and Fogelman, A. M. (2001). Human paraoxonase-3 is an HDL-associated enzyme with biological activity similar to paraoxonase-1 protein but is not regulated by oxidized lipids. *Arterioscler Thromb Vasc Biol* 21, 542–7.

Rosenblat, M., Draganov, D., Watson, C. E., Bisagier, C. L., La Du, B. N. and Aviram, M. (2003). Mouse macrophage paraoxonase 2 activity is increased whereas cellular paraoxonase 3 activity is decreased under oxidative stress. *Arterioscler. Throm. Vasc. Biol.* 23, 468–474.

Rosenblat, M., Hayek, T., Hussein, K. and Aviram, M. (2004). Decreased macrophage Paraoxonase 2 Expression in Patients With Hypercholesterolemia Is the Result of Their Increased Cellular Cholesterol Content: Effect of Atorvastatin Therapy. *Arterioscler. Thromb. Vasc. Biol.* 24, 175–180.

Shih, D. M., Gu, L., Hama, S., Xia, Y.-R., Navab, M., Fogelman, A. M. and Lusis, A. J. (1996). Genetic-dietary regulation of serum paraoxonase expression and its role in atherogenesis in a mouse model. *J Clin Invest* 97, 1630–9.

Shih, D. M., Kast-Woelbern, H. R., Wong, J., Xia, Y.-R., Edwards, P. A. and Lusis, A. J. (2006). A role for FXR and human FGF-19 in the repression of paraoxonase-1 gene expression by bile acids. *J. Lipid Res.* 47, 384–392.

Shiner, M., Fuhrman, B. and Aviram, M. (2004). Paraoxonase 2 (PON2) expression is upregulated via a reduced-nicotinamide-adenine-dinucleotide-phosphate (NADPH)-oxidase-dependent mechanism during monocytes differentiation into macrophages. *Free Radic. Biol. Med.* 37, 2052–2063.

Suehiro, T., Nakamura, T., Inoue, M., Shiinoki, T., Ikeda, Y., Kumon, Y., Shindo, M., Tanaka, H. and Hashimoto, K. (2000). A polymorphism upstream from the human paraoxonase (PON1) gene and its association with PON1 expression. *Atherosclerosis* 150, 295–8.

Van Lenten, B. J., Reddy, S. T., Navab, M. and Fogelman, A. M. (2006). Understanding Changes in High Density Lipoproteins During the Acute Phase Response. *Aterioscler. Thromb. Vasc. Biol.* 26, 1687–1688.

Van Lenten, B. J., Wagner, A. C., Navab, M. and Fogelman, A. M. (2001). Oxidized phospholipids induce changes in hepatic paraoxonase and ApoJ but not monocyte chemoattractant protein-1 via interleukin-6. *J Biol Chem* 276, 1923–9.

CHAPTER 16

EFFECT OF LIPID LOWERING MEDICATIONS ON PON1

G. PARAGH, M. HARANGI AND I. SERES

1st Department of Medicine, Medical and Health Science Center, University of Debrecen, Hungary
Email: paragh@internal.med.unideb.hu

Abstract: Lipid abnormalities are among the most important risk factors of the development of atherosclerosis. Inhibitors of 3-hydroxy-methyglutaryl-coenzyme A reductase (statins) and peroxisome proliferator-activated receptor (PPAR) alpha ligand (fibrates) are widely used as lipid lowering drugs in patients with disturbed lipid metabolism. Apart from reducing plasma lipid levels, these agents have additional beneficial effects on other processes involved in the atherosclerotic process. In the last few years the favourable effect of these drugs on human paraoxonase-1 activity was extensively investigated both in in vitro and in vivo studies. This chapter summarizes the results of in vivo and in vitro studies, and the putative mechanisms of altered PON1 activity caused by statin and fibrate administration. The possible causes of discrepancies in currently available data will be also discussed

Keywords: statin, fibrate, lipid lowering

1.   LIPID LOWERING MEDICATION – NEW GUIDELINES: AGGRESSIVE MANAGEMENT

Cardiovascular disease (CVD) is currently the leading cause of morbidity and mortality worldwide and its incidence is likely to increase. Multiple risk factors contribute to CVD. Elevated LDL-cholesterol (LDL-C) and triglyceride levels, and low HDL-cholesterol (HDL-C) levels are key modifiable risk factors (Ballantyne et al., 2005). In the last decades various therapeutic strategies that target lowering of LDL-C or augmentation of HDL-C have been employed to prevent atherogenesis (Saini et al., 2005). Of among these strategies fibrates and statins are the clinically most important and widely used agents. Primary and secondary prevention trials with lipid lowering agents in a wide variety of populations have demonstrated that lowering plasma LDL-C levels retards the progression of atherogenesis and reduces the risk of coronary events.

251

*B. Mackness et al. (eds.), The Paraoxonases: Their Role in Disease Development
and Xenobiotic Metabolism, 251–266.*
© 2008 *Springer.*

It must be noted that none of these trials have achieved the optimal reduction in LDL-cholesterol concentration for high-risk patients as recommended by the third report of the National Cholesterol Education Program (NCEP) Adult Treatment Panel III (ATP III) guidelines. The recent update of the NCEP is the most aggressive approach to date for reducing LDL-cholesterol. A basic element of the update is the modification of LDL-cholesterol goal in very high risk patients to 70 mg/dl (Grundy et al., 2004). Overall, the trend continues towards a more aggressive management of patients with atherogenic lipid profile, particularly if they fall with in the high risk or very high risk catergories. These studies have been carried out using 3-hydroxy-3-methylglutaryl coenzyme A (HMG-CoA) reductase inhibitors (statins). It has been proved that high doses of statins are needed for aggressive lipid lowering but these were found to have a greater risk of causing adverse effects (Grundy et al., 2004; Saini et al., 2005).

On the other hand, previous experimental and clinical evidence demonstrates that the antiatherogenicity of statins also includes cholesterol-independent pleiotropic effects. Such effects include improvement of endothelial dysfunction, increased nitric oxide bioavailability, antioxidant effects, antiinflammatory properties, stabilization of atherosclerotic plaques, the ability to recruit endothelial progenitor cells, immunosuppressive activity, and inhibition of cardiac hypertrophy. Many of these pleiotropic effects operate independently of LDL-cholesterol reduction, correlate poorly or not at all with LDL-cholesterol changes (Davidson, 2005).

The dyslipidemia associated with obesity, metabolic syndrome, insulin resistance, and type 2 diabetes consists of increased triglyceride, reduced high-density lipoprotein (HDL) cholesterol, and increased numbers of small, dense low-density lipoprotein (LDL) particles (Krauss and Siri, 2004). The inverse relationship between plasma levels of HDL and coronary heart disease has been demonstrated in a number of observational epidemiological studies, as well as in several interventional studies, which showed that increased HDL concentrations independently predicted lowered risk of coronary artery disease (Ballantyne et al., 2005). Fibrates have been shown to activate peroxisome proliferator-activated receptor alpha (PPARα)/ retinoid X receptor (RXR) signal transduction pathway and cause a significant decrease in serum triglyceride and increase in HDL-cholesterol levels. Results of several clinical trials showed that fibrates attenuate atherosclerosis and reduce the incidence of cardiovascular death, myocardial infarction and stroke in patients with coronary artery disease (CAD). In addition it has been demonstrated that fibrates prevent progression of CAD in diabetic patients (Israelian-Konaraki and Reaven, 2005).

In the last few years, because of the limited effectivity and the risk of adverse effects, targets other than LDL-cholesterol and triglyceride lowering ones have been suggested to increase HDL-cholesterol levels, including the cholesterol absorption inhibitor ezetimibe, acyl-cholesterol acyl transferase (ACAT) inhibitors and cholesterol ester transfer protein (CETP) inhibitors. Furthermore, the multicausal nature of atherosclerotic diseases including oxidative processes, inflammation, endothelial dysfunction, immune processes and other cellular events indicates new therapeutical possibilities for the prevention of atherogenesis, especially agents associated with antiinflammatory and antioxidant effects (Saini et al., 2005).

2. HUMAN PARAOXONASE 1 (PON1) ACTIVITY AND LIPID LOWERING MEDICATIONS – MORE QUESTIONS THAN ANSWERS

Although the clinical trials have provided strong evidence that lowering plasma LDL-cholesterol and raising HDL-cholesterol reduce the risk for cardiovascular events, information on the antioxidant activity of common lipid lowering drugs is suprisingly scanty. Since PON1 is the most potent HDL-associated antioxiant enzyme, and PON1 activity has been shown to correlate negatively to cardiovascular risk, it is necessary to clarify the effect of these drugs on PON1 activity. In recent years numerous clinical studies have been published about the effects of statins and fibrates on PON1 activity. In what follows we summarize the most important data of the available studies on lipid lowerings and PON1 activity.

3. EFFECT OF STATINS ON PON1 ACTIVITY

3.1. Statins: The Gold Standard of LDL Cholesterol Lowering

The HMG-CoA reductase inhibitors (statins) were developed to compete with HMG-CoA for binding at the catalytic site of HMG-CoA reductase and thereby reduce the synthesis of mevalonate. The plasma cholesterol lowering effect of these drugs results mainly from the enhanced receptor-mediated uptake of LDL in the liver via the upregulation of LDL-receptor (LDL-R). The newer statins, such as atorvastatin and rosuvastatin are more effective and are able to reduce plasma cholesterol by up to 60% (Kleemann and Kooistra, 2005). The efficacy of HMG-CoA reductase inhibitors in lowering serum cholesterol levels is well documented. In addition to their lipid-lowering effect, statins and their metabolites may have an antioxidant potential (Aviram and Rosenblat, 2005).

There is extensive evidence that links hypercholesterolemia with increased lipid peroxidation and increased oxidative stress (Morrow, 2005). The oxidative modification of lipoproteins – particularly LDL – has emerged as a fundamental process in the development of atherosclerosis. Oxidatively modified LDL that has been heralded as an initiating factor in atherogenesis, possesses numerous unfavourable biological effects, including induction of endothelial dysfunction, activation of endothelial adhesiveness, monocyte differentiation and adhesion, and smooth muscle cell proliferation. Several studies suggest a relationship between oxidized LDL and severity of atherosclerosis in coronary arteries (Heinecke, 2006).

Only a few studies have shown the ability of statins to reduce the levels of circulating oxidized LDL or other measures of LDL oxidation, such as circulating conjugated dienes or malondialdehyde (MDA). The pioneer work by Aviram et al. in an in vitro study using several oxidation systems on isolated lipoproteins demonstrated that oxidized metabolites of atorvastatin (5–50 μM) but not the parent compound exerts inhibitory effect on lipoprotein oxidation (Aviram et al., 1998). Several in vitro studies proved that statins show a dose-dependent decrease in free radical production and its effect can be reversed by mevalonate (Giroux et al., 1993). The majority of these studies were done on animal models and their results are contradictory.

Fuhrman et al. studied the effect of one month atorvastatin therapy on isolated monocytes of hypercholesterolemic patients. Atorvastatin was found to increase serum paraoxonase activity, and reduce free radical-induced lipid peroxidation and scavenger receptor expression (Fuhrman et al., 2002).

The question is how can statins affect paraoxonase activity and serum paraoxonase enzyme concentrations? What is the effect of statins on HDL level and paraoxonase activity?

3.2.   Clinical Studies: Data on Simvastatin and Atorvastatin

Interestingly, to date only simvastatin and atorvastatin have been used in clinical studies (Table 1.).

All the studies demonstrated significant LDL-C lowering effect of statins. In each study the HDL-C level was increased. Only Kural and coworkers demonstrated that 10 mg atorvastatin significantly increased the HDL-C level unlike the other studies which could not demonstrate significant HDL-C elevation (Kural et al., 2004). Tomás et al. proved that a 4-month simvastatin treatment could increase PON1 activity, thereby improving the antiatherogenic effect of this statin (Tomás et al., 2000). However, Balogh et al. found a nonsignificant decrease in PON1 activity after one month of simvastatin treatment in patients with types IIa and IIb hyperlipidemia (Balogh et al., 2001a). Paragh et al investigated the effects of a 3-month treatment with simvastatin and atorvastatin on PON1 activity in hyperlipidemic patients. They found that a short term administration of simvastatin did not influence the activity of PON1, while atorvastatin significantly increased it (Paragh et al., 2004). Tsimihodimos et al. studied the effect of atorvastatin therapy on the activities of serum platelet-activating factor acetylhydrolase (PAF-AH) and PON1 in patients with dyslipidemia of types IIa and IIb. They found that atorvastatin did not affect PON1 activities either towards paraoxon or phenylacetate in both patient groups (Tsimihodimos et al., 2002). Harangi et al. found that atorvastatin was able to improve serum PON1 activity and PON1/HDL ratio in type IIa hyperlipidemic patients, while the serum PON1 concentration was not significantly decreased after atorvastatin therapy (Harangi et al., 2004). Kassai et al. reported a significant increase in PON1 activity after 3 months of atorvastatin administration in patients with dyslipidemia of types IIa and IIb (Kassai et al., 2006). The paraoxonase activity changes were between 12.3–37.9%. The reasons for these differences could be attributed to the diverse study populations, different types of hyperlipoproteinaemia and various types of statins and their dosages. To assess whether the observed rise in paraoxonase activity was due to HDL elevation the PON/HDL ratio was calculated. Statin effect on paraoxonase activity seems to be independent of HDL elevation.

Table 1. Clinical studies on the effect of statins on human paraoxonase-1 (PON1) activity

Author, year of publication	Statin (dose)	Patients (number)	Treatment period	PON1 activity Δ % (p)	PON1/HDL Δ % (p)
Statins					
Tomás et al. 2000.	Simvastatin (20 mg/d)	familial hyperc-holesterinemia (64)	4 months	↑12.3% (p = 0.005)	↑ 6.1% (p < 0.05)
Balogh et al. 2001a.	Simvastatin (20 mg/d)	IIa and IIb hypleripidemia (112)	4 weeks	↓ 8.65% (n.s.)	↓ 10.9% (n.s.)
Tsimihodimos et al. 2002.	Atorvastatin (20 mg/d)	IIa hyperlipidemia (40)	4 months	↓ 0.5% (n.s.)	↑ 1.5% (n.s.)
,,	,,	IIb hyperlipidemia (36)	,,	↑ 3.8% (n.s.)	↓ 3.99% (n.s.)
Deakin et al. 2003.	Simvastatin (10/20 mg/d)	hypercholesterolemia, CHD (21)	6.7 weeks	↑ 22.8% (p < 0.01)	↑ 12.8% (p < 0.05)
Kural et al. 2004.	Atorvastatin (10 mg/d)	mixed hyperlipidemia (40)	no data	↑ 23.6% (p < 0.05)	↑ 19.1% (p < 0.05)
Paragh et al. 2004.	Simvastatin (20 mg/d)	IIa és IIb hyperlipidemia (49)	8 weeks	↑ 25.1% (p < 0.05)	↑ 14.4% (p < 0.05)
	Atorvastatin (10 mg/d)	,,	8 weeks	↑ 3.6% (n.s.)	↑ 1.8% (n.s.)
Harangi et al. 2004.	Atorvastatin (10 mg/d)	IIa hyperlipidaemia (13)	6 months	↑ 37.9% (p < 0.05)	↑ 8.7% (p < 0.05)
Kassai et al. 2006.	Atorvastatin ((20 mg/d)	and IIb hypleripidemia (33)	3 months	↑ 21.2% (p < 0.001)	↑ 30.7% (p < 0.01)

Kural et al. showed that PON1 activity in serum and isolated HDL was significantly increased after atorvastatin treatment in dyslipidemic patients (Kural et al., 2004). A recent study on 134 patients with familial hypercholesterinaemia demonstrated that PON1 modifies the HDL-C increment during statin therapy (Himbergen et al., 2005)

3.2.1. Pon1 lactonase activity and statins

Although PON1 catalyzes the hydrolysis of multiple substrates it has been shown that PON1 and the other PONs are in fact lactonases, catalyzing both the hydrolysis and formation of a variety of lactones (Draganov et al., 2005; Khersonsky and Tawfik, 2006). Beltowski et al investigated the effect of a natural statin, pravastatin and that of a synthetic one, fluvastatin on plasma PON1 lactonase activity. Pravastatin tended to decrease the gamma-decanolactone hydrolizing activity of plasma, though the effect was not significant. Fluvastatin administration (2 and 20 mg/kg/day) lowered gamma-decanolactonase activity in a dose-dependent manner (Beltowski et al., 2002b). Further studies will be necessary to elucidate the effect of statin on lactonase activity of PON1.

3.3. Animal Studies

Beltowski et al. investigated the effect of a natural statin, pravastatin, and a synthetic one, fluvastatin on plasma PON1 activity in normolipidemic Wistar rats. They found that fluvastatin (20 mg/kg/day for 3 weeks) reduced both plasma and liver PON1 activities, while a lower dose (2 mg/kg/day) decreased only liver PON1 activity. On the other hand, pravastatin (4 mg or 40 mg/kg/day) had no significant effect on PON1 activity. They concluded that natural statins have no effect on PON1 whereas synthetic ones decrease its activity independently of their effect on plasma lipids (Beltowski et al., 2002a).

The effect of cerivastatin on plasma PON1 activity was investigated. The adult male Wistar rats received cerivastatin at a dose of either 0.03 or 0.3 mg/kg/day for 3 weeks. PON1 activity on paraoxon was decreased by 16.1% and 11.6% in low- and high-dose groups, respectively. PON/arylesterase ratio was raised in the cerivastatin-treated rats by 227.4% and 328.2% (Beltowsky et al., 2002a).

The effect of statins on paraoxonase activity in different species can be varied. The question is: how do statins influence PON1 activity in human cell lines and what is the putative mechanism of statin effect on PON1 activity?

3.4. Possible Mechanisms by Which Statins Increase PON1 Activity

3.4.1. Statins, prenylated proteins and PON1

Gouédard et al. showed that in vitro Hu H7 human hepatoma cell culture PON1 activity and mRNA levels were significantly decreased by pravastatin, simvastatin and fluvastatin (10–100 μM). Moreover, statins suppressed the activity of the

PON1 gene promoter in transient and stable transfection assays. Data from our paper suggest that the effect of statins on PON1 activity may be mediated by the inhibition of HMG-CoA reductase, and that modifying the levels of downstream products of the mevalonate pathway could account for these effects (Gouédard et al., 2003). This pathway produces numerous bioactive signaling molecules including farnesyl pyrophosphate (FPP), and geranylgeranyl pyrophosphate (GGPP), which regulate transcriptional and posttranscriptional events that affect various biological processes. FPP is the major branch point in the mevalonate pathway and in addition to cholesterol is incorporated into prenylated proteins. Members of the Ras and Rho GTPase family are major substrates for posttranslational modification by farnesylation and geranylgeranylation, a process essential for their proper membrane localization and activation. GTPases function as molecular switches, cycling between an active GTP-bound and an inactive GDP-bound state thereby mediating cellular responses through their association with numerous effector molecules including kinases. It has been demonstrated that HMG-CoA reductase inhibition, through reduced RhoA activation, results in decreased phosphorylation and enhanced activity of PPARα. Furthermore, it has been demonstrated that the non-sterol mevalonate intermediate, GGPP, can directly antagonize LXR activity as well as indirectly inhibit expression of LXR-responsive genes, by reducing Rho protein activation. Therefore, mevalonate metabolites derived from the cholesterol biosynthetic pathway may represent an additional endogenous mechanism for regulation of cellular cholesterol homeostasis through modulation of LXR-responsive genes (Argmann et al., 2005).

3.4.2.   Possible role of srebp-2 pathway

Deakin et al. examined the influence of simvastatin (1.5–25 μg/ml) on PON1 gene expression in human hepatic HepG2 cells. They found that simvastatin was able to modulate in vitro the expression of PON1 regulated by sterol regulatory element-binding protein-2 (SREBP-2) and increase serum PON1 concentration and activity (Deakin et al., 2003). SREBPs are important transcriptional regulators of intracellular cholesterol homeostasis located at the endoplasmatic reticulum (ER) membrane. When intracellular cholesterol levels are decreased, SERBP-1 and SREBP-2 are transported from the ER to the Golgi by an SREBP-cleavage activating protein (SCAP), which contains a cholesterol-sensing domain, where two proteases, site 1 and site 2 protease cleave the SREBPs. The active SREBPs are released into the cytosol and enter the nucleus where they bind to a sterol regulatory element-1 (SRE-1) in the promoter regions of genes required for cholesterol biosynthesis, including HMG-CoA reductase-, fatty acid synthesis and genes mediating inflammatory reactions (Weber et al., 2004). Data from this study suggest that PON1 gene is a member of this wide range of genes regulated by SREBPs (Fig. 1.).

When intracellular cholesterol is abundant, SREBP2 remains in the endoplasmatic reticulum (ER) associated with the escort protein SREBP cleavage activating protein (SCAP) and the ER retention protein Insig. Low cholesterol causes a conformational change in the sterol-sensing domain of SCAP, dissociating Insig and allowing

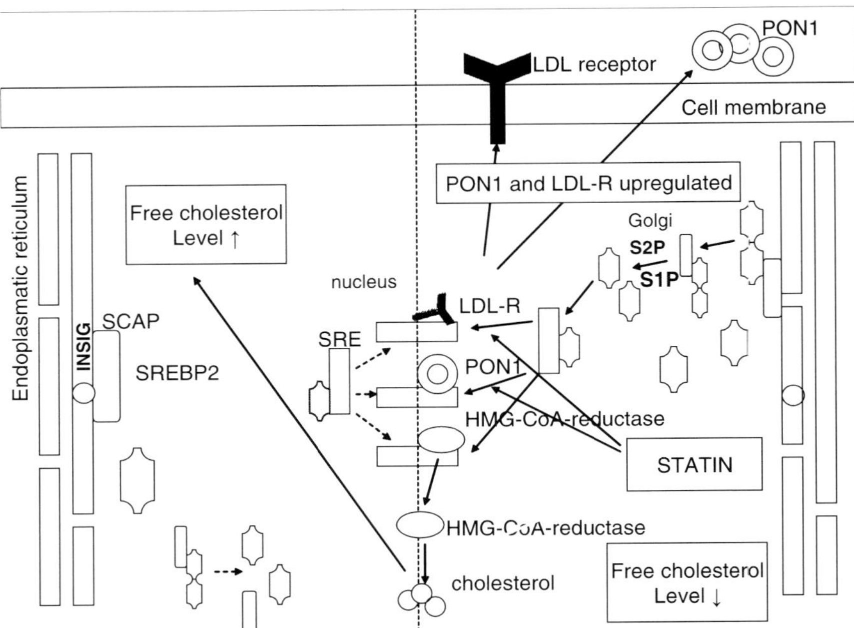

Figure 1. Possible mechanism of the effect of statins on PON1 activity via the SREBP2 regulatory pathway

SREBP-SCAP to reach the Golgi. Two proteases, S1P and S2P in the Golgi release the active form of SREBP, which translocates to the nucleus to activate transcription of target genes: LDL-receptor, HMG-CoA reductase and PON1 via binding to the sterol regulatory element (SRE) in the promoter regions of these genes. Cholesterol synthesis is also regulated posttranscriptionally: high cholesterol accelerates degradation of HMG-CoA reductase, the rate-limiting enzyme in cholesterol synthesis, by promoting association of its sterol-sensing domain with Insig (Soccio and Breslow, 2004).

Recently, Ota et al. found that pitavastatin significantly increased promoter activity in HEK293 cells 24 hours after PON1 (587/-6) plasmid transfection. Because the PON1 gene promoter activity was also increased by two other fat-soluble statins (atorvastatin and simvastatin), it was concluded that the transactivation was not specific to pitavastatin, but rather a general effect of statins. Their results also suggested that statin effect on PON1 activity may have occurred through the mevalonic acid-derived farnesyl pyrophosphate pathway. Although the exact mechanism of pitavastatin effect on PON1 promoter activity is not known, some other mechanisms are speculated. Since the PON1 gene promoter was activated by Sp1 and the increased transactivation by pitavastatin was completely suppressed by an Sp1 inhibitor, they theorised that the effect of pitavastatin on PON1 was an action in conjugation with Sp1. They also found 4 sequences of the SREBP-1 binding site in the PON1 promoter region, which underlines the possible role of SREBPs on PON1 activation (Ota et al., 2005).

The controversial results of in vitro studies on human cell lines may be due to differences in the cell lines used, as was suggested in a recent rewiew by Aviram et al. (Aviram and Rosenblat, 2005).

4. EFFECT OF FIBRATES ON PON1 ACTIVITY

4.1. Fibrates and Pparα

Fibrates belong to the commonly used lipid lowering drugs acting on the Peroxisome proliferator-activated receptor (PPAR) alpha. PPARs are ligand-activated transcription factors belonging to the nuclear receptor superfamily, which also includes the steroid and thyroid hormone receptors. Upon activation by their ligands, PPARs form heterodimers with the nuclear receptor RXR and bind to specific PPAR response elements (PPREs) in the promoter region of their target genes. PPARs can also repress gene expression in a DNA-binding–independent manner by interfering with other signaling pathways as well as in a DNA binding–dependent way through the recruitment of corepressors to unliganded PPARs (Marx et al., 2004)). PPARα can be activated by certain polyunsaturated fatty acids, by oxidized phospholipids, by lipoprotein lipolytic products, and by fibrates. PPARα regulates several genes that are involved in lipid and lipoprotein metabolism. PPARα activators induce the expression of apolipoprotein CIII (apoCIII), which is the main apolipoprotein of the LDL particle. Additionally, it induces the expression of the HDL apolipoprotein AI (apoAI) and HDL receptor CLA-1/SRB-I and ABCA1, a transporter controlling apoAI–mediated cholesterol efflux. Moreover, PPARα ligands downregulate the expression of the apoB48-remnant receptor in differentiated macrophages and reduce the uptake of glycated LDL and triglyceride-rich remnant lipoproteins. Conflicting results exist on the role of PPARs on the expression and activation of lipoprotein lipase enzyme, with data demonstrating a decreased secretion and activity of LPL on treatment with PPARα ligands. On the whole, the most important clinical effects of fibrates are the marked reduction in serum triglyceride level and an increase in HDL cholesterol level (Li and Glass, 2004).

Additionally. PPARα activators act on a variety of vascular cells. In human ECs, PPARα activators interfere with processes involved in leukocyte recruitment and cell adhesion. PPARα agonists, such as fenofibrate, have been shown to enhance endothelial NO synthase expression and NO release, and diminish thrombin-induced and oxidized low-density lipoprotein (LDL)-induced expression of ET-1 (Marx et al., 2004).

The effect of fibrates on PON1 activity is still largely unclear.

4.2. Clinical Studies: Conflicting Results

In the last few years several clinical studies have been published (Table 2.).

Fibrates significantly decreased plasma triglyceride levels in all studies, and increased plasma HDL-cholesterol levels except in one study. Durrington et al. described how an 8 weeks therapy with bezafibrate and gemfibrozil failed

Table 2. Clinical studies on the effect of fibrates medication on human paraoxonase-1 (PON1) activity

Author, year of public.	Fibrate (dose)	Patients (number)	Treatment period	PON1 activity $\Delta\%$ (p)	PON1/HDL $\Delta\%$ (p)
Fibrate Durrington et al. 1998.	Bezafibrate (400 mg/d)	IIb hyperlipidemia (29)	8 weeks	↑ 10.6% (n.s.)	↑ 1.35% (n.s.)
Fenofibrate (2 × 600 mg/d)	”	”	8 weeks	↑ 7.1% (n.s.)	↑ 2.29% (n.s.)
Paragh et al. 2000.	Gemfibrozil (2 × 600 mg/d)	hypertrigliceridemia (57)	3 months	↑15%↑ (p < 0.001)	↑ 3.7% (p < 0.01)
Turay et al. 2000.	Ciprofibrate (100 mg/d)	combined fam. hypelipopr. (15)	no data	↓ 56.1% (n.s.)	↓ 68.6% (n.s.)
Balogh et al. 2001b.	Gemfibrozil (2 × 600 mg/d)	type 2 diabetes mellitus (56)	3 months	↑ 18.5% (p < 0.001)	↑ 6.93% (p < 0.05)
Paragh et al. 2003.	Fenofibrate (200 mg/d)	IIb hyperlipidemia, IHD (52)	3 months	↑ 37.6% (p < 0.05)	↑ 12% (p < 0.05)
Tsimihodimos et al. 2003.	Fenofibrate (200 mg/d)	IIa, hyperlipidemia (18)	3 months	↓ 9.05% (n.s.)	↑ 11.2% (n.s.)
”	”	IIb hyperlipidemia (23)	”	↓ 2.26% (n.s.)	↓ 14.4% (n.s.)
”	”	IV hyperlipidemia (30)	”	↑ 9.8% (n.s.)	↓ 55% (p < 0.05)

to influence the activity of paraoxonase in type IIb hyperlipidaemic patients (Durrington et al., 1998). Tsimihodimos et al. found that 3 months of micronised fenofibrate administration did not influence serum PON1 and arylesterase activity in type IIa, IIb and IV dyslipidemic patients (Tsimihodimos et al., 2003). Turay et al. investigated the effect of ciprofibrate on PON1 activity in patients with familial combined hyperlipoproteinemia. They found a non-significant decrease in PON1 activity (Turay et al., 2000). Previously we investigated the effect of micronised fenofibrate in patients with coronary heart disease. After a three-month treatment period PON1 activity increased significantly (Paragh et al., 2003). We found out, following a three-month treatment period, that gemfibrozil was able to increase the activity of paraoxonase in patients with hypertriglyceridemia (Paragh et al., 2000). A three-month gemfibrozil treatment exerted an effect on PON1 activity in type 2 diabetic patients with associated hypertriglyceridemia. Since fibrates are able to activate peroxisome-activated receptors (PPARs), the authors hypothesized that increased PON1 activity observed after gemfibrozil therapy was caused by the elevation cf HDL and ApoA1 via activation of hepatic PPARα (Balogh et al., 2001b) (Fig. 2.).

The PPARα forms a heterodimer with the retinoid X receptor (RXR) in the presence of their ligands: fibrates and many other molecules. The resultant heterodimer binds to the PPAR-response elements (PPRE) in the promoter regions

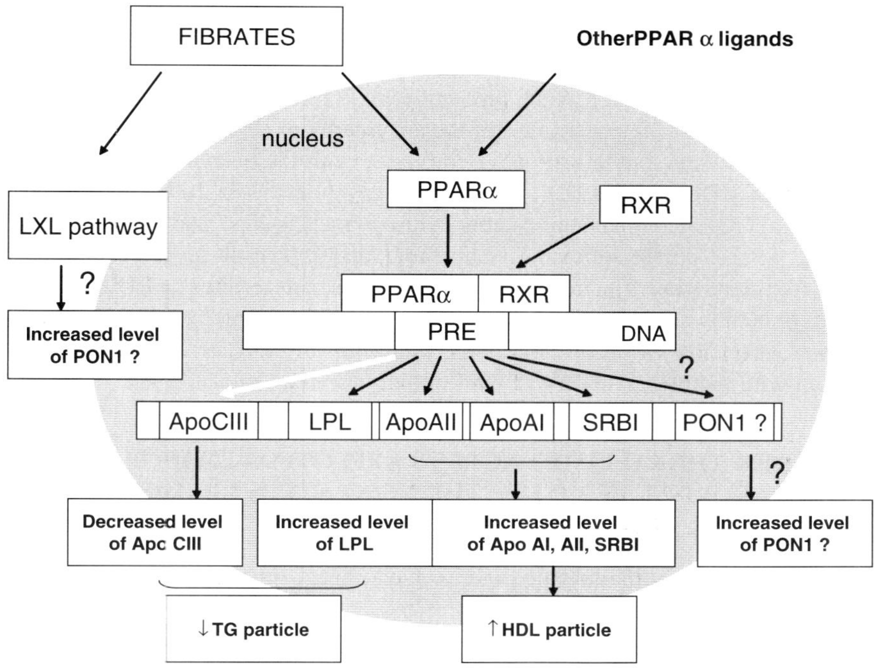

Figure 2. Theoretical mechanism of fibrates on PON1 activity by the PPARα/RXR regulatory pathway

of PPAR-driven genes through DNA-binding domains of PPARα and RXR, leading to the modulation of the transcription of PPAR target genes: upregulation of apolipoprotein AI (ApoAI) and AII (ApoAII), lipoprotein lipase (LPL), scavenger receptor class B type I (SRBI) and probably PON1, and downregulation of apolipoprotein CIII (ApoCIII) resulting in increased HDL and decreased triglyceride (Tg) particle size. On the other hand a recent study hypothesized the role of liver X receptor (LXL) pathway in fibrate induced PON1 activity modulation.

Altough there are clinical data on the effect of fibrates on PON1 activity, the possible mechanism remained to be clarified. Is there a role of PPARα regulatory pathway?

4.3.   In Vitro Studies – Data Against Pparα Activation

Gouédard et al. demonstrated in an in vitro study that fenofibrate resulted in an increase in PON1 activity and mRNA levels in Hu H7 human hepatoma cell cultures and slightly increased its promoter activity. This effect seemed to be dose-dependent for the 1-kb PON1 promoter. Different fibrates were tested. Fenofibrate provided the most efficient induction, while clofibrate and gemfibrozil did not increase the PON1 gene promoter activity. The involvement of PPARα in the fenofibrate-elicited induction of the PON1 gene expression was also investigated. Unexpectedly, it was observed that PPARα expression prevented the induction mediated by fenofibrate, suggesting that the inducing effect does not involve this receptor. They hypothesized that fibrates can alter gene expression through alternative mechanisms, the LXR pathway being one of them (Gouedard et al., 2003) (Fig. 2.).

Finally, it must be noted, that the protection by PON1 against LDL oxidation is accompained by a partial inactivation of the enzyme by oxidized lipids. The degradation of PON1 paraoxonase activity is correlated with the decrease in the antioxidant action of HDL towards LDL oxidation (Nguyen and Sok, 2003). Previously, Aviram et al. reported that atorvastatin hydroxy metabolites, and a gemfibrozil metabolite possess potent antioxidative potential, and as a result protect lipoprotein particles from oxidation. The inhibitory effects of these metabolites on LDL, VLDL, and HDL oxidation could be related to their free radical scavenging activity, as well as to their metal ion chelation capacities. Inhibition of HDL oxidation has been associated with the preservation of PON1 activity (Aviram et al., 1998).

5.   OTHER THERAPEUTIC AGENTS AFFECTING LIPID METABOLISM AND PON1: THE FIRST STEPS LEADING TO OTHER DIRECTIONS

Recently, some other therapeutic agents have been investigated as potential modulators of PON1 activity.

Rosiglitazone is a PPAR-alpha receptor activator antidiabetic agent with potent anti-oxidant and lipid lowering properties. In a randomized, cross-over, placebo-controlled, double-bind clinical trial rosiglitasone effectively increased fasting

PON1 activity by 9.4%, and attenuated the post-prandial fall in PON1 activity by 39.8% in 19 type 2 diabetic patients, with the PON1 mass not changed significantly (van Wijk et al., 2006). Glitazones are one of the most promising oral antidiabetics of recent years showing pleiotropic effects over their serum glucose lowering effect. Since diabetes is one of the pathological conditions with enhanced atheroslerosis, the improvement in PON1 activity by glitazone administration may underline the significance of this group of antidiabetics.

Enhanced atherogenesis and increased systemic oxidative stress have been demonstrated in obese patients (Couillard et al., 2005). Orlistat is a gastroin-testinal lipase enzyme inhibitor that enhances weight reduction in obese subjects. In addition to weight loss, the reduction of dietary fat intake has an effect on serum lipid levels and its composition. Its cholesterol reducing effect is independent of its weight reducing action, while controversial data exists on its effect on HDL levels. Recently, in a longitudinal, multicenter, randomized study orlistat adminis-tration significantly improved PON1 activity by 22% in 78 obese subjects. Also, PON1/HDL ratio was significantly increased (8.5%) after orlistat treatment. The exact mechanism has not yet been elucidated, but the significant weight reduction may contribute to the increase in PON1 activity (Audikovszky et al., 2006).

6.   FUTURE DIRECTIONS

We have no data on the effects of other newly developed lipid lowering agents such as cholesterol absorption inhibitor ezetimibe, ACAT inhibitor avasimibe, CETP inhibitor torcetrapib phospholipase A2 antagonists, Liver X receptor ligands or recombinant HDL ApoA1 Milano/phospholipid complex particles, etc. Especially exciting is the effect of LXR ligands on PON1, since fibrate studies have highlighted the role of this pathway.

7.   CONCLUSIONS

It is remarkable that there were significant differences between the results of the clinical studies. In some cases the data were contradictory, thus it was not easy to discern the final conclusions. Unfortunately, there were marked differences in sample size, selection bias in patients, age, ethnicity and environmental background of the studied populations discussed above. Besides the different efficacies of these agents on altering PON1 activity these factors may be responsible for the contradictory results. In the future further well organized, multicentre studies on large patient populations will be necessary to elucidate the effect of lipid lowering medications on PON1 activity.

ACKNOWLEDGEMENTS

The authors work is supported by the Hungarian Scientific Research Fund (K63025, and Hungarian Health Science Council (ETT 243/2006).

REFERENCES

Argmann CA, Edwards JY, Sawyez CG, O'Neil CH, Hegele RA, Pickering JG, Huff MW. 2005, Regulation of macrophage cholesterol efflux through hydroxymethylglutaryl-CoA reductase inhibition: a role for RhoA in ABCA1-mediated cholesterol efflux. *J. Biol. Chem.* 230:22212–22221

Audikovszky M, Pados G, Seres I, Harangi M, Fulop P, Katona E, Illyes L, Winkler G, Katona EM, Paragh G. 2006, Orlistat increases serum paraoxonase activity in obese patients. *Nutr. Metab. Cardiovasc Dis.* [Epub ahead of print]

Aviram M, Rosenblat M. 2005, Paraoxonases and cardiovascular diseases: pharmacological and nutritional influences. *Curr. Opin. Lipidol.* 16:393–399

Aviram M, Rosenblat M, Bisgaier CL, Newton RS. 1998, Atorvastatin and gemfibrozil metabolites, but not parent drugs, are potent antioxidants against lipoprotein oxidation. *Atherosclerosis* 138:271–80

Ballantyne C, Arroll B, Shepherd J. 2005, Lipids and CVD management: towards a global consensus. *Eur. Heart. J.* 26(21):2224–31

Balogh Z, Fülöp P, Seres I, Harangi M, Katona E, Kosztáczky B, Paragh GY. 2001a, Effects of simvastatin on serum paraoxonase activity. *Clin. Drug. Invest.* 21:505–10

Balogh Z, Seres I, Harangi M, Kovács P, Kakuk G, Paragh G. 2001b, Gemfibrozil increases paraoxonase activity in type 2 diabetic patients. A new hypothesis of beneficial action of fibrates? *Diab. Met.* 27:604–10

Beltowski J, Wojcicka G, Jamroz A. 2002a, Differential effect of 3-hydroxy-3-methylglutaryl coemzyme A reductase inhibitors on plasma paraoxonase 1 activity in the rat. *Pol. J. Pharmacol.* 54:661–71

Beltowski J, Wojcicka G, Mydlarczyk M, Jamroz A. 2002b, The effect of peroxisome proliferator-activated receptors α (PPARα) agonist, fenofibrate, on lipid peroxidation, total antioxidant capacity, and plasma paraoxonase 1 (PON1) activity. *J. Physiol. Pharmacol.* 53:463–75

Couillard C, Ruel G, Archer WR, Pomerleau S, Bergeron J, Couture P, Lamarche B, Bergeron N. 2005, Circulating levels of oxidative stress markers and endothelial adhesion molecules in men with abdominal obesity. *J. Clin. Endocrinol. Metab.* 90(12):6454–9

Davidson M.H. 2005, Clinical significance of statin pleiotropic effects: hypotheses versus evidence. *Circulation* 111(18):2280–1

Deakin S, Leviev I, Guernier S, James RW. 2003, Simvastatin modulates expression of the PON1 gene and increases serum paraoxonase. A role for sterol reulatory element-binding protein-2. *Arterioscler. Thromb. Vasc. Biol.* 23:2083–89

Draganov DI, Teiber JF, Speelman A, Osawa Y, Sunahara R, La Du BN. 2005, Human paraoxonases (PON1, PON2, and PON3) are lactonases with overlapping and distinct substrate specificities. *J. Lipid. Res.* 46(6):1239–47

Durrington PN, Mackness MI, Bhatnagar D, Julier K, Prais H, Arrol S, Morgan J, Wood GN. 1998, Effects of two different fibric acid derivates on lipoproteins, cholesteryl ester transfer, fibrinogen, plasminogen activator inhibitor and paraoxonase activity in type Iib hyperlipoproteinaemia. *Atherosclerosis* 138:217–25

Fuhrman B, Koren L, Volkova N, Keidar S, Hayek T, Aviram M. 2002, Atorvastatin therapy in hypercholesterolemic patients suppressed cellular uptake of oxidized-LDL by differenticting monocytes. *Atherosclerosis* 64:179–85

Giroux LM, Davignon J, Naruszewicz M. 1993, Simvastatin inhibits the oxidation of low-density lipoproteins by activated human monocyte-derived macrophages. *Biochim. Biophys. Acta.* 1165(3):335–8

Gouédard C, Koum-Besson N, Barouki R, Morel Y. 2003, Opposite regulation of the human paraoxonase-1 gene PON-1 by fenofibrate and statins. *Mol. Pharmacol.* 64(4):945–56

Grundy SM, Cleeman JI, Merz CN, Brewer HB Jr, Clark LT, Hunninghake DB, Pasternak RC, Smith SC Jr, Stone NJ. 2004, Coordinating Committee of the National Cholesterol Education Program. Implications of recent clinical trials for the National Cholesterol Education Program Adult Treatment Panel III Guidelines. *J. Am. Coll. Cardiol.* 44:720–732

Harangi M, Seres I, Varga Z, Emri G, Szilvassy Z, Paragh G, Remenyik E. 2004, Atorvastatin effect on HDL associated paraoxonase activity and oxidative DNA damage. *Eur. J. Clin. Pharm.* 60:685–91

Heinecke JW. 2006, Lipoprotein oxidation in cardiovascular disease: chief culprit or innocent bystander? *J. Exp. Med.* 203(4):813–6

Himbergen TM, van Tits LJ, Voorbij HA, de Graaf J, Stalenhoef AF, Roest M. 2005, The effect of statin therapy on plasma high-density lipoprotein cholesterol levels is modified by paraoxonase-1 in patients with familial hypercholesterolaemia. *J. Intern. Med.* 258(5):442–9

Israelian-Konaraki Z, Reaven PD. 2005, Peroxisome proliferator-activated receptor-alpha and atherosclerosis:from basic mechanisms to clinical implications. *Cardiology* 103:1–9

Kassai A, Illyes L, Mirdamadi HZ, Seres I, Kalmar T, Audikovszky M, Paragh G. 2006, The effect of atorvastatin therapy on lecithin:cholesterol acyltransferase, cholesteryl ester transfer protein and the antioxidant paraoxonase. *Clin. Biochem.* [Epub ahead of print]

Khersonsky O, Tawfik DS. 2006, The histidine 115-histidine 134 dyad mediates the lactonase activity of mammalian serum paraoxonases. *J. Biol. Chem.* 281(11):7649–56

Kleemann M, Kooistra T. 2005, HMG-CoA reductase inhibitors: effects on chronic subacute inflammation and onset of atherosclerosis induced by dietary cholesterol. *Curr. Drug Targets Cardiovasc. Haematol. Disord.* 5:441–453

Krauss RM, Siri PW. 2004, Dyslipidemia in type 2 diabetes. *Med Clin North Am.* 88: 897–909

Kural BV, Orem C, Uydu HA, Alver A, Orem A. 2004, The effects of lipid-lowering therapy on paraoxonase activities and their relationships with the antioxidant system in patients with dyslipidemia. *Coron. Artery. Dis.* 15:277–83

Li AC, Glass CK. 2004, PPAR- and LXR-dependent pathways controlling lipid metabolism and the development of atherosclerosis. *J. Lipid. Res.* 45(12):2161–73

Marx N, Duez H, Fruchart JC, Staels B. 2004, Peroxisome proliferator-activated receptors and atherogenesis: regulators of gene expression in vascular cells. *Circ. Res.* 94(9):1168–78

Morrow JD. 2005. Quantification of isoprostanes as indices of oxidant stress and the risk of atherosclerosis in humans. *Arterioscler. Thromb. Vasc. Biol.* 25(2):279–86

Nguyen SD, Sok D. 2003, Benefitial effect of oleoylated lipids on paraoxonase 1: protection against oxidative inactvation and stabilisation. *Biochem. J.* 375:257–85

Ota K, Suehiro T, Arii K, Ikeda Y, Kumon Y, Osaki F, Hashimoto K. 2005, Effect of pitavastatin on transactivation of human serum paraoxonase 1 gene. *Metabolism* 54:142–50

Paragh G, Balogh Z, Seres I, Harangi M, Boda J, Kovacs P. 2000, Effect of gemfibrozil on HDL-associated serum paraoxonase activity and lipoprotein profile in patients with hyperlipidaemia. *Clin. Drug. Invest.* 19:277–82

Paragh G, Seres I, Balogh Z, Harangi M, Illyés L, Boda J, Varga ZS, Kovács P. 2003, The effect of micronised fenofibrate on serum paraoxonase activity in patients with coronary heart disease. *Diab. Met.* 29:613–18

Paragh G, Töröcsk D, Seres I, Harangi M. 2004, Effect of short term treatment with simvastatin and atorvastatin on lipids and paraoxonase activity in patients with hyperlipoproteinaemia. *Curr. Med. Res. Opin.* 20:1321–27

Saini HK, Xu YJ, Arneja AS, Tappia PS, Dhalla NS. 2005,Pharmacological basis of different targets for the treatment of atherosclerosis. *J. Cell. Mol. Med.* 9:818–839

Soccio RE, Breslow JL. 2004, Intracellular cholesterol transport. *Arterioscler. Thromb. Vasc. Biol.* 24(7):1150–60

Tomás M, Sentí M, García-Faria F, Vila J, Rorrents A, Covas M, Marrugat J. 2000, Effect of simvastatin therapy on paraoxonase activity and related lipoproteins in familial hypercholesterolemic patiens. *Arterioscler. Thromb. Vasc. Biol.* 20:2113–19

Tsimihodimos V, Kakafika A, Tambaki AP, Bairaktari E, Chapman MJ, Elisaf M, Tselepis AD. 2003, Fenofibrate induces HDL-associated PAF-AH but attenuates enzyme activity associated with apoB-containing lipoproteins. *J. Lipid. Res.* 44:927–34

Tsimihodimos V, Karabina SAP, Tambaki AP, Bairaktari E, Goudevenos JA, Chapman MJ, Elisaf M, Tselepis AD. 2002, Atorvastatin preferentially reduces LDL-associated platelet-activating factor acetylhydrolase activity in dyslipidemias of type IIa and type IIb. *Arterioscler. Throm. Vasc. Biol.* 22:306–11

Turay J, Grnaková V, Valka J. 2000, Changes in paraoxonase and apolipoprotein A-I, B, C-III and E in subjects with combined familial hyperlipoproteinaemia treated with ciprofibrate. *Drug. Exp. Clin. Res.* 26:83–88

van Wijk J, Coll B, Cabezas MC, Koning E, Camps J, Mackness B, Joven J. 2006, Rosiglitazone modulates fasting and post-prandial paraoxonase 1 activity in type 2 diabetic patients. *Clin. Exp. Pharmacol. Physiol.* 33:1134–1137

Weber LW, Boll M, Stampfl A. 2004, Maintaining cholesterol homeostasis: sterol regulatory element-binding proteins. *World. J. Gastroenterol.* 10:3081–7

CHAPTER 17

THE FUNCTIONAL CONSEQUENCES
OF POLYMORPHISMS IN THE HUMAN PON1 GENE

C.E. FURLONG,[1,2] R.J RICHTER[1,2], W.-F. LI*, V.H. BROPHY[1,2**],
C. CARLSON[2†], M. RIEDER[2], D. NICKERSON[2], L.G. COSTA[3,4],
J. RANCHALIS[1,2], A.J. LUSIS[5], D.M. SHIH[5], A. TWARD[5‡]
AND G.P. JARVIK[1,2]

[1] *Departments of Medicine (Div. Medical Genetics),*
[2] *Genome Sciences, and*
[3] *Environmental and Occupational Health Sciences, University of Washington, Seattle, WA 98195, USA*
[4] *Dept. of Human Anatomy, Pharmacology and Forensic Medicine, University of Parma, Italy*
[5] *Department of Medicine, UCLA, Los Angeles, CA 90095-1697*
* *Present address: Division of Environmental Health and Occupational Medicine, National Health Research Institutes, Zhunan, Taiwan*
** *Present address: Roche Diagnostics, Alameda, CA*
† *Present address, The Fred Hutchinson Cancer Research Center, Seattle, WA*
‡ *Present address, University of California, San Francisco, CA*

Abstract: Early research on population distributions of plasma PON1 paraoxonase activity indicated a polymorphic distribution with high, intermediate and low metabolizers. Cloning and characterization of the cDNA encoding human PON1 and follow-on experiments demonstrated that the molecular basis of the activity polymorphism (PM) was a Q192R PM with $PON1_{R192}$ specifying high paraoxonase activity. Further research demonstrated that the $PON1_{192}$ polymorphism had little effect on the catalytic efficiencies of hydrolysis of phenylacetate and diazoxon (DZO), but did affect the efficiencies of hydrolysis of chlorpyrifos oxon (CPO), soman and sarin, with $PON1_{R192}$ having a higher efficiency of CPO hydrolysis and $PON1_{Q192}$ having higher rates of hydrolysis of soman and sarin. Plots of rates of DZO hydrolysis (at a salt concentration that differentially inhibited $PON1_{R192}$) vs. paraoxon hydrolysis clearly separated the three $PON1_{192}$ phenotypes (QQ, QR, RR) and also showed a wide range of activity among individuals with the same $PON1_{192}$ genotype. The term PON1 status was introduced to include both $PON1_{192}$ functional genotype and plasma PON1 level, both important in determining risk for either exposure to specific organophosphorus compounds (OPs) or disease. Characterization of 5 promoter-region polymorphisms by several groups indicated that an Sp1 binding site was responsible for significant ($\sim 30\%$) variation in plasma PON1 levels. Re-sequencing of the PON1 genes of 47 individuals (24 African-American/23 European) revealed an additional 180 polymorphisms in 27 kb of the PON1 genomic DNA including 8 more 5′ regulatory region PMs, 1 coding region polymorphism (W194X), 162 additional intronic PMs and 9

B. Mackness et al. (eds.), The Paraoxonases: Their Role in Disease Development and Xenobiotic Metabolism, 267–281.

additional 3' UTR PMs. The generation of PON1 null mice and "PON1 humanized mice" expressing either tgHuPON1$_{R192}$ or tgHuPON1$_{Q192}$ at the same levels on the PON1$^{-/-}$ background allowed for a functional analysis of the Q192R PM under physiological conditions. Toxicology experiments with the PON1 humanized mice and the PON1 null mice injected with purified human PON1$_{192}$ alloforms clearly demonstrated that the catalytic efficiency of substrate hydrolysis is important in determining whether PON1 is able to protect against a given OP exposure. HuPON1$_{R192}$ protects well against CPO and DZO exposure, but HuPON1$_{Q192}$ does not protect well against CPO exposure and neither protects against PO exposure. Studies on PON1 status and carotid artery disease show that low PON1 levels are a risk factor. The effects of PON1$_{192}$ alloforms on rates of hydrolysis of quorum sensing factors are not yet known. Taken together, these data along with those of the leading researchers in the PON1 field indicate that it is important to measure PON1 levels/activities in any epidemiological study. SNP analysis alone is inadequate for epidemiological studies, due to the wide variability of PON1 levels within the three PON1$_{192}$ genotypes Q/Q, Q/R R/R). Even the most comprehensive PON1 SNP analyses are unable to accurately predict PON1 levels. PON1 activity or level accurately predicts CHD risk, while genotype does not

Keywords: paraoxonase, PON1, organophosphate, organophosphorus compounds (OPs), chlorpyrifos, chlorpyrifos oxon, diazinon, diazoxon, arylesterase, regulation of gene expression, nerve agents, single nucleotide polymorphisms (SNPs), PON1 expression, Sp1 transcription factor, quorum sensing, quorum sensing factor, carotid artery disease (CAAD), coronary heart disease (CHD)

1.   INTRODUCTION

The early history of PON1 research is outlined in Chap. 1 of this volume. This chapter will focus on the research carried out following the cloning and characterization of the human PON1 cDNA. As noted in Chap. 1, sequencing the initial cDNA isolates of human *PON1* revealed the two common coding polymorphisms L55M and Q192R (Hassett et al., 1991) that have since been studied extensively. More recent studies have identified many new polymorphisms including those found in the promoter region, coding region, introns and 3' UTR as well as a number of mutations in the PON1 gene.

2.   FUNCTIONAL CONSEQUENCES OF THE PON1 POLYMORPHISMS

2.1.   Coding Polymorphisms and Paraoxon Hydrolysis

Following the characterization of PON1 cDNAs clones in 1991 that revealed the two common polymorphisms L55M and Q192R, Humbert et al. (1993) and Adkins et al. (1993) reported that the Q192R polymorphism was responsible for the high vs. low paraoxonase activity of PON1. The early research that examined levels of paraoxonase in population studies all found a large variability in levels among individuals (reviewed in: Chap. 1).

2.2. Promoter Polymorphisms

One explanation for the large variability in PON1 levels among individuals emerged from three independent studies that examined the effects of promoter region polymorphisms on PON1 expression (Brophy et al., 2001a, b; Leviev and James, 2000; Suehiro et al., 2000). All three groups made use of reporter gene constructs in cultured cells. The most significant contribution to variability in expression in the reporter gene systems as well as in plasma levels was the C-107/108T polymorphism (referred to here as -108) that accounted for up to 30% of the variability of PON1 serum levels. Suehiro et al. (2000) examined the C(-108)T, G(-126)C and G(-160)A polymorphisms; Leviev and James (2000) the C(-107)T, G(-824)A and G(-907)C polymorphisms and Brophy the C(-108)T, C(-126G), A(-162)G, A(-832)G and G(-909)C polymorphisms. The C(-108)T polymorphism occurs at an Sp1 binding site and the A(-162)G polymorphism at a potential NF-1 transcription factor binding site when A is present at this position. Leviev and James (2000) noted that earlier reports showed that mutations in Sp1 binding sites affected promoter activity (Saikawa et al., 1995).

A report by Blatter Garin et al. (1997) found an association between M55 and lower serum PON1 concentrations and Leviev et al. (1997) found that the *M55* allele produced lower levels of mRNA. Both reports by Leviev and James (2000) and Brophy et al. (2001b) noted that part of this difference in expression was related to a strong linkage disequilibrium between L55M and C(-108)T, with M55 being in linkage disequilibrium with the inefficient promoter polymorphism T(-108). However, not all of the difference in plasma PON1 levels could be accounted for by the association of M55 with T(-108) (Deakin and James, 2004). Leviev et al. (2001) noted that the $PON1_{M55}$ alloform was less stable than $PON1_{L54}$. Two recent reports support this conclusion. The crystal structure of a "hybrid PON1" supports an important role of L55 in packing of PON1 (Harel et al., 2004) and a recent study of PON1 genotypes in 1527 subjects [Q192R, T(-107C) and L55M] reported an independent contribution of $PON1_{L55M}$ to plasma PON1 levels with M55 having lower activity. Further support for the inefficiency of the T(-107) allele came from experiments reported by Deakin et al. (2003) where they demonstrated a better affinity of the C(-108) sequence for binding nuclear factors and demonstrated a difference in affinity for an Sp1 binding factor using anti Sp1 antibodies and polymorphic fragments representing the two different -108 promoter PM alleles.

In a follow-up study, Deakin et al. (2003) found that simvastatin modulated the expression of the PON1 gene in a dose dependent manner and appeared to act in concert with Sp1 to up regulate expression of PON1. Intermediates of cholesterol biosynthesis blocked the stimulation, particularly mevalonate and farnesyl pyrophosphate. More recent work, reported at this conference (see Chap. 15), showed that only the C(-108) allele was responsive to simvastatin upregulation, suggesting that patients with the C(-108) allele would perhaps benefit more from simvastatin therapy than those with the T(-108) allele. Presumably heterozygotes would exhibit an intermediate response.

2.3.   Polymorphisms and OP Detoxication

This section will deal primarily with the effects of PON1 polymorphisms on OP detoxication. Some of the background material has been covered in Chaps. 1 and 13. Figure 1 shows the OP detoxication pathways that involve PON1.

The insecticides parathion, chlorpyrifos and diazinon are manufactured as organophosphorothioates, compounds thought to be relatively safe to humans. These compounds are "bioactivated" in the liver by cytochromes P450s to their highly toxic oxon forms that can be hydrolyzed by PON1. They also react with B esterases stoichiometrically with one enzyme molecule inactivating one OP molecule. None the less, the B esterases are capable of providing some protection against OP exposure (Chambers et al., 1994; Pond et al., 1995). The P450s also directly detoxify a percentage of a given organophosphorothioates instead of bioactivating it with the percentage dependent on the P450 (Jokanovic, 2001).

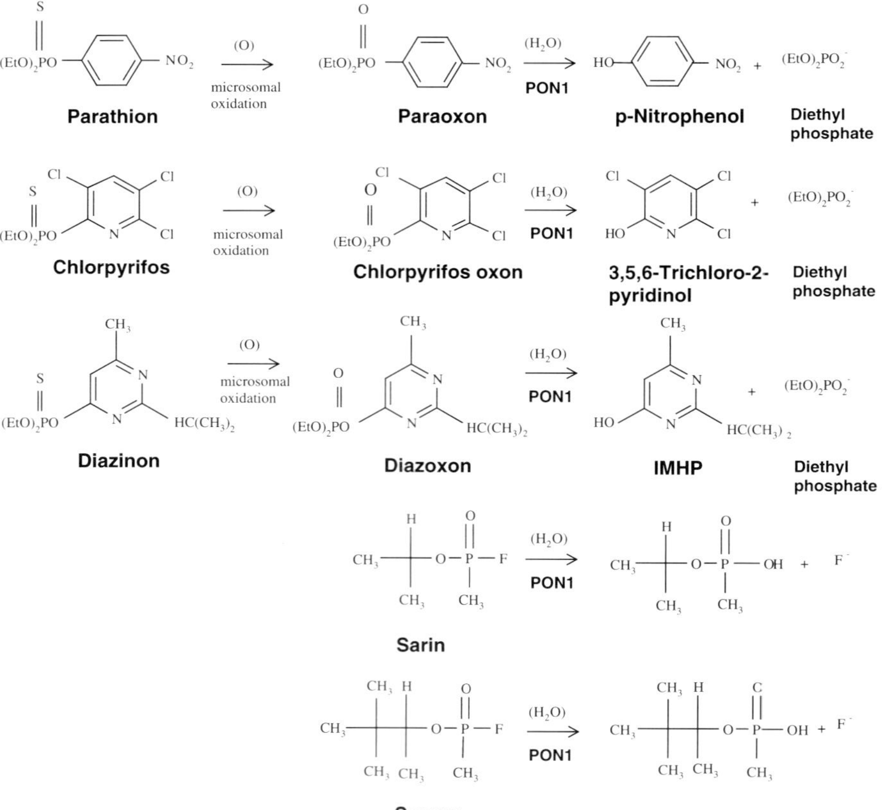

Figure 1. Proposed P450/PON1 pathway for detoxication of some PON1 substrates. Reproduced from Davies et al. 1996 with permission

While polymorphisms are known for the P450s involved in bioactivation, there has not been sufficient research to know the effects of these polymorphisms on sensitivity to OP exposure. What has been reported recently, however, is that the organophosphorothioates are suicide substrates for the P450s, e.g., P450 3A4, and inactivate them during the bioactivation reaction (Usami et al., 2003, 2006). This would suggest a possible concern of the effects of exposure to parent compounds on reproductive health, for example.

One major point regarding the effects of the PON1 polymorphisms on OP sensitivity is that while all of the safety testing of the OP compounds was carried out with highly pure parent organophosphorothioates, an actual exposure may include rather high levels of the highly toxic oxon forms (Table 1). Since the oxons inactivate acetylcholinesterase at approximately 1000 times the rate of the parent compound (Huff and Abou-Donia, 1995), a small percentage of oxon residue in an exposure is significant. The importance of this point will be obvious from what follows.

As noted in Chaps. 1 and 13, experiments with genetically modified mice have clearly demonstrated that the PON1 polymorphisms are most important in the detoxication of the oxon component of an exposure. The experiments that involved reconstituting $PON1^{-/-}$ mice with purified human $PON1_{Q192}$ or $PON1_{R192}$ demonstrated that it was the catalytic efficiency of hydrolysis of a given OP that determined whether

Table 1. Oxon levels in total pesticide residues taken from dislodgeable leaf foliar residue and dermal exposure studies. Reproduced from Yuknavage et al., 1997 with permission

	Pesticide (units)	Oxon[a]	Thioate	Total OP	Oxon (%)
Ralls et al. (1966)	Diazinon (ppm[d])	0.05	0.25	0.3	17
Kansoug and Hopkins (1968)	Diazinon[b]	ND[c]	–	–	ND
Wolfe et al. (1975)	Parathion (ng/cm^2)[d]	8	106	114	7
Kraus et al. (1977)	Azinphosmethyl (%)[d]	0.05	99.95	100	0.05
Nigg et al. (1977)	Ethion (ng/cm^2)[d]	42	285	327	13
Spear et al. (1977a)	Parathion (ng/cm^2)[d]	84	29	113	74
	Parathion (μg)[e]	145	39	184	79
Spear et al. (1977b)	Parathion (ng/cm^2)[d]	229	8	237	97
Maddy and Meinders (1987)	Azinphosmethyl (μg)[e]	ND	–	–	ND
Costello et al. (1989)	Malathion (μg)[e]	659	2301	2960	22
Schneider et al. (1990)	Azinphosmethyl (ng/cm^2)[d]	0.008	0.31	0.32	2.5
	Azinphosmethyl (μg)[e]	272	1450	1722	16
Spencer et al. (1991)	Azinphosmethyl (%)[d]	15	85	100	15
McCurdy et al. (1994)	Azinphosmethyl[b]	–	–	–	2.3

[a] Based on the highest value reported in study.

[b] Units or values not given in study.

[c] ND, none detected.

[d] Foliar residue measurement.

[e] Dermal monitoring measurement.

or not a given $PON1_{192}$ alloform was able to provide protection against exposure to that OP. Figure 2 shows the dose response curves for exposure of wild type and $PON1^{-/-}$ mice to chlorpyrifos oxon (CPO), diazoxon (DZO) and paraoxon (PO). The genetic dissection of the PON1 gene from the mice rendered them dramatically more sensitive to CPO (Shih et al., 1998) and DZO (Li et al., 2000), but not PO. These experiments demonstrated that PON1 does play an important role in protecting against OP exposure, providing that the catalytic efficiency is high enough.

The data in Fig. 2 show the effects of the PON1 Q192R polymorphism on detoxication of CPO, DZO and PO in PON1 reconstituted *PON1$^{-/-}$* mice. There was an excellent correlation between the catalytic efficiency of OP hydrolysis (Vm/Km) of each PON1 alloform and the protection afforded by injecting that purified human $PON1_{192}$ alloform into $PON1^{-/-}$ mice. $PON1_{R192}$ has a better efficiency of chlorpyrifos oxon hydrolysis than $PON1_{Q192}$ and provided better protection. Both alloforms have nearly equivalent catalytic efficiencies for hydrolyzing DZO and provided nearly equivalent protection against dermal DZO exposure. While $PON1_{R192}$ had a much higher catalytic efficiency for PO hydrolysis, (i.e., it was this feature that originally defined the polymorphic distribution of PON in human populations) it is still too low to provide significant protection against PO dermal exposure (Li et al., 2000) (Fig. 3).

These observations provide a good framework for thinking about *PON1* polymorphisms and risk of disease or exposure. If both alloforms metabolize biological toxicants (e.g., oxidized lipids) with the same efficiency, then it will be the level of PON1 that will be important. If there is a difference in catalytic efficiency of detoxication between the two $PON1_{192}$ alloforms, as in the case of CPO, then it will be both the level and the position 192 amino acid that will be important. However, in no case will PON1 levels be unimportant (see also Chap. 1). The L55M polymorphism has a minor influence on PON1 levels (Leviev et al., 1997, 2001; Roest et al., 2007), however, this is measured when PON1 status is determined.

2.4.   PON1 and Vascular Disease

Several observations directed early attention to the possibility of genetic variability of PON1 affecting the risk of vascular disease. These were: 1) the description of the PON1 coding region polymorphisms (L55M and Q192R) by Hasset et al. (1991); 2) the report in by Mackness et al. (1991), that PON1 prevented the accumulation of lipoperoxides in low-density lipoprotein and 3) the findings that the Q192R polymorphism of PON1 was responsible for determining catalytic efficiency for hydrolysis of some substrates (Adkins et al., 1993; Davies et al., 1996; Humbert et al., 1993). Following these observations, many studies were carried out that looked for an effect of either L55M or Q192R on vascular disease (summarized below). Following the description of polymorphisms in the promoter region of PON1 (Brophy et al., 2001a, b; Leviev and James, 2000; Suehiro et al., 2000) and the finding that the C(-108)T polymorphism had a significant effect on levels of PON1 expression, this polymorphism was also included in some of the studies that followed.

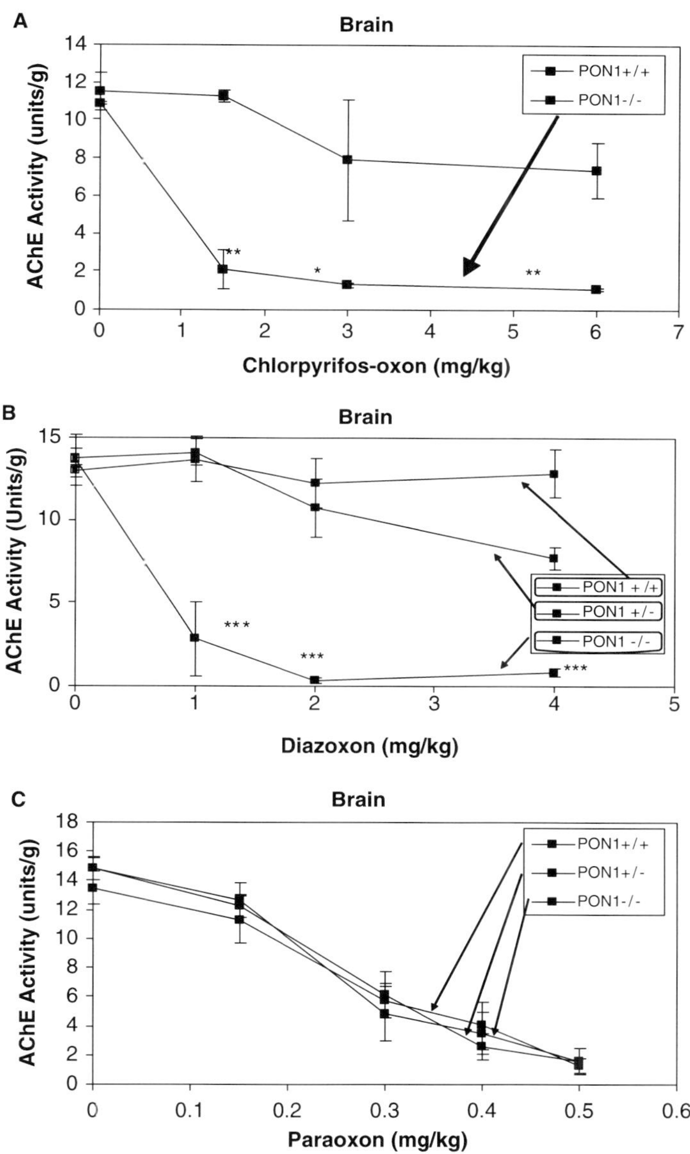

Figure 2. A) dose response of dermal CPO exposure for wild type and PON1$^{-/-}$ mice (data from Shih et al., 1998); B), dose response of dermal DZO exposure for wild type, hemizygous and PON1$^{-/-}$ mice (data from Li et al., 2000); C), dose response for wild type, hemizygous and PON1$^{-/-}$ mice (data from Li et al., 2000). Reproduced with permission

Catalytic Efficiency of PON1$_{Q192}$ & PON1$_{R192}$

Protection afforded PON1$^{-/-}$ mice by injection of hPON1$_{Q192}$ or hPON1$_{R192}$

Chlorpyrifos-oxon Hydrolysis

	PON1Q192	PON1R192
Km (mM)	0.54	0.25
Vmax (units/mg)	82	64
Vmax/Km	152	256

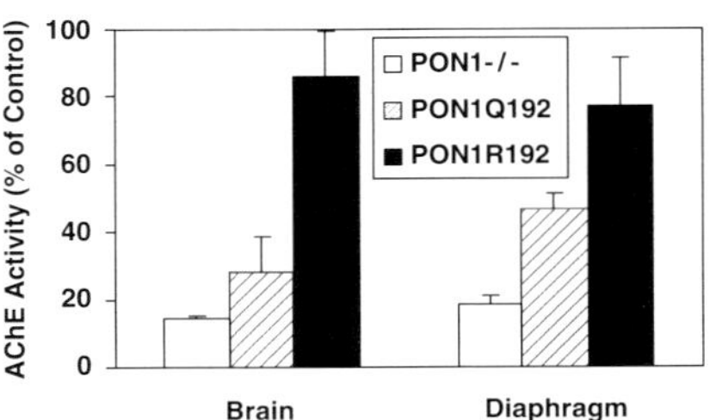

Diazoxon Hydrolysis

	PON1Q192	PON1R192
Km (mM)	2.98	1.02
Vmax (units/mg)	222	79
Vmax/Km	75	77

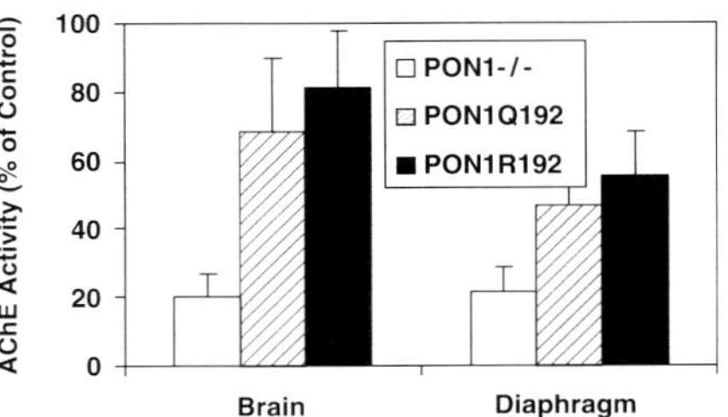

Paraoxon Hydrolysis

	PON1Q192	PON1R192
Km (mM)	0.81	0.52
Vmax (units/mg)	0.57	3.26
Vmax/Km	0.71	6.27

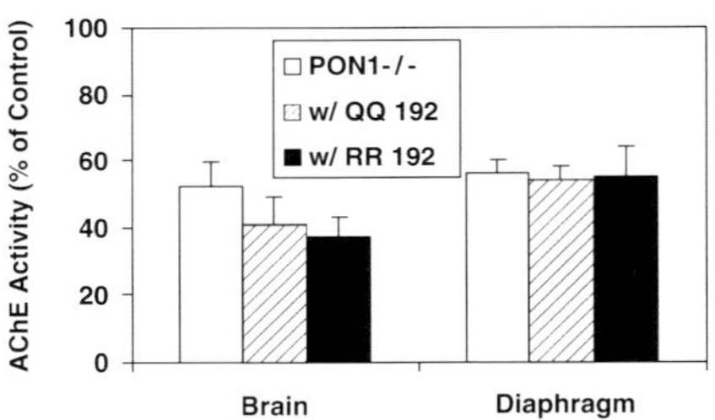

Figure 3. Effects of the PON1 Q192R polymorphism on OP detoxication. Adapted from Li et al. 2000 with permission

A few studies on the possible relationship of PON1 genetic variability to coronary heart disease (CHD) are worth mentioning. Mackness et al. (2001) examined both PON1-55 and -192 polymorphisms, plasma protein levels and POase activity. In addition, they carried out a meta analysis of the results of the previous 18 studies, only three of which involved a measurement of PON1 activities or levels. They found that both PON1 activity (POase) and concentration were lower in subjects compared with controls. POase in patients= 122.8 [16 to 527.4] vs. 214.6 [26.3 to 620.8] nmol min^{-1} mL^{-1} in controls, and plasma PON1 concentration (by ELISA) in patients= 71.6 [11.4 to 489.3 vs. 89.1 [16.8 to 527.5] μg mL^{-1} in controls. Their meta analysis of PON1-192 genotype and CHD noted heterogeneity between studies, but noted that the R allele was associated with a modest increase in risk for CHD. At the same time, they noted that the majority of studies did not measure the activity or quantity of PON1, both of which were lower in CHD patients than controls in their study. They concluded that "… it is likely that PON1 activity and mass are more important determinants of susceptibility to CHD than are the PON1-55 and PON1-192 genotypes." They recommended that *We, along with other authors, would strongly suggest that all further epidemiological studies into the role of PON1 and disease should include a measurement of the enzyme itself in addition to the genetic polymorphisms.*"

A second study (Lawlor et al., 2004) that included a prospective analysis of the association of the PON1 Q192R PM with CHD in a cohort (N = 3, 266) combined with a meta analysis of 38 other published studies (10,738 case and 17,068 controls) concluded that *"There is no robust evidence that the PON1 Q192R polymorphism is associated with CHD risk in Caucasian women or men."* This study, as well as many others, suffered from the serious flaw of not including measures of PON1 levels.

A third meta analysis published the same year (Wheeler et al., 2004) examined 4 *PON1* PMs and one *PON2* PM for association with CHD. Their analysis included 43 genetic association studies with more than 11,000 cases and nearly 13,000 controls. Their analysis yielded a relative risk for the R allele under a dominant genetic model of 1.15. Restricting the analysis to the 5 largest studies generated a relative risk of only 1.05. The authors note that, *"The findings of these genetic association meta-analyses are rather sobering, suggesting important limitations in the assumptions under which several dozen investigations of paraoxonase polymorphisms and CHD have been conducted during the past decade."* They go on to list a number of what they feel to be the main limitations. However, they completely missed the main limitation of the studies, which is not including a measurement and analysis of PON1 levels. The SNPs are interesting with respect to any affect that they might have on PON1 levels, but in the end, it is the PON1 level that matters and possibly the position 192 amino acid if there is a difference in catalytic efficiency for the metabolite(s) involved in generating the risk of CHD. A PON1 status analysis provides both of these crucial bits of information.

The study by Jarvik et al. (2003) provides an excellent framework for evaluating the risk of genetic variability of PON1 for disease or exposure (see also Chap. 1). PON1 levels were compared between patients and controls within each $PON1_{192}$ genotype (Q/Q, Q/R, R/R). Low PON1 levels were clearly a risk for carotid artery disease (CAAD) in Q/Q homozygotes and Q/R heterozygotes. There were too few R/R homozygotes for a statistically significant comparison. Similar studies need to be carried out with populations with higher gene frequencies for $PON1_{R192}$.

2.5. Recently Discovered PON1 Polymorphisms

Figure 4 shows the positions and gene frequencies for the 180 new PON1 SNPs that were revealed in the PON1 re-sequencing efforts of Nickerson and co-workers (Seattle SNPs http://pga.gs.washington.edu). Since PON1 appears to have an

Additional *PON1* SNPs discovered by SeattleSNPs
(NIEHS Environmental Genome Program)

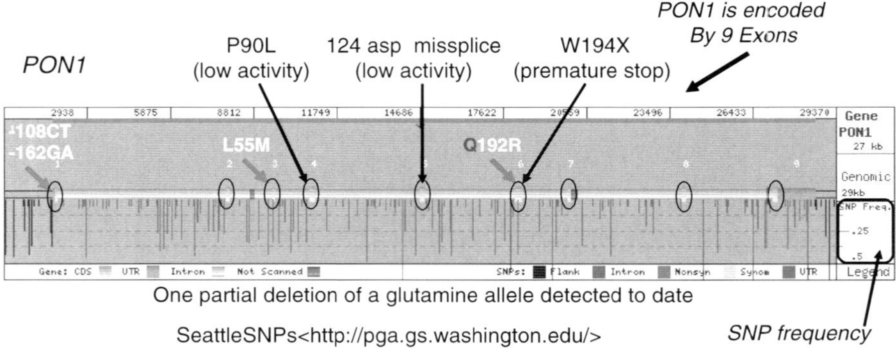

Figure 4. The human PON1 gene with recently discovered PON1 polymorphisms and mutations

important role as a risk factor for vascular disease, the entire PON1 genes of 47 individuals (24 African-Americans and 23 Europeans) were re-sequenced. This effort revealed 8 new promoter region PMs that have yet to be characterized, 1 new coding PM (W194X), 162 additional intronic PMs, and 9 additional 3′ UTR PMs. The elucidation of the W194X prompted us to examine a previously studied set of samples where there was a disagreement between PCR determined genotype and the functional phenotype assay (Jarvik et al., 2003). Some individuals genotyped as heterozygotes and by the functional assay were shown to produce only one PON1 alloform. Sequencing of the entire PON1 genes of several individuals who exhibited discrepancies between genotype and phenotype revealed one additional individual with the W194X mutation, one with a P90L mutation, one with an Asp124 splice site mutation, and one individual with a probable partial deletion of their $PON1_{Q192}$ allele. These results demonstrate the ability of the functional assay to identify heterozygous individuals with mutations in one of their PON1 genes. Individuals who are homozygous for $PON1_{Q192}$ or $PON1_{R192}$ would be identified as possibly having a PON1 mutation only by having low levels of PON1. It would require a sophisticated sequencing effort to identify such individuals.

Jarvik et al. (2003) examined the comparison of haplotype associations with risk for CAAD with measures of PON1 activity. The PMs included A(-162)G, C(-108)T, L55M, and Q192R. The results showed that PON1 did not have useful linkage disequilibrium across the gene, although there were several discrete regions of the PON1 gene with strong local LD (Fig. 5). They concluded that PON1 activity predicted CAAD, whereas four functional polymorphisms did not.

The relationship between PON1 genetic variability and other diseases is not reviewed here (see Li et al., 2003; Mackness et al., 2002). However, it is important to note that the same considerations that apply to analyzing PON1 genetic variability

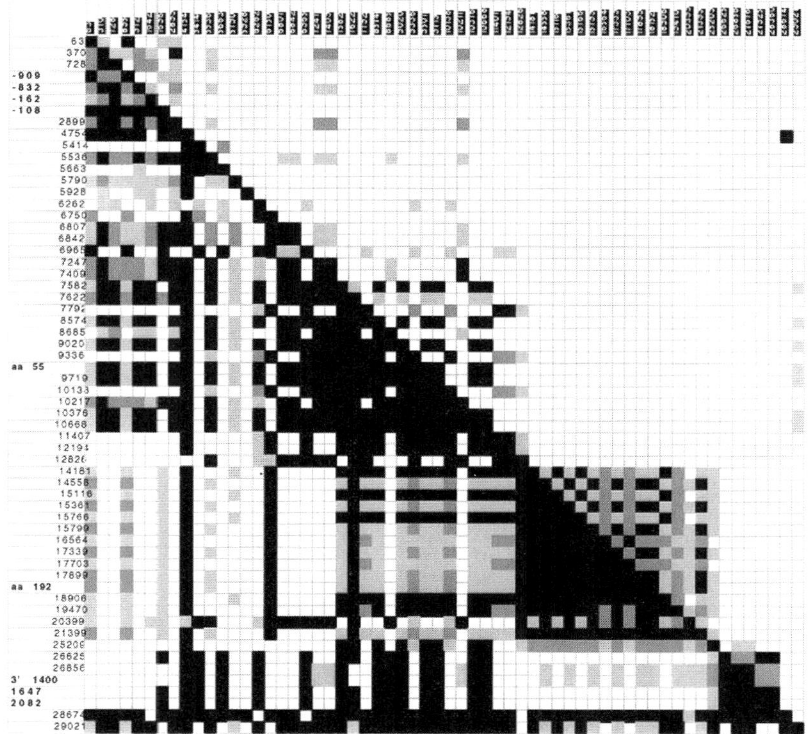

Figure 5. Associations between PON1 region SNPs with a minor allele frequency of ≥ 0.1 r2 is shown above the diagonal and D′ below. Sixty SNPs with a minor allele frequency greater than 5% were detected in the 46 chromosome SNP discovery panel. Each cell represents the observed LD value for a pair of SNPs; r2 is shown above the diagonal and D′ below the diagonal. Each cell is shaded by value: values below 0.5 are white, values between 0.5 and 0.75 are light grey, values between 0.75 and 1.0 are dark grey, and values of 1.0 are black. D′ is useful in detecting nonrecombinant regions; D′ is 1 for all pairwise comparisons within a nonrecombinant region, so these regions can be seen as dark triangles below the diagonal, as in the region between 6807 and 12194. Reproduced from Jarvik et al. 2003 with permission

for risk of OP exposure or risk for vascular disease apply to the possible relationship of PON1 genetic variability to other diseases.

3.   FUTURE RESEARCH

3.1.   Regulation of PON1 Levels

There is a lot of research to carry out in the future. With respect to regulation of PON1 expression, it appears that only about 30% of the variability of plasma PON1 levels can be explained by known polymorphisms. The possibility of the role of DNA methylation as a regulatory component has not yet been explored (see, e.g., Hantusch et al., 2007) nor have any intronic or exonic enhancer elements yet been identified. The possibility of interchromosomal regulation (see, e.g., Ling and Hoffman, 2007) has not yet been explored. Environmental regulation of PON1 has been reviewed recently and will not be covered here (Costa et al., 2005) except to note that the factors that modulate the activity of plasma PON1 appear to have only a modest effect on levels and over time, PON1 levels for a given individual appear to be quite stable over time (Zech and Zurcher, 1974).

Two recent studies by Winnier et al. examined the effects of other loci on PON1 expression. They found significant evidence for one locus on chromosome 12 (LOD = 3.56) that influenced PON1 levels with suggestive evidence for two other *PON*-related quantitative trait loci (QTLs) on chromosomes 17 and 19 (Winnier et al., 2006). In a follow up study, they found evidence for a sex-specific QTL on chromosome 17 (LOD = 2.32, P = 0.0003) that accounted for 6% of the additive genetic variance in males and 20% in females (Winnier et al., 2007).

3.2.   Epidemiological Studies

It should be clear from the studies reviewed in this chapter and Chap. 1 that it is critical to measure PON1 activities/levels in any study that explores the relationship of genetic variability of PON1 with risk of exposure or disease – to fail to do so will provide inadequate data from which to draw meaningful conclusions. If you plan to carry out an epidemiological study and feel that you can estimate levels of plasma PON1 accurately, try with a small subset of samples to characterize all 180 plus PON1 SNPs to predict PON1 levels and compare the prediction with actual measurements of initial rates of hydrolysis useful substrates such as phenyl acetate that allows comparison of PON1 levels across genotypes (Furlong et al., 2006). It is quite certain that you will not be able to accurately predict this most important parameter.

In a recent review, Deakin and James (2004) state, *"Thus when carrying out studies into the association of PON1 with CHD and other diseases, it is essential that PON1 serum concentration and/or activity are measured. The fact that the vast majority of published studies do not take into account that individual serum PON1 levels may be responsible for the lack of conclusive epidemiological evidence for a link between PON1 and CHD."*

La Du (2003) notes, *"The conclusion in the article by Jarvik et al. (2003) that the level of PON1 activity is the best predictor of vascular disease should encourage other workers in this field to continue in this direction, rather than search for key SNPs to advance the field, at this time."*

3.3.   Significant Substrates of PON1

Much more needs to be learned about the physiologically significant substrates for PON1, including the family of quorum sensing factors, lipid substrates and drugs as well as the effect of the Q192R polymorphism on their metabolism.

The genetically modified mice generated by Shih and colleagues at UCLA will provide an excellent system with which to address these questions, as they have for the research described here, in Chap. 1, and elsewhere in this volume. The role of all three PONs in modulating oxidative stress is an important area of future research.

3.4.   Structure-function Studies

The only crystal structure obtained to date for a PON, is from a domain shuffled, evolved PON. It will be useful to have a solution structure for $PON1_{R192}$, $PON1_{Q192}$ and the L55M variants of each.

ACKNOWLEDGEMENTS

The following grants provided support for this effort: ES09883, ES07033, ES-09601/EPA-R826886, ES04696 and P30ES07033.

REFERENCE

Adkins, S., Gan, K.N., Mody, M., La Du, B.N., 1993, Molecular basis for the polymorphic forms of human serum paraoxonase/arylesterase: Glutamine or arginine at position 191 for the respective A or B allozymes, Am. J Hum. Gene. 52:598–608

Blatter Garin, M.-C., James, R.W., Dussoix, P., et al., 1997, Paraoxonase polymorphism Met-Leu54 is associated with modified serum concentrations of the enzyme. A possible link between the paraoxonase gene and increased risk of cardiovascular disease in diabetes. J. Clin. Invest. 99:62–66

Brophy, V.H., Hastings, M.D., Clendenning, J.B., Richter, R.J., Jarvik, G.P., Furlong, C.E., 2001a, Polymorphisms in the human paraoxonase (PON1) promoter. Pharmacogenetics 11:77–84

Brophy, V.H., Jampsa, R.L., Clendenning, J.B., Jarvik, G.P., Furlong, C.E., (2001b), Promoter polymorphisms affect paraoxonase (PON1) expression. Am. J. Hum. Genet. 68:1428–1436

Chambers, J.E., Ma, R., Boone, J.S., Chambers, H.W., 1994, Role of detoxication pathways in acute toxicity of phosphorothioate insecticides in the rat. Life Sci. 54:1357–1364

Costa, L.G., Vitalone, A., Cole, T.B., Furlong, C.E., 2005, Modulation of paraoxonase (PON1) activity. Biochem Pharmacol., 15,69(4):541–550

Deakin, S.P., James, R.W., 2004, Genetic and environmental factors modulating serum concentrations and activities of the antioxidant enzyme paraoxonase-1. Clin Sci (Lond). 107(5):435–447

Deakin, S., Leviev, I., Brulhart-Meynet, M.C., James, R.W., 2003, Paraoxonase-1 promoter haplotypes and serum paraoxonase: a predominant role for polymorphic position - 107, implicating the Sp1 transcription factor. Biochem, J. 372(Pt 2):643–649

Davies, H., Richter, R.J., Keifer, M., Broomfield, C., Sowalla, J., Furlong, C.E., 1996, The effect of the human serum paraoxonase polymorphism is reversed with diazoxon, soman and sarin. Nature Genet. 14:334–336

Furlong, C., Holland, N., Richter, R., Bradman, A., Ho, A., Eskenazi, B., 2006, PON1 status of farmworker mothers and children as a predictor of organophosphate sensitivity. Pharmacogenetics and Genomics. 16:183–190

Hantusch, B., Kalt, R., Krieger, S., Puri, C., Kerjaschki, D., 2007, Sp1/Sp3 and DNA-methylation contribute to basal transcriptional activation of human podoplanin in MG63 versus Saos-2 osteoblastic cells. BMC Mol Biol., 8:20

Harel, M., Aharoni, A., Gaidukov, L., Brumshtein, B., Khersonsky, O., Meged, R., Dvir, H., Ravelli, R.B.G., McCarthy, A., Toker, L., Silman, I., Sussman, J.L., Tawfik, D.S., 2004, Structure and evolution of the serum paraoxonase family of detoxifying and anti-atherosclerotic enzymes. Nature Structural & Molecular Biology 11:412–419

Hassett, C., Richter, R.J., Humbert, R., Chapline, C., Crabb, J.W., Omiecinski, C.J., Furlong, C.E., 1991, Characterization of cDNA clones encoding rabbit and human serum paraoxonase: the mature protein retains its signal sequence. Biochemistry 30:10141–10149

Huff, R., Abou-Donia, M.B., 1995, In vitro effect of chlorpyrifos oxon on muscarinic receptors and adenylate cyclase. Neurotoxicology 16:281–290

Humbert, R., Adler, D.A., Disteche, C.M., Hassett, C., Omiecinski, C.J., Furlong, C.E., 1993, The molecular basis of the human serum paraoxonase activity polymorphism. Nat. Genet., 3:73–76

Jarvik, G.P., Jampsa, R., Richter, R.J., Carlson, C., Rieder, M., Nickerson, D., Furlong, C.E., 2003, Novel Paraoxonase (PON1) nonsense and missense mutations predicted by functional genomic assay of PON1 status. Pharmacogenetics 13:291–295

Jokanovic, M., 2001, Biotransformation of organophosphorus compounds. Toxicology 166:139–160

La Du, BN., 2003, Future studies of low-activity PON1 phenotype subjects may reveal how PON1 protects against cardiovascular disease. Arterioscler. Thromb. Vasc. Biol., 23(8):1317–1318

Lawlor, D.A., Day, I.N., Gaunt, T.R., Hinks, L.J., Briggs, P.J., Kiessling, M., Timpson, N., Smith, G.D., Ebrahim, S., 2004, The association of the PON1 Q192R polymorphism with coronary heart disease: findings from the British Women's Heart and Health cohort study and a meta-analysis. BMC Genet., 5:17

Leviev, I., Deakin, S., James, R.W., 2001, Decreased stability of the M54 isoform of paraoxonase as a contributory factor to variations in human serum paraoxonase concentrations. J. Lipid Res. 42:528–535

Leviev, I., James, R.W., 2000, Promoter polymorphisms of the human paraoxonase PON1 gene and serum paraoxonase activities and concentrations. Arterioscler. Thromb. Vasc. Biol. 20:516–552

Leviev, I., Negro, F., James, R.W., 1997, Two alleles of the human paraoxonase gene produce different amounts of mRNA: an explanation for differences in serum concentrations of paraoxonase associated with the (Leu-Met54) polymorphism. Arterioscler. Thromb. Vasc. Biol. 17:3935–3939

Li, H.-L., Liu, D.-P., Liang, C.-H., 2003, Paraoxonase gene polymprphisms, oxidative stress and diseases. J. Mol. Med, 81:766–779

Li, W.-F., Costa, L.G., Richter, R.J., Hagen, T., Shih, D.M., Tward, A., Lusis A.J., Furlong, C.E., 2000, Catalytic efficiency determines the in vivo efficacy of PON1 for detoxifying organophosphates. Pharmacogenetics, 10:767–780

Ling, J.Q., Hoffman, A.R., 2007, Epigenetics of long-range chromatin interactions. Pediatr Res. 2007 Mar 15; [Epub ahead of print]

Mackness, B., Davie, G.K., Turkie, W., Lee, E., Roberts, D.H., Hill, E., Roberts, C., Durrington, P.N., Mackness, M.I., 2001, Paraoxonase status in coronary heart disease: are activity and concentration more important than genotype?. Arterioscler. Thromb. Vasc. Biol., 21(9):1451–1457

Mackness, B., Durrington, P.N., Mackness, M., 2002, PON1 in other diseases. In *Paraoxonase (PON1) in Health and Disease: Basic and Clinical Aspects* (L.G. Costa and C.E. Furlong, eds.), Kluwer Academic Publishers, Boston, USA, pp. 185–195

Mackness, M., Arrol, S., Durrington, P.N., 1991, Paraoxonase prevents accumulation of lipoperoxides in low-density lipoprotein. FEBS Lett., 286:152–154

Pond, A.L., Chambers, H.W., Chambers, J.E., 1995, Organophosphate detoxication potential of various rat tissues via A-esterase and aliesterase activities. Toxicol. Lett. 70:245–252

Roest, M., van Himbergen, T.M., Barendrecht, A.B., Peeters, P.H., van der Schouw, Y.T., Voorbij, H.A., 2007, Genetic and environmental determinants of the PON-1 phenotype. Eur, J Clin, Invest., 37(3):187–196

Saikawa, Y., Price, K., Hance, K.W., Chen, T.Y., Elwood, P.C., 1995, Structural and functional analysis of the human KB cell folate receptor gene P4 promoter: cooperation of 3 clustered Sp1-binding sites with the initiator region for basal promoter activity. Biochemistry 34:995–9961

Shih, D.M., Gu, L., Xia, Y.-R., Navab, M., Li, W.-F., Hama, S., Castellani, L.W., Furlong, C.E., Costa, L.G., Fogelman, A.M., Lusis, A.J., 1998, Mice lacking serum paraoxonase are susceptible to organophosphate toxicity and atherosclerosis. Nature 394:284–287.

Suehiro, T.. Nakamura, T., Inoue, M., Shiinoki, T., Ikeda, Y., Kumon, Y., Shindo, M., et al., 2000, A polymorphism upstream from the human paraoxonase (PON1) gene and its association with PON1 expression. Atherosclerosis 150:295–298

Usmani, K.A., Cho, T.M., Rose, R.L., Hodgson, E., 2006, Inhibition of the human liver microsomal and human cytochrome P450 1A2 and 3A4 metabolism of estradiol by deployment-related and other chemicals. Drug Metab. Dispos., 34(9):1606–1614

Usmani, K.A., Rose, R.L., Hodgson, E., 2003, Inhibition and activation of the human liver microsomal and human cytochrome P450 3A4 metabolism of testosterone by deployment-related chemicals. Drug Metab. Dispos., 31(4):384–91.

Wheeler, J.G., Keavney, B.D., Watkins, H., Collins, R., Danesh, J., 2004, Four paraoxonase gene polymorphisms in 11212 cases of coronary heart disease and 12786 controls: meta-analysis of 43 studies. Lancet, 363(9410):689–695

Winnier, D.A, Rainwater, D.L., Cole, S.A., Dyer, T.D., Blangero, J., Maccluer, J.W., Mahaney, M.C., 2006, Multiple QTLs influence variation in paraoxonase 1 activity in Mexican. Americans. Hum. Biol., 78(3):341–352

Winnier, D.A., Rainwater, D.L., Cole, S.A., Williams, J.T., Dyer, T.D., Blangero, J., MacCluer, J.W., Mahaney, M.C., 2007, Sex-specific QTL effects on variation in paraoxonase 1 (PON1) activity in Mexican Americans. Genet. Epidemiol, 31(1):66–74

Yuknavage, K.L., Fenske, R.A., Kalman, D.A., Keifer, M.C., Furlong, C.E., 1997, Simulated dermal contamination with capillary samples and field cholinesterase biomonitoring. J. Toxicol. and Env. Health 51:35–55

Zech, R., Zurcher, K., 1974, Organophosphate splitting serum enzymes in different mammals. Comp. Biochem. Physiol. 48B:427–433

CHAPTER 18

DIETARY MODULATION OF PARAOXONASE-1 ACTIVITY AND CONCENTRATION IN HUMANS

C. DALGÅRD

Institute of Public Health – Environmental Medicine, University of Southern Denmark, Odense, Denmark

Abstract: Due to the assumed anti-atherogenic properties of PON1, it is of potential interest to identify pharmacological and environmental modulators of PON1 activity and concentration. Diet may be one of these environmental factors and this chapter focuses on dietary factors that may affect serum PON1 activity and concentration in humans. In general, the effect of most of the various dietary factors tested on enzyme activity is modest. However, a few studies indicate that the genetic variation in *PON1* may contribute to individual variation in response to nutrient intake. Further studies including larger sample sizes are needed to clarify this important public health issue

Keywords: Diet, fruit and vegetables, pomegranate juice, oils, dietary fatty acids, postprandial effects, nutrient-gene interaction

1.   FRUITS AND VEGETABLES – EPIDEMIOLOGIC STUDIES

Several epidemiological studies report that specific dietary patterns, i.e. diets rich in fruits, vegetables and whole grains are protective against cardiovascular disease (Hu and Willett, 2002; Hu et al., 2000) and randomised trials testing whole diets support these observations (de Lorgeril et al., 1999; Sacks et al., 2001). These diets are rich in specific vitamins, minerals and phytochemicals. Many of the phytochemicals, like flavonoids, have potent biological activities in mammalian systems, including antioxidative and anti-inflammatory effects. As discussed thoroughly throughout this book PON1 possess antioxidative and anti-inflammatory properties, thus it is possible that these diets, foods or food components modulate PON1 activity and concentration.

A few epidemiologic studies have examined the effects of fruit and vegetables on PON1 activity; however the results are not consistent. In a small-sized community based cross-sectional survey performed three times after an education programme

B. Mackness et al. (eds.), The Paraoxonases: Their Role in Disease Development and Xenobiotic Metabolism, 283–293.

was initiated in the community demonstrated increased intake of fruit and vegetables (Rowley et al., 2001) and PON1 activity (Cohen et al., 2002) over the years. Regrettably, it is not possible to make valid conclusion as the dietary intake was estimated at the community level and the PON1 activity was measured at the individual level. The intake of vegetables, fruits and berries were negatively correlated with enzyme activity in Finish women with adequate nutrient intakes but not in men (Kleemola et al., 2002). In Spanish subjects no associations were found between the intake of vitamin C and E (Ferre et al., 2003) in contrast to the positive association observed in elderly male Americans (Jarvik et al., 2002). Finally, an Australian study observed a positive correlation in Greek migrants between PON1 activity and plasma concentrations of carotenoids (Lee et al., 2005), which reflects an increased intake of vegetables and fruits (Al Delaimy et al., 2005).

These ambiguous results could be caused by the differences in the methods used for assessment of dietary intakes, since food frequency questionnaires may discriminates intakes among persons differently than 3-d food records do. Moreover, little variation in intakes among persons within the study populations may typically result in no association observed and in addition, the association between foods (or nutrients/phytochemicals) and paraoxonase activity may well be non-linear. Finally, the study populations differ in age and health status and all are of small size. These are all factors that may confound the association together with the fact that high vitamin intake are generally associated with a healthy lifestyle.

Larger studies using validated methods for diet assessment at the individual level that include biomarkers of intakes are warranted.

2. FRUIT AND VEGETABLES – INTERVENTION STUDIES

Due to the risk of confounding factors in the above mentioned observational studies, the strongest evidence of a causal relationship between PON1 and the intake of fruit and vegetables may come from intervention studies. Based on the published human intervention studies the pomegranate juice is so far the most promising fruit-and-vegetable-modulator. Pomegranate juice has a high content of flavonoids including quercetin and ellagitannins of which punicalagin is responsible for more than half of the antioxidant activity. One liter of pomegranate juice contains more than 2 grams of punicalagin (Adams et al., 2006).

Administration of 50 mL pomegranate juice for two weeks to 13 healthy young men resulted in an increased PON1 activity (Aviram et al., 2000). Also, 19 elderly patients with carotid artery stenosis was randomised to 50 mL pomegranate each day for a year (n = 10) or nothing (control, n = 9) which resulted in an increased PON1 activity by 83% in line with a decreased oxidised LDL antigen level already after three months. Furthermore, the pomegranate consumption resulted in a decreased mean intima (IMT) thickness by 35% in contrast to the patients that did not drink pomegranate juice who increased the IMT by 9% (Aviram et al., 2004). There was no additional effect of pomegranate juice consumption for 3 years in the five patients that agreed to continue the trial (Aviram et al., 2004). In 10 male type

II diabetics consumption of 50 mL pomegranate juice for 3 months resulted in increased PON1 activity by 24% as well as decreased levels of lipid peroxides and TBARS (Rosenblat et al., 2006). Although pomegranate juice thus seems to increase PON1 activity, the above described studies have limited power of evidence due to certain methodological issues, e.g. treatments are not allocated by randomisation (Aviram et al., 2000; Rosenblat et al., 2006), control groups are not treated like the treated group in all other aspects but the intervention regime (Aviram et al., 2000; Aviram et al., 2004), the studies do not describe whether they are (investigator) blinded and finally, the studies do not control for or measure usual diet. Nonetheless, the studies are supported by evidence from experimental studies in mice consistent with the idea that pomegranate juice is a modulator of PON1 activity (Kaplan et al., 2001).

Other juices made on fruits or vegetables are not able to modify PON1 activity. Administration of 330 mL tomato juice daily for 8 weeks increased PON1 activity in healthy elderly but so did mineral water in the control group (Bub et al., 2002). In healthy young subjects neither 330 mL carrot or tomato juice affected PON1 activity after 2 weeks of supplementation (Bub et al., 2005). Finally, we observed no changes in PON1 activity after 4 weeks of supplementation with 250 mL orange and 250 mL black currant juice daily in patients with peripheral arterial disease (unpublished results). Other drinks rich in polyphenols have been tested, for instance red wine increased PON1 paraoxonase activity in healthy men, but so did the beer and spirits, thus it may have been an alcohol-induced increase and not a result of an increased antioxidant intake (van der Gaag et al., 1999). Finally, a well controlled 6 weeks intervention study in 77 Finnish healthy subjects tested four different diets; high in linoleic acid and either low in vegetables or high in vegetables, berries and apples, or high in oleic acid and either low in vegetables or high in vegetables, berries and apples in a parallel design. PON1 activity decreased in all treatment groups (Freese et al., 2002). Thus the study indicates no difference in PON1 activity during either high or low intake of various dietary vitamins and phytochemicals or monounsaturated or polyunsaturated fat.

The inconsistencies among these studies may be explained by the beverage (or the fruits and vegetables) choice or amount of beverage provided as it is reasonable to suggest that type of flavonoids ingested as well as plasma concentration obtained after consumption is important with respect to their physiologic functions. But also choice of enzyme substrate, study populations and length of supplement period may influence study results.

Clearly, it is difficult at the present time to depict with certainty the general effect of fruit and vegetables – and the juices – on PON1 activity or concentration. Further studies are warranted to clarify the role in humans of fruit and vegetables as well as specific dietary (antioxidant) vitamins and phytochemicals as modulators of the enzyme activity or concentration. Before any safe recommendations can be made regarding for instance the use of pomegranate juice to boost the antioxidative system and health, studies are needed that clarify the impact of the putative drug interactive effects. Pomegranate juice contains, like grapefruit juice, inhibitors of intestinal CYP which may affect statin metabolism (Sorokin et al., 2006) and a case of rhabdomyolysis during rosuvastatin treatment suggestively caused by sudden increased intake of pomegranate juice (200 mL twice weekly for 3 weeks) has been reported (Sorokin et al., 2006).

3. MECHANISMS

Still, several plausible mechanisms for the possible PON1 modulator effect at both the expression and posttranslational level of fruit and vegetables (or the specific food constituents herein, e.g. phytochemicals or vitamins C and E) have been suggested. Human PON1 is inactivated *in vitro* by oxidative stress (Aviram et al., 1999). It is therefore possible that PON1 activity is protected secondary to the antioxidative effects of the vitamins and flavonoids, i.e. scavenging reactive oxygen species or peroxyl radicals, which reduce the levels of oxidised lipoproteins. Although this may be true for some of the flavonoids (Aviram et al., 1999) vitamin E, which decrease the susceptibility of LDL and HDL to oxidation is not able to maintain or increase PON1 activity either *in vitro* (Aviram et al., 1999) or *in vivo* (Arrol et al., 2000). Thus the putative beneficial effects of fruit and vegetables may also be explained by recent work, which suggests that especially some of the flavonoids may influence PON1 mass and activity at the gene expression level as demonstrated in a human cell line (Gouedard et al., 2004a; Gouedard et al., 2004b). Specific flavonoids interact with the transcription factor aryl hydrocarbon receptor, AhR, which together with the AhR nuclear translocator binds to the xenobiotic responsive element (XREs) in target gene promoters. The PON1 gene promoter contains sequences similar to XRE (Gouedard et al., 2004a).

4. OILS, FATS AND VARIOUS FATTY ACIDS – EPIDEMIOLOGIC STUDIES

PON1 is produced in the hepatocytes and released into the blood stream mediated by high density lipoprotein, HDL. Evidence suggests that HDL size as well as composition is important for PON1 activity (James and Deakin, 2004). Consequently the effects on HDL of various oils like olive oil, rapeseed oil and sunflower oil or dietary fatty acids like saturated fatty acids and *trans* fatty acids are likely to influence PON1 concentration and activity.

Only a few epidemiologic studies are published investigating this relationship. A cross –sectional study including 654 men showed that a high intake of oleic acid, which is the dominant oil in olive oil, correlated with a high HDL concentration as well as a high PON1 activity but only in PON1 192 RR homozygous (Tomas et al., 2001). In Spanish subjects no associations were found between the intake of saturated fatty acids and tertiles of PON1 activity (Ferre et al., 2003). Although the diet has been assessed in other epidemiological studies for instance by using food-frequency questionnaires, these studies do not report on the relationship between fatty acid intake and PON1 activity albeit reporting on for instance vitamins C and E intakes (Jarvik et al., 2002; Lee et al., 2005).

5.   OILS, FATS AND VARIOUS FATTY ACIDS – INTERVENTION STUDIES

More intervention studies report results regarding the effects of various oils on PON1 activity or mass. In 14 patients with familial combined hyperlipidemia administration of 4 grams of Omacor, containing 92% n-3 fatty acids, per day for eight weeks caused a 10% increase in PON1 concentration compared with a placebo, containing corn oil. The effects are likely explained by the paralleled changes in the HDL subfractions, as the n-3 fatty acids increased the HDL_2 and decreased the HDL_3 subfractions (Calabresi et al., 2004). Evidence support that PON1 is mainly associated with HDL_2 (James and Deakin, 2004).

A cross-over design was used to study the effects of two types of fat in 33 subjects provided with diets rich in either *trans* fat (9.3 energy percent) or saturated fat (22.3 energy percent) on PON1 activity. After 4 weeks on the *trans* fat diet PON1 activity was lower by 6% compared with the diet rich in saturated fat (De Roos et al., 2002a). *Trans* fat are known to decrease HDL concentration, which may have caused the lowered PON1 activity. Unfortunately, there was no baseline measures thus it is not possible to say whether the saturated or the *trans* fat enriched diets actually resulted in increased or decreased PON1 activities *per se* – although less so after the *trans* fat.

In a well-controlled cross-over study providing 20% of energy from palm oil, partially hydrogenated soybean oil, soybean oil or canola oil for five weeks in 15 subjects no effects on PON1 activity were observed (Vega-Lopez et al., 2006). This is in contrast with the significant differences in HDL subfractions that appeared after the diets, probably due to the different fatty acid profiles of the oils, which in other studies have been demonstrated to be associated with changes in PON1 activity (Calabresi et al., 2004).

In 60 normolipidemic Moroccan men diets enriched with 25 g argan oil per day (n = 30) or 25 grams of olive oil daily for three weeks increased PON1 activity during both diets (Cherki et al., 2005). The argan oil enriched diets have a lower MUFA, lower phenolic and a higher PUFA content. The study does not report on HDL concentrations during diets.

Eating introduces pertubations of lipoprotein metabolism, and these dynamic postprandial changes in HDL concentration as well as subfractions distribution may influence PON1 concentration and activity. In humans, a few studies have been performed also testing the postprandial effects of fats on PON1 mass and activity. A breakfast rich in either saturated fatty acids or *trans* fatty acids resulted in an increased PON1 activity by less than 3% three hours after consumption of the meal in a study including 21 males who received both test meals (De Roos et al., 2002b). There was no difference in the PON1 changes between the two meals.

One study tested the postprandial effect of an eu-caloric Mediterranean style meal vs. a Western style meal with comparable total fat content in 10 young healthy males (Blum et al., 2006). The important differences between the two meals were the high content of monounsaturated fatty acids (MUFA) and low saturated fatty acid and the inclusion of vegetables and orange fruit in the Mediterranean meal compared with the Western meal. Two hours after the ingestion of either meal PON1 activity was decreased, which is likely to be a consequence of the decreased HDL level observed, however after four hours PON1 activity increased.

In 77 type II diabetics, 10 glucose intolerant patients as well as 38 controls, consumption of a high fat, high carbohydrate meal resulted in a moderately decreased PON1 activity during the postprandial phase (Beer et al., 2006), probably as a results of the normal postprandial HDL metabolism.

Also, the influence of two types of thermally stressed oils that differed in their content of MUFA or polyunsaturated fatty acids (PUFA) (olive oil vs. safflower oil) on PON1 activity was measured 4 h after consumption in 14 patients with type II diabetes (Wallace et al., 2001). Oils rich in PUFA are more susceptible to oxidation when heated and animal studies suggest that these lipid peroxidation products are taken up from the diet through the gut (Grootveld et al., 1998). It could therefore be hypothesised that these oxidised lipids would inactivate PON1 in the circulation. However, the thermally stressed safflower oil did not change PON1 maybe because no increase in circulating lipid oxidations products was observed after either meal. In contrast, the thermally stressed olive oil increased PON1 activity in parallel with an increase in postprandial lipaemia, but only in the female patients (n = 8) (Wallace et al., 2001).

Another study have demonstrated a decreased PON1 activity 4 h after consumption of used fat (similar to the thermally stressed oil, although the heating period was much longer) in contrast to an increased activity 4 h after consumption of the corresponding non-used oil in 12 healthy males (Sutherland et al., 1999).

Several factors, such as a large variety of different feeding regimes, different study populations, and different study periods or time schedules in the postprandial studies, complicate the interpretation of the above described results. Furthermore, some results seem to contradict each other. However, some prudent general conclusions may be made; type and amount of fat intake have an effect on PON1 activity, which is likely caused by the influence on HDL although direct effects on PON1 expression cannot be ruled out.

6.   MECHANISMS

As thoroughly reviewed by James and Deakin (James and Deakin, 2004), HDL is of major importance for PON1. PON1 is transferred from the cell membrane to HDL as HDL anchors transiently to the cell membrane by a specific receptor. The lipid component of HDL seems to be of importance in this secretary process (James and Deakin, 2004). Furthermore, PON1 seems to be found primarily in HDL_2 (James and Deakin, 2004) although this is a matter of dispute (Kontush et al., 2003). It is well-known that higher intakes of most types of fat increase HDL cholesterol concentration, the major exception is *trans* fatty acids that decrease HDL (Denke, 2006). Accordingly, the results from the above described studies could be explained by the effects of increased fat intake on HDL cholesterol, i.e. the more HDL lipid the more PON1 is excreted to circulation and as a consequence higher enzyme activity can be demonstrated. However, another possible mechanism can be suggested to explain an effect of various fatty acids on PON1 activity and mass. Fatty acids and, in particular, PUFAs are good activators of peroxisome proliferator activated receptors (PPARs) which upregulate the transcription of apo AI. This protein component of HDL is important as it stabilises the enzyme and increases the activity (Deakin et al., 2002).

7.   GENE-DIET INTERACTION IN MODULATION OF PON1 ACTIVITY AND CONCENTRATION

As described elsewhere in this book, PON1 genotype is of major importance for mass and activity and a recent review demonstrated a small increased risk of cardiovascular disease for carriers of the 192R-allele. Is has been demonstrated that the 192R-isoform of the enzyme has decreased ability top prevent the oxidation of LDL and hydrolyse lipid-peroxide as well as is more sensitive to inhibition by oxidative stress (Aviram et al., 1998; Mackness et al., 1998). This may lead to speculations whether certain specific genotypes are more responsive to dietary components than others.

Only a few studies have tested the interaction between *PON1* gene variations and diet on PON1 activity or mass in humans. For instance, no influence on PON1 activity of Q192R during tomato juice consumption could be demonstrated in healthy subjects (Bub et al., 2005). Nor did this genotype influence the PON1 activity response to tomato juice in elderly subjects, although subjects carrying the 192R-allele had a reduced lipid peroxidation level after treatment (Stanger et al., 2002). In addition the decrease in PON1 activity after a diet high in vegetables was highest in the women carrying the high activity genotypes (Rantala et al., 2002). Our results suggest that there may be a moderate interaction between orange and black currant supplementation and the L55M genotypes, such that only in L allele carriers PON1 activity increased significantly while in the MM homozygotes a decrease was observed (unpublished results). Significant interactions have been described between the intake of specific PUFAs and *PPAR* gene variations and as well as *APO AI* gene variations (Ordovas, 2006). Thus, presence of these polymorphisms may therefore indirectly influence the PON1 activity response to PUFA's. So far, this has not been investigated.

However, the sample sizes are often small in the above described studies for making valid interferences and results should be interpreted with caution. Even so, the results illustrate that genetic variation may contribute to individual variation in response to nutrient intake. Further studies including larger sample sizes are needed to clarify this important public health issue.

8.   CONCLUDING REMARKS

In general, from the above narrative review of the literature on diet and PON1 activity and concentration it has become clear that diet does affect PON1 activity (less evidence is present for PON1 concentration), but the effect is mostly modest – with pomegranate juice as an exception. However, results are too inconsistent to make clear inferences or even recommendations.

Other vitamins, which may directly or indirectly influence PON1 activity or mass, have had minimal focus in the literature. For instance, it could be speculated that niacin, which inhibits the apo AI catabolism and thus is one of the best known therapies to increase HDL (Meyers et al., 2004), could increase PON1 activity. Also, an increased intake of folic acid and/or vitamin B_{12} is associated with lower blood homocysteine concentrations. A recent study suggests that homocysteine thiolactone, which blood concentration is proportional to the homocysteine level (Jakubowski, 2000), decreases PON1 activity (Ferretti et al., 2003). (Hydrolysis of lactones appears to be the primary function of PON1 (Khersonsky and Tawfik, 2006)). It could therefore be speculated, that an increased intake of folic acid would increase PON1 activity. However, studies testing these hypothesises have not been performed or published yet.

REFERENCES

Adams, L. S., Seeram, N. P., Aggarwal, B. B., Takada, Y., Sand, D., and Heber, D. (2006). Pomegranate juice, total pomegranate ellagitannins, and punicalagin suppress inflammatory cell signaling in colon cancer cells. *J Agric. Food Chem* **54**, 980–985.

Al Delaimy, W. K., Ferrari, P., Slimani, N., Pala, V., Johansson, I., Nilsson, S., Mattisson, I., Wirfalt, E., Galasso, R., Palli, D., Vineis, P., Tumino, R., Dorronsoro, M., Pera, G., Ock+®, M. C., Bueno-de-Mesquita, H. B., Overvad, K., Chirlaque, M. D., Trichopoulou, A., and Naska, A. (2005). Plasma carotenoids as biomarkers of intake of fruits and vegetables: individual-level correlations in the European Prospective Investigation into Cancer and Nutrition (EPIC). *European Journal of Clinical Nutrition* **59**, 1387–1396.

Arrol, S., Mackness, M. I., and Durrington, P. N. (2000). Vitamin E supplementation increases the resistance of both LDL and HDL to oxidation and increases cholesteryl ester transfer activity. *Atherosclerosis* **150**, 129–134.

Aviram, M., Billecke, S., Sorenson, R., Bisgaier, C., Newton, R., Rosenblat, M., Erogul, J., Hsu, C., Dunlop, C., and La Du, B. (1998). Paraoxonase Active Site Required for Protection Against LDL Oxidation Involves Its Free Sulfhydryl Group and Is Different From That Required for Its Arylesterase/Paraoxonase Activities: Selective Action of Human Paraoxonase Allozymes Q and R. *Arteriosclerosis, Thrombosis, and Vascular Biology* **18**, 1617–1624.

Aviram, M., Dornfeld, L., Rosenblat, M., Volkova, N., Kaplan, M., Coleman, R., Hayek, T., Presser, D., and Fuhrman, B. (2000). Pomegranate juice consumption reduces oxidative stress, atherogenic modifications to LDL, and platelet aggregation: studies in humans and in atherosclerotic apolipoprotein E-deficient mice. *Am.J.Clin.Nutr.* **71**, 1062–1076.

Aviram, M., Rosenblat, M., Billecke, S., Erogul, J., Sorenson, R., Bisgaier, C. L., Newton, R. S., and La Du, B. (1999). Human serum paraoxonase (PON 1) is inactivated by oxidized low density lipoprotein and preserved by antioxidants. *Free Radic.Biol.Med.* **26**, 892–904.

Aviram, M., Rosenblat, M., Gaitini, D., Nitecki, S., Hoffman, A., Dornfeld, L., Volkova, N., Presser, D., Attias, J., Liker, H., and Hayek, T. (2004). Pomegranate juice consumption for 3 years by patients with carotid artery stenosis reduces common carotid intima-media thickness, blood pressure and LDL oxidation. *Clin Nutr* **23**, 423–433.

Beer, S., Moren, X., Ruiz, J., and James, R. W. (2006). Postprandial modulation of serum paraoxonase activity and concentration in diabetic and non-diabetic subjects. *Nutrition, Metabolism and Cardiovascular Diseases* **16**, 457–465.

Blum, S., Aviram, M., Ben Amotz, A., and Levy, Y. (2006). Effect of a Mediterranean meal on postprandial carotenoids, paraoxonase activity and C-reactive protein levels. *Ann.Nutr Metab* **50**, 20–24.

Bub, A., Barth, S., Watzl, B., Briviba, K., Herbert, B. M., Luhrmann, P. M., Neuhauser-Berthold, M., and Rechkemmer, G. (2002). Paraoxonase 1 Q192R (PON1-192) polymorphism is associated with reduced lipid peroxidation in R-allele-carrier but not in QQ homozygous elderly subjects on a tomato-rich diet. *Eur.J Nutr.* **41**, 237–243.

Bub, A., Barth, S. W., Watzl, B., Briviba, K., and Rechkemmer, G. (2005). Paraoxonase 1 Q192R (PON1-192) polymorphism is associated with reduced lipid peroxidation in healthy young men on a low-carotenoid diet supplemented with tomato juice. *Br.J Nutr* **93**, 291–297.

Calabresi, L., Villa, B., Canavesi, M., Sirtori, C. R., James, R. W., Bernini, F., and Franceschini, G. (2004). An [omega]-3 polyunsaturated fatty acid concentrate increases plasma high-density lipoprotein 2 cholesterol and paraoxonase levels in patients with familial combined hyperlipidemia*1. *Metabolism* **53**, 153–158.

Cherki, M., Derouiche, A., Drissi, A., El Messal, M., Bamou, Y., Idrissi-Ouadghiri, A., Khalil, A., and Adlouni, A. (2005). Consumption of argan oil may have an antiatherogenic effect by improving paraoxonase activities and antioxidant status: Intervention study in healthy men. *Nutrition, Metabolism and Cardiovascular Diseases* **15**, 352–360.

Cohen, J., Jenkins. A. J., Karschimkus, C., Qing, S., Lee, C. T., O'Dea, K., Best, J. D., and Rowley, K. G. (2002). Paraoxonase and other coronary risk factors in a community-based cohort. *Redox.Rep.* **7**, 304–307.

Deakin, S., Leviev, I., Gomaraschi, M., Calabresi, L., Franceschini, G., and James, R. W. (2002). Enzymatically Active Paraoxonase-1 Is Located at the External Membrane of Producing Cells and Released by a High Affinity, Saturable, Desorption Mechanism. *Journal of Biological Chemistry* **277**, 4301–4308.

Denke, M. A. (2006). Dietary fats, fatty acids, and their effects on lipoproteins. *Curr.Atheroscler.Rep.* **8**, 466–471.

Durrington, P. N., Mackness, B., and Mackness, M. I. (2002). The Hunt for Nutritional and Pharmacological Modulators of Paraoxonase. *Arteriosclerosis, Thrombosis, and Vascular Biology* **22**, 1248–1250.

de Lorgeril, M., Salen, P., Martin, J. L., Monjaud, I., Delaye, J., and Mamelle, N. (1999). Mediterranean diet, traditional risk factors, and the rate of cardiovascular complications after myocardial infarction: final report of the Lyon Diet Heart Study [see comments]. *Circulation* **99**, 779–785.

De Roos, N. M., Schouten, E. G., Scheek, L. M., van Tol, A., and Katan, M. B. (2002a). Replacement of dietary saturated fat with trans fat reduces serum paraoxonase activity in healthy men and women. *Metabolism* **51**, 1534–1537.

De Roos, N. M., Siebelink, E., Bots, M. L., van Tol, A., Schouten, E. G., and Katan, M. B. (2002b). Trans monounsaturated fatty acids and saturated fatty acids have similar effects on postprandial flow-mediated vasodilation. *Eur.J Clin Nutr* **56**, 674–679.

Ferre, N., Camps, J., Fernandez-Ballart, J., Arija, V., Murphy, M. M., Ceruelo, S., Biarnes, E., Vilella, E., Tous, M., and Joven, J. (2003). Regulation of Serum Paraoxonase Activity by Genetic, Nutritional, and Lifestyle Factors in the General Population. *Clinical Chemistry* **49**, 1491–1497.

Ferretti, G., Bacchetti, T., Marotti, E., and Curatola, G. (2003). Effect of homocysteinylation on human high-density lipoproteins: A correlation with paraoxonase activity*1. *Metabolism* **52**, 146–151.

Freese, R., Alfthan, G., Jauhiainen, M., Basu, S., Erlund, I., Salminen, I., Aro, A., and Mutanen, M. (2002). High intakes of vegetables, berries, and apples combined with a high intake of linoleic or oleic acid only slightly affect markers of lipid peroxidation and lipoprotein metabolism in healthy subjects. *American Journal of Clinical Nutrition* **76**, 950–960.

Gouedard, C., Barouki, R., and Morel, Y. (2004a). Dietary polyphenols increase paraoxonase 1 gene expression by an aryl hydrocarbon receptor-dependent mechanism. *Mol.Cell Biol* **24**, 5209–5222.

Gouedard, C., Barouki, R., and Morel, Y. (2004b). Induction of the Paraoxonase-1 Gene Expression by Resveratrol. *Arteriosclerosis, Thrombosis, and Vascular Biology* **24**, 2378–2383.

Grootveld, M., Atherton, M. D., Sheerin, A. N., Hawkes, J., Blake, D. R., Richens, T. E., Silwood, C. J. L., Lynch, E., and Claxson, A. W. D. (1998). In Vivo Absorption, Metabolism, and Urinary Excretion of alpha, beta -Unsaturated Aldehydes in Experimental Animals. Relevance to the Development of

Cardiovascular Diseases by the Dietary Ingestion of Thermally Stressed Polyunsaturate-rich Culinary Oils. *Journal of Clinical Investigation* **101**, 1210–1218.

Hu, F. B., Rimm, E. B., Stampfer, M. J., Ascherio, A., Spiegelman, D., and Willett, W. C. (2000). Prospective study of major dietary patterns and risk of coronary heart disease in men. *Am.J.Clin.Nutr.* **72**, 912–921.

Hu, F. B., and Willett, W. C. (2002). Optimal diets for prevention of coronary heart disease. *JAMA* **288**, 2569–2578.

Jakubowski, H. (2000). Homocysteine Thiolactone: Metabolic Origin and Protein Homocysteinylation in Humans. *Journal of Nutrition* **130**, 377.

James, R. W., and Deakin, S. P. (2004). The importance of high-density lipoproteins for paraoxonase-1 secretion, stability, and activity. *Free Radic.Biol Med* **37**, 1986–1994.

Jarvik, G. P., Rozek, L. S., Brophy, V. H., Hatsukami, T. S., Richter, R. J., Schellenberg, G. D., and Furlong, C. E. (2000). Paraoxonase (PON1) phenotype is a better predictor of vascular disease than is PON1(192) or PON1(55) genotype. *Arteriosclerosis, Thrombosis, and Vascular Biology* **20**, 2441–2447.

Jarvik, G. P., Tsai, N. T., McKinstry, L. A., Wani, R., Brophy, V. H., Richter, R. J., Schellenberg, G. D., Heagerty, P. J., Hatsukami, T. S., and Furlong, C. E. (2002). Vitamin C and E Intake Is Associated With Increased Paraoxonase Activity. *Arteriosclerosis, Thrombosis, and Vascular Biology* **22**, 1329–1333.

Kaplan, M., Hayek, T., Raz, A., Coleman, R., Dornfeld, L., Vaya, J., and Aviram, M. (2001). Pomegranate Juice Supplementation to Atherosclerotic Mice Reduces Macrophage Lipid Peroxidation, Cellular Cholesterol Accumulation and Development of Atherosclerosis. *Journal of Nutrition* **131**, 2082–2089.

Khersonsky, O., and Tawfik, D. S. (2006). The Histidine 115-Histidine 134 Dyad Mediates the Lactonase Activity of Mammalian Serum Paraoxonases. *Journal of Biological Chemistry* **281**, 7649–7656.

Kleemola, P., Freese, R., Jauhiainen, M., Pahlman, R., Alfthan, G., and Mutanen, M. (2002). Dietary determinants of serum paraoxonase activity in healthy humans. *Atherosclerosis* **160**, 425–432.

Kontush, A., Chantepie, S., and Chapman, M. J. (2003). Small, Dense HDL Particles Exert Potent Protection of Atherogenic LDL Against Oxidative Stress. *Arteriosclerosis, Thrombosis, and Vascular Biology* **23**, 1881–1888.

Lee, C. T., Rowley, K., Jenkins, A. J., O'Dea, K., Itsiopoulos, C., Stoney, R. M., Su, Q., Giles, G. G., and Best, J. D. (2005). Paraoxonase activity in Greek migrants and Anglo-Celtic persons in the Melbourne Collaborative Cohort Study: relationship to dietary markers. *Eur J Nutr* **44**, 223–230.

Mackness, B., Durrington, P., McElduff, P., Yarnell, J., Azam, N., Watt, M., and Mackness, M. (2003). Low Paraoxonase Activity Predicts Coronary Events in the Caerphilly Prospective Study. *Circulation* **107**, 2775–2779.

Mackness, B., Mackness, M. I., Arrol, S., Turkie, W., and Durrington, P. N. (1998). Effect of the human serum paraoxonase 55 and 192 genetic polymorphisms on the protection by high density lipoprotein against low density lipoprotein oxidative modification. *Febs Letters* **423**, 57–60.

Meyers, C. D., Kamanna, V. S., and Kashyap, M. L. (2004). Niacin therapy in atherosclerosis. *Curr.Opin.Lipidol.* **15**, 659–665.

Ng, C. J., Shih, D. M., Hama, S. Y., Villa, N., Navab, M., and Reddy, S. T. (2005). The paraoxonase gene family and atherosclerosis[star, open]. *Free Radical Biology and Medicine* **38**, 153–163.

Ordovas, J. M. (2006). Genetic interactions with diet influence the risk of cardiovascular disease. *American Journal of Clinical Nutrition* **83**, 443S–4446.

Rantala, M., Silaste, M. L., Tuominen, A., Kaikkonen, J., Salonen, J. T., Alfthan, G., Aro, A., and Kesaniemi, Y. A. (2002). Dietary modifications and gene polymorphisms alter serum paraoxonase activity in healthy women. *J Nutr* **132**, 3012–3017.

Rosenblat, M., Hayek, T., and Aviram, M. (2006). Anti-oxidative effects of pomegranate juice (PJ) consumption by diabetic patients on serum and on macrophages. *Atherosclerosis* **187**, 363–371.

Rowley, K. G., Su, Q., Cincotta, M., Skinner, M., Skinner, K., Pindan, B., White, G. A., and O'Dea, K. (2001). Improvements in circulating cholesterol, antioxidants, and homocysteine after dietary intervention in an Australian Aboriginal community. *American Journal of Clinical Nutrition* **74**, 442–448.

Sacks, F. M., Svetkey, L. P., Vollmer, W. M., Appel, L. J., Bray, G. A., Harsha, D., Obarzanek, E., Conlin, P. R., Miller, E. R., Simons-Morton, D. G., Karanja, N., Lin, P. H., and The DASH-Sodium Collaborative Research Group (2001). Effects on Blood Pressure of Reduced Dietary Sodium and the Dietary Approaches to Stop Hypertension (DASH) Diet. *The New England Journal of Medicine* **344**, 3–10.

Shih, D. M., Gu, L., Xia, Y. R., Navab, M., Li, W. F., Hama, S., Castellani, L. W., Furlong, C. E., Costa, L. G., Fogelman, A. M., and Lusis, A. J. (1998). Mice lacking serum paraoxonase are susceptible to organophosphate toxicity and atherosclerosis. *Nature* **394**, 284–287.

Shih, D. M., Xia, Y. R., Wang, X. P., Miller, E., Castellani, L. W., Subbanagounder, G., Cheroutre, H., Faull, K. F., Berliner, J. A., Witztum, J. L., and Lusis, A. J. (2000). Combined serum paraoxonase knockout/apolipoprotein E knockout mice exhibit increased lipoprotein oxidation and atherosclerosis. *J Biol Chem* **275**, 17527–17535.

Sorokin, A. V., Duncan, B., Panetta, R., and Thompson, P. D. (2006). Rhabdomyolysis Associated With Pomegranate Juice Consumption. *The American Journal of Cardiology* **98**, 705–706.

Stanger, O., Semmelrock, H. J., Wonisch, W., Bos, U., Pabst, E., and Wascher, T. C. (2002). Effects of Folate Treatment and Homocysteine Lowering on Resistance Vessel Reactivity in Atherosclerotic Subjects. *Journal of Pharmacology And Experimental Therapeutics* **303**, 158–162.

Sutherland, W. H., Walker, R. J., de Jong, S. A., van Rij, A. M., Phillips, V., and Walker, H. L. (1999). Reduced postprandial serum paraoxonase activity after a meal rich in used cooking fat. *Arterioscler.Thromb.Vasc.Biol.* **19**, 1340–1347.

Tomas, M., Senti, M., Elosua, R., Vila, J., Sala, J., Masia, R., and Marrugat, J. (2001). Interaction between the Gln-Arg 192 variants of the paraoxonase gene and oleic acid intake as a determinant of high-density lipoprotein cholesterol and paraoxonase activity. *European Journal of Pharmacology* **432**, 121–128.

Tward, A., Xia, Y. R., Wang, X. P., Shi, Y. S., Park, C., Castellani, L. W., Lusis, A. J., and Shih, D. M. (2002). Decreased Atherosclerotic Lesion Formation in Human Serum Paraoxonase Transgenic Mice. *Circulation* **106**, 484–490.

van der Gaag, M. S., van Tol, A., Scheek, L. M., James, R. W., Urgert, R., Schaafsma, G., and Hendriks, H. F. J. (1999). Daily moderate alcohol consumption increases serum paraoxonase activity; a diet-controlled, randomised intervention study in middle-aged men. *Atherosclerosis* **147**, 405–410.

Vega-Lopez, S., Ausman, L. M., Jalbert, S. M., Erkkila, A. T., and Lichtenstein, A. H. (2006). Palm and partially hydrogenated soybean oils adversely alter lipoprotein profiles compared with soybean and canola oils in moderately hyperlipidemic subjects. *American Journal of Clinical Nutrition* **84**, 54–62.

Wallace, A. J., Sutherland, W. H., Mann, J. I., and Williams, S. M. (2001). The effect of meals rich in thermally stressed olive and safflower oils on postprandial serum paraoxonase activity in patients with diabetes. *Eur.J Clin Nutr* **55**, 951–958.

PART 7

PONs BIOCHEMISTRY

CHAPTER 19

PONS' NATURAL SUBSTRATES – THE KEY FOR THEIR PHYSIOLOGICAL ROLES

D.I. DRAGANOV AND J.F. TEIBER

Department of Metabolism, WIL Research Laboratories, LLC, Ashland, OH 44 805, USA
Department of Internal Medicine, Division of Epidemiology, The University of Texas
Southwestern Medical Center, Dallas, TX 75390, USA

Abstract: Paraoxonase family members, PON1, PON2 and PON3, have different cell and tissue distribution as well as different regulation of expression suggesting distinct physiological roles for each of them. These roles, however, remain largely unknown. Phylogenetic, structural and biochemical data demonstrate that all three PONs are primarily lactone hydrolyzing enzymes, albeit they do have different substrate specificity. Lactones are ubiquitous in nature and have the ability to affect prokaryotic and eukaryotic cellular signaling, growth and differentiation. Therefore, it is very likely that PONs exert their physiological roles by metabolizing and thus altering the biological activity and/or distribution of endogenous and exogenous lactones. The synthesis and biological activity of known or putative lactone substrates for the PONs are reviewed

Keywords: lactones, lactonase, 5-HETEL, acylhomoserine lactones

1. EVOLUTION OF THE PON ENZYMES

In mammals, the paraoxonase gene family consists of three members: PON1, PON2 and PON3 (Primo-Parmo et al., 1996). The three genes are well conserved: at the amino acid level, the orthologs share 79–95% and the paralogs about 65% identities (Draganov and La Du, 2004; Primo-Parmo et al., 1996). In other vertebrates such as birds, frogs and fishes, it appears that only one PON gene is present which shares highest percent identity with PON2. PON-like genes have been identified in the worm *C. elegans*, in plants, bacteria and fungi. Phylogenetic analysis demonstrated that PON2 is the oldest member of the family from which PON3 and next PON1 arose (Draganov and La Du, 2004). We studied the conservation and variation in the amino acid sequences of 21 mammalian PONs (5 PON1, 8 PON2 and 8 PON3) from mice, rat, rabbit, dog, cattle, swine, macaques, chimpanzee and human and

B. Mackness et al. (eds.), The Paraoxonases: Their Role in Disease Development
and Xenobiotic Metabolism, 297–305.

followed the location of the highly conserved and variable strings on the template of the solved PON1 crystal structure (Harel et al., 2004). In general, the residues comprising the β-sheets of the propeller were highly conserved among all PON sequences, while the majority of the variation among the paralogs was located primarily in the 3 helices forming a unique lid over the PON's active site. The N-terminal sequence (helix 1) is the main determinant for PONs' cell distribution, translocation and secretion (Draganov et al., 2002); however, the other two helices are also likely to have a profound effect on their protein-lipid and protein-protein interactions.

Lactonase activity is the plausible common enzymatic activity preserved through the evolution of the PON proteins. For example, a lactone hydolase from the fungus *Fusarium oxisporum* has appreciable percent similarity with PON1 and shares a number of lactone substrates with it, but has no paraxonase activity (Kobayashi et al., 1998). Recent comprehensive structure-activity studies with PON1 and PON1 variants generated through directed evolution have demonstrated that PON1's native activity is that of a lactonase and that its arylesterase and phosphotriesterase activities are promiscuous (Khersonsky and Tawfik, 2005). Furthermore, binding of PON1 to HDL particles, i.e. the natural carrier of PON1 in the blood, was shown to greatly stimulate its lactonase activity but only to a limited extent its arylesterase or phosphotriesterase activities (Gaidukov and Tawfik D, 2005). We characterized the enzymatic activities of recombinant human PON1, PON2 and PON3 towards a broad range of substrates (Draganov et al., 2005). The phosphotriesterase activity was almost exclusive to PON1. All three PONs exhibited arylesterase activity, although, with the exception of PON1 phenyl acetate hydrolysis, these activities were very low. Lactone hydrolysis/lactonization was catalyzed by all three PONs. PON1 and PON3 hydrolyzed over twenty aromatic and aliphatic lactones with a high degree of overlapping substrate specificity, whereas PON2's lactonase activity was much more restricted. Since PON2 is the oldest member of the PON family and by analogy to the directed evolution experiments described above, we hypothesized that PON2 posses the primary enzymatic activity for which the PON enzymes have evolved and that this activity would be preserved to some extent in PON1 and PON3 as well. Based on the substrate specificity determined with the purified recombinant PONs (Draganov et al., 2005) it appears, that PON1 evolved towards a smaller active site compared to PON2 and is able to catalyze non-substituted and short chain-substituted lactones, whereas PON3 evolved towards a bigger than PON2's active site, capable of accommodating bulky substrates like the statin lactones and spironolactone.

2.   NATURAL LACTONE SUBSTRATES FOR PONs

Lactones are ubiquitous in nature and are produced by organisms in all five kingdoms (*Monera, Protista, Plantae, Fungi* and *Animalia*). Natural lactones display a broad phylogenetic diversity. They can act as simple alkylating agents and are considered generic enzyme inhibitors (Konaklieva and Plotkin, 2005). Many lactones have the ability to affect prokaryotic and eukaryotic cellular signaling, growth and differentiation. Therefore, it is conceivable that PONs exert their physiological roles by metabolizing and thus altering the biological activity and/or distribution of endogenous and exogenous lactones. Lactone metabolites of polyunsaturated fatty acids (PUFA) were fond to be substrates for all three human PONs (Draganov et al., 2005; Teiber et al., 2004).

The synthesis and biological activity of arachidonic acid metabolites which are known or putative lactone substrates for the PONs are briefly reviewed below.

5,6-dihydroxytrienoic acid-1,5 lactone (5,6-DHETL). Cytochrome P_{450} epoxygenases (CYPs 1A2, 2C9, 2C19, 2D6) convert arachidonic acid into 4 epoxytrienoic acid (EET) regioisomers, 5,6-EET, 8,9-EET, 11,12-EET, and 14,15-EET, corresponding to the carbons where the epoxide group is formed (Oltman et al., 1998). Cytosolic epoxide hydrolases rapidly converted the last three EETs to respective dihydroxyeicosatrienoic acids (DHETs), however, 5,6-EET did not appear to be a substrate for them (Zeldin et al., 1993). EETs are very potent vasoactive compounds with constriction or dilation effects depending on the vessel type and experimental animal species. 5,6-EET can spontaneously re-arrange into 5,6-DHETL. Both eicosanoids produced extremely potent vasodilation in canine coronary microcirculation where 5,6-DHETL (EC50 values ranging from −15.8 to −13.1 log [M]) was about 100-fold more potent than 5,6-EET (Oltman et al., 1998). The vasodilatory effect was shown to be mediated by activation of K_{Ca} channels. We found that PON1 purified from human or rabbit serum rapidly hydrolyzed 5,6-DHETL to the corresponding 5,6-dihydroxyeicosatrienoic (5,6-DHET), but had no effect the rate of 5,6-EET conversion to 5,6-DHETL (Draganov and Stetson unpublished results). The biological activity of the 5,6-DHET was not evaluated by Oltman et al. (1998), thus it is unclear what the physiological effect of 5,6-DHETL hydrolysis, if catalyzed by the PONs *in vivo*, will be. Nevertheless, 5,6-DHETL was the first lactone metabolite of a PUFA identified as a PON substrate (Fig. 1).

5-hydroxy-eicosatetraenoic acid lactone (5-HETEL). 5-Hydroxy-6,8,11,15-eicosatetraenoic acid (5-HETE) and 5-HETEL are synthesized by 5-lipoxigenase (5-LO) in B-lymphocytes and in uterine and extrauterine tissues during pregnancy (Saeed and Mitchell, 1983; Schulam and Shearer, 1990). Schulam et al. demonstrated that 5-HETE could not be converted into its intramolecular ester 5-HETEL during extraction and sample preparation unless cells were present, suggesting that 5-HETEL is of cellular origin (Schulam and Shearer, 1990). 5-HETE and 5-HETEL inhibited platelet neutrophil phospholipase A_2 (PLA_2) and peritoneal macrophage cycloxygenase (COX) and thus leukotrien B_4, tromboxane B_2 and prostaglandin E_2 synthesis (Chang et al., 1985a, b). 5-HETEL was found to be the best substrate for all three PONs identified so far (Draganov et al., 2005; Gaidukov and Tawfik, 2005; Teiber et al., 2004). The catalytic efficiency (k_{cat}/K_M) of PON1 with 5-HETEL was determined to be in the 10^{-6} to 10^{-7} $M^{-1}s^{-1}$ range with apparent K_M in the low micromolecular range (Gaidukov and Tawfik, 2005). These values are characteristic of a highly efficient enzyme acting on its natural/physiological substrate. For comparison, most of PON1 substrates have K_M values in the low milimolar range.

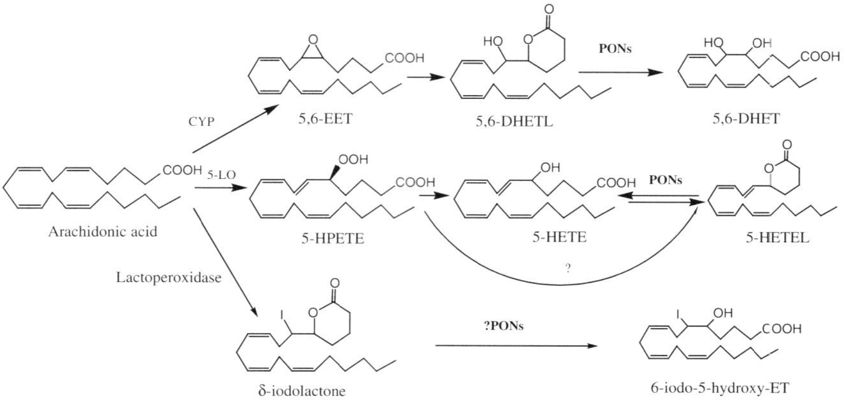

Figure 1. Structure of hydroxy-acids and lactone metabolites of arachidonic acid. See text for explanation of the abbreviations used

PON3 and PON2 hydrolyze 5-HETEL at about 3- and 10-fold slower rates than PON1, respectively. Although the physiological consequences of 5-HETEL hydrolysis remain unclear at present, 5-HETEL is a likely endogenous substrate for the PONs.

6-Iodo-5-hydroxy-8,11,14-eicosatrienoic trienoic acid δ-lactone (δ-iodolactone). Thyroid cells were found to iodinate fatty acids to form iodolactones (reviewed in Dugrillon, 1996). δ-Iodolactone was shown to inhibit signal transduction pathways induced by local growth factors such as epidermal growth factor and fibroblast growth factor. δ-Iodolactone appeared to act as a mediator of iodine in the autoregulation of the cAMP-independent thyroid cell proliferation (Dugrillion, 1996). Interestingly, the open acid form of the δ-iodolactone, 6-iodo-5-hydroxy-ET, and 4-hydroxy-7,10,13,16,19-docosapentaenoic acid γ-lactone (γ-iodolactone) had no effect on thyroid cell proliferation (Dugrillion et al., 1994). δ-Iodolactone is very similar structurally to 5,6-DHETL and 5-HETEL and thus is a putative PON substrate. Biological effects of δ-iodolactone on thyroid cell growth are relatively well characterized and it is appealing to study if PON inhibition or overexpression in these cells will modulate the δ-iodolactone activity.

Oxidative metabolites of archidonic acid such as epoxides and hydroxy-carboxylic acids do get incorporated in phospholipids in sn_2-position and it is possible that corresponding lactones are formed during their cleavage by phospholipases A_2. Oxidative metabolites of other PUFA such as eicosapentaenoic and docoahexaenoic acids may also result in formation of lactone substrates for the PONs.

Acylhomoserine lactones (AHLs). AHLs are quorum sensing factors in gram-negative bacteria. In the opportunistic pathogen *Pseudomonas aeruginosa*, N-3-oxodocanoyl-homoserine lactone (3O-C12HSL) is the primary signal molecule that regulates the expression of genes controlling virulence and biofilm formation. 3O-C12HSL and other long chain AHLs have been shown to affect gene expression in mammalian cells, including upregulation of proinflammatory mediators and induction of apoptosis (for a recent reviews see Pritchard, 2006; Shiner et al., 2005). The endogenous mammalian AHL receptors have not been identified. 3O-C12HSL is a fatty acid-based molecule that can diffuse through cell membranes, thus molecular targets for the AHLs in eukaryotic cells can be located in the cytoplasm, nucleus or at the cell surface (Kravchenko et al., 2006; Shiner et al., 2006; Zimmermann et al., 2006). We and others have demonstrated the ability of the PONs and especially of PON2 to hydrolyze and thereby inactivate AHLs (Draganov et al., 2005; Ozer et al., 2005; Yang et al., 2005), Fig. 2. We proposed that PON2 could play a dual role disrupting the AHL signaling in bacteria (and thus attenuating their virulence) and modulating the host response to the infection (Draganov et al., 2005).

Recently it was demonstrated that indeed PON2 deficiency enhanced *Pseudomonas aeruginosa* quorum sensing in murine tracheal epithelia (Stoltz et al., 2007). PON1 also hydrolyses AHLs albeit less efficiently than PON2. Paradoxically, *PON1*-knock-out mice were more protected in *Pseudomonas aeruginosa* peritonitis/sepsis model than wild type mice (Ozer et al., 2005). This was likely due to a compensatory overexpression of PON2 in the epithelia of the *PON1*-knockout mice (Ozer et al., 2005).

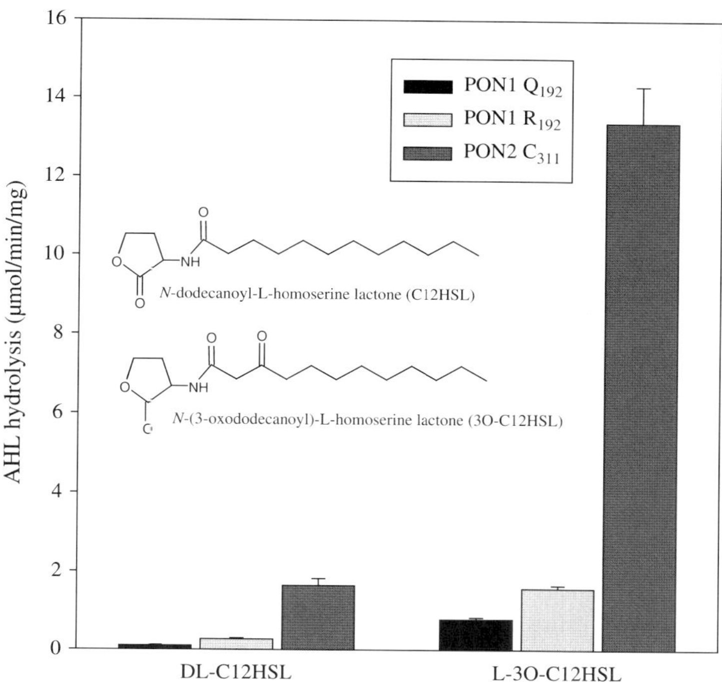

Figure 2. AHL hydrolysis by purified recombinant PON1 and PON2. Reaction conditions and analysis were as described in Draganov et al., 2005

Quorum sensing in bacteria is similar in principle to the hormone signaling in eukaryotes. AHLs exhibit structural similarities to some mammalian hormones and the later can also interact with components of the bacterial quorum sensing machinery (Shiner et al., 2005). Bacteria and eukaryotes have co-existed for millions if not billions of years and it is not surprising that the ability to decipher cross-kingdom signals has evolved (Shiner et al., 2005). Interference with quorum sensing in pathogenic bacteria would provide an evolutionary advantage for the host organism. Thus we hypothesize that the fitness for AHL hydrolysis was preserved during the evolution of the PON proteins, exemplified by PON2, the oldest PON member among the mammalian PON enzymes. Furthermore, there is evidence for the existence of a eukaryotic autocrine/paracrine system utilizing AHL-like mediator(s). The major protein product of the *rTS* gene (*ENOSF1*), rTSβ, produces signal molecules that share biological and chemical properties with AHLs and downregulate the enzyme thymidylate synthase (TS) (Dolnick, 2005; Dolnick et al., 2003). The rTSβ-overproducing cell line H630-1 was shown to secrete significant amounts of lipophilic metabolites derived from methionine that can downregulate TS protein in other cells without direct cellular contact (Dolnick et al., 2003), and this effect could be reproduced with 3O-12CHSL (Dolnick, 2005). The chemical identity of the rTSβ products remains to be established and we are currently investigating if these compounds are PON substrates.

3.   FROM A PHYSIOLOGICAL ROLE TO A SUBSTRATE

As outlined above, endogenous and exogenous lactones with distinct biological activities were found to be PONs' substrates *in vitro*, however, the physiological relevance or significance of their metabolism by the PONs *in vivo* remains to be determined. On the other hand, it could be argued, that if a physiological role for a PON enzyme has been defined, there should be a (lactone) substrate mediating that role. Here is an example of such speculation.

High-density lipoprotein-associated PON1 has been shown to reduce oxidative stress in macrophages by modulating cholesterol influx, cholesterol biosynthesis and cholesterol efflux from macrophages (reviewed in Aviram and Rosenblat, 2004). It was demonstrated that PON1 stimulates HDL binding and HDL-mediated cholesterol efflux, where both were found to involve the ABCA1 transporter (Rosenblat et al., 2005). Furthermore, PON1 stimulated HDL-induced lysophosphosphatidylcholine (lyso-PC) formation in the macrophages and loading of macrophages with lyso-PC stimulated apolipoprotein A-I mediated cholesterol efflux (Rosenblat et al., 2005). PON1's enzymatic activity (Rosenblat et al., 2005) and more specifically, its lactonase activity (Rosenblat et al., 2006) was essential for the stimulation of the macrophage cholesterol efflux. It was proposed that PON1 hydrolyses certain oxidized lipids in macrophages to generate lyso-PC (Rosenblat et al., 2006), Fig. 3, panel A. We have the following critiques to that model: 1) the model implies that all the lyso-PC generated will originate from 5-OH-fatty acid in *sn2*-postion of PC which is highly unlikely mass-balance wise; and 2) highly purified serum

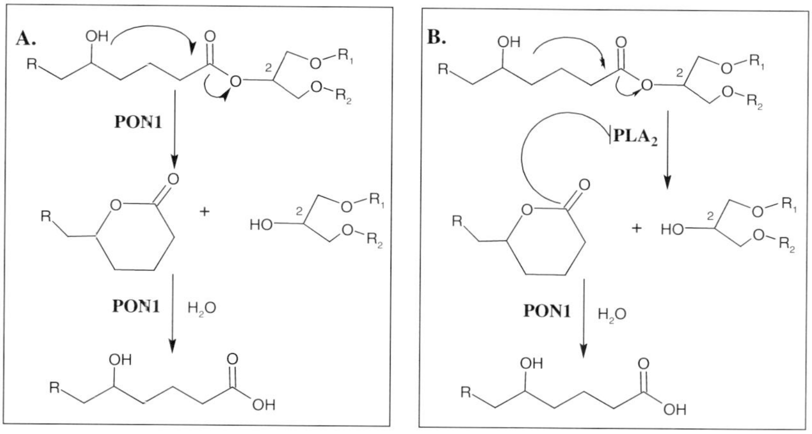

Figure 3. Hypothesized mechanisms for PON1's role in lyso-PC generation in macrophages. A, (modified from Rosenblat et al., 2006). B., an alternative model proposing a role for δ-valerolactone derivatives in regulation of PLA$_2$ activity

PON1 and recombinant PON1 did not possess phospholipase A$_2$ (PLA$_2$) activity (Connelly et al., 2005; Draganov et al., 2005; Marathe et al., 2003). Hence, we are proposing an alternative model (Fig. 3, panel B.) where lyso-PC is generated by PLA$_2$ which will be inhibited by the δ-valerolactone derivatives putatively formed during the hydrolysis of 5-6-epoxy or 5-OH-fatty acids in *sn*2-postion of the PC, unless PON1 is present. This model requires a close proximity of the PLA$_2$ and PON1 enzymes which may both be located on the HDL particle or alternatively the PLA$_2$ located in the cell membrane will get activated upon HDL binding to the ABCA1 transporter. In the absence of PON1, PLA$_2$ activity will be reversely proportional to the abundance of oxidized PC because their hydrolysis may generate lactones inhibitors for the enzyme. In the presence of PON1 such lactones will be hydrolyzed thereby prolonging PLA$_2$ activity. Noteworthy, PON1 hydrolyzes bromoenol lactone which is a classic inhibitor of γ-(Ca-independent)-PLA$_2$ (Teiber and Draganov unpublished results).

4. CONCLUDING REMARKS

There is ample evidence that PON enzymes evolved as lactonases and therefore lactones are likely to be their natural substrates. It has been demonstrated for PON1 and most likely holds for PON2 and PON3 as well, that the protein and phospholipid environment has a deep impact on PON1's stability and enzymatic activity. Therefore the "true" enzymatic activities of the PONs should be considered in the temporal and spatial constrains of their protein-protein and protein-lipid interactions. The notion that PONs' natural substrates modulate intracellular biochemical reactions and regulate gene expression necessitates the use of cell-based test systems in the study of PONs' physiological roles.

REFERENCES

Aviram, M., and Rosenblat, M., 2004, Paraoxonases 1, 2, and 3, oxidative stress, and macrophage foam cell formation during atherosclerosis development. *Free Radic. Biol. Med.* 37, 1304–1316

Chang, J., Blazek, E., Kreft, A.F., and Lewis, A.J., 1985a, Inhibition of platelet and neutrophil phospholipase A2 by hydroxyeicosatetraenoic acids (HETES). A novel pharmacological mechanism for regulating free fatty acid release. *Biochem. Pharmacol.* 34, 1571–1575

Chang, J., Lamb, B., Marinari, L., Kreft, A.F., and Lewis, A.J., 1985b, Modulation by hydroxyeicosatetraenoic acids (HETEs) of arachidonic acid metabolism in mouse resident peritoneal macrophages. *Eur. J. Pharmacol.* 107, 215–222

Connelly, P.W., Draganov, D., and Maguire, G.F., 2005, Paraoxonase-1 does not reduce or modify oxidation of phospholipids by peroxynitrite. *Free Radic. Biol. Med.* 38, 164–174

Dolnick, B.J., 2005, The rTS signaling pathway as a target for drug development. *Clin. Colorectal. Cancer* 5, 57–60

Dolnick, B.J., Angelino, N.J., Dolnick, R., and Sufrin, J.R., 2003, A novel function for the rTS gene. *Cancer Biol. Ther.* 2, 364–369

Draganov, D.I., and La Du, B.N., 2004, Pharmacogenetics of paraoxonases: a brief review. *Naunyn Schmiedebergs Arch. Pharmacol.* 369, 78–88

Draganov, D.I., Sass, K.M., Watson, C.E., Bisgaier, C.L., Reddy, S.T., Teiber, J.F., and La Du, B.N., 2002, The N-terminal sequences of human paraoxonase-1 (PON1) and paraoxonase-3 (PON3) are responsible for their different translocation and secretion. *Circulation* 106 (19), II–123.

Draganov, D.I., Teiber, J.F., Speelman, A., Osawa, Y., Sunahara, R., and La Du, B.N., 2005, Human paraoxonases (PON1, PON2, and PON3) are lactonases with overlapping and distinct substrate specificities. *J. Lipid Res.* 46, 1239–1247

Dugrillon, A., 1996, Iodolactones and iodoaldehydes–mediators of iodine in thyroid autoregulation. *Exp. Clin. Endocrinol. Diabetes.* 104 Suppl 4, 41–45

Dugrillon, A., Uedelhoven, W.M., Pisarev, M.A., Bechtner, G., and Gartner, R., 1994, Identification of delta-iodolactone in iodide treated human goiter and its inhibitory effect on proliferation of human thyroid follicles. *Horm. Metab. Res.* 26, 465–469

Gaidukov, L., and Tawfik, D.S., 2005, High affinity, stability, and lactonase activity of serum paraoxonase PON1 anchored on HDL with ApoA-I. *Biochemistry* 44, 11843–11854

Harel, M., Aharoni, A., Gaidukov, L., Brumshtein, B., Khersonsky, O., Meged, R., Dvir, H., Ravelli, R.B., McCarthy, A., Toker, L., Silman, I., Sussman, J.L., and Tawfik, D.S., 2004,Structure and evolution of the serum paraoxonase family of detoxifying and anti-atherosclerotic enzymes. *Nat. Struct. Mol. Biol.* 11, 412–419

Khersonsky, O., and Tawfik, D.S., 2005, Structure-reactivity studies of serum paraoxonase PON1 suggest that its native activity is lactonase. *Biochemistry* 44, 6371–6382

Kobayashi, M., Shinohara, M., Sakoh, C., Kataoka, M., and Shimizu, S., 1998, Lactone-ring-cleaving enzyme: genetic analysis, novel RNA editing, and evolutionary implications. *Proc. Natl. Acad. Sci. U S.* 95, 12787–12792

Konaklieva, M.I., and Plotkin, B.J., 2005, Lactones: generic inhibitors of enzymes?. *Mini Rev. Med. Chem.* 5, 73–95

Kravchenko, V.V., Kaufmann, G.F., Mathison, J.C., Scott, D.A., Katz, A.Z., Wood, M.R., Brogan, A.P., Lehmann, M., Mee, J.M., Iwata, K., Pan, Q., Fearns, C., Knaus, U.G., Meijler, M.M., Janda, K.D., and Ulevitch, R.J., 2006, N-(3-oxo-acyl)homoserine lactones signal cell activation through a mechanism distinct from the canonical pathogen-associated molecular pattern recognition receptor pathways. *J. Biol. Chem.* 281, 28822–28830

Marathe, G.K., Zimmerman, G.A., and McIntyre, T.M., 2003, Platelet-activating factor acetylhydrolase, and not paraoxonase-1, is the oxidized phospholipid hydrolase of high density lipoprotein particles. *J. Biol. Chem.* 278, 3937–3947

Oltman, C.L, Weintraub, N.L, VanRollins, M., and Dellsperger, K.C., 1998, Epoxyeicosatrienoic acids and dihydroxyeicosatrienoic acids are potent vasodilators in the canine coronary microcirculation. *Circ. Res.* 83, 932–939

Ozer, E.A., Pezzulo, A., Shih, D.M., Chun, C., Furlong, C., Lusis, A.J., Greenberg, E.P., and Zabner, J., 2005, Human and murine paraoxonase 1 are host modulators of Pseudomonas aeruginosa quorum-sensing. *FEMS Microbiol. Lett.* 253, 29–37

Primo-Parmo, S.L., Sorenson, R.C., Teiber, J., and La Du, B.N., 1996, The human serum paraox-onase/arylesterase gene (PON1) is one member of a multigene family. *Genomics* 33, 498–507

Pritchard, D.I., 2006, Immune modulation by Pseudomonas aeruginosa quorum-sensing signal molecules. *Int. J. Med. Microbiol.* 296, 111–116

Rosenblat, M., Vaya, J., Shih, D., and Aviram, M., 2005, Paraoxonase 1 (PON1) enhances HDL-mediated macrophage cholesterol efflux via the ABCA1 transporter in association with increased HDL binding to the cells: a possible role for lysophosphatidylcholine. *Atherosclerosis* 179, 69–77

Rosenblat. M., Gaidukov, L., Khersonsky, O., Vaya, J., Oren, R., Tawfik, D.S., and Aviram, M., 2006, The Catalytic Histidine Dyad of High Density Lipoprotein-associated Serum Paraoxonase-1 (PON1) Is Essential for PON1-mediated Inhibition of Low Density Lipoprotein Oxidation and Stimulation of Macrophage Cholesterol Efflux. *J. Biol. Chem.* 281, 7657–7665.

Saeed, S.A., and Mitchell, M.D., 1983, Conversion of arachidonic acid to lipoxygenase products by human fetal tissues. *Biochem Med.* 30, 322–327

Schulam, P.G., Shearer, W.T., 1990, Evidence for 5-lipoxygenase activity in human B cell lines. A possible role for arachidonic acid metabolites during B cell signal transduction. *J. Immunol.* 144, 2696–2701

Shiner, E.K., Rumbaugh, K.P., and Williams, S.C., 2005, Inter-kingdom signaling: deciphering the language of acyl homoserine lactones. *FEMS Microbiol. Rev.* 29, 935–947

Shiner, E.K., Terentyev, D., Bryan, A., Sennoune, S., Martinez-Zaguilan, R., Li, G., Gyorke, S., Williams, S.C., and Rumbaugh, K.P., 2006, Pseudomonas aeruginosa autoinducer modulates host cell responses through calcium signalling. *Cell Microbiol.* 8, 1601–1610

Stoltz, D.A., Ozer, E.A., Ng, C.J., Yu, J., Reddy, S.T., Lusis, A.J., Bourquard, N., Parsek, M.R., Zabner, J., Shih, D.M., 2007, Paraoxonase-2 Deficiency Enhances *Pseudomonas aeruginosa* Quorum Sensing in Murine Tracheal Epithelia. *Am. J. Physiol. Lung Cell Mol. Physiol.* 292, 852–860

Teiber, J.F., Draganov, D.I., and La Du, B.N., 2004, Lactonase and lactonizing activities of human serum paraoxonase (PON1) and rabbit serum PON3. *Biochem. Pharmacol.* 66, 887–896

Yang, F., Wang, L.H., Wang, J., Dong, Y.H., Hu, J.Y., and Zhang, L.H., 2005, Quorum quenching enzyme activity is widely conserved in the sera of mammalian species. *FEBS Lett.* 579, 3713–3717

Zeldin, D.C., Kobayashi, J., Falck, J.R., Winder, B.S., Hammock, B.D., Snapper, J.R., and Capdevila, J.H., 1993, Regio- and enantiofacial selectivity of epoxyeicosatrienoic acid hydration by cytosolic epoxide hydrolase. *J. Biol. Chem.* 268, 6402–6407

Zimmermann, S., Wagner, C., Muller, W., Brenner-Weiss, G., Hug, F., Prior, B., Obst, U., and Hansch. G.M., 2006, Induction of neutrophil chemotaxis by the quorum-sensing molecule N-(3-oxodocecanoyl)-L-homoserine lactone. *Infect. Immun.* 74, 5687–5692

CHAPTER 20

PARAOXONASES, QUORUM SENSING, AND *PSEUDOMONAS AERUGINOSA*

D.A. STOLTZ, E.A. OZER AND J. ZABNER

Department of Internal Medicine, University of Iowa, Iowa City, IA 52242, USA

Abstract: *Pseudomonas aeruginosa* is a major cause of pulmonary infections in hospitalized, immunocompromised, and chronic lung disease patients, such as those with cystic fibrosis (CF). CF affects multiple organ systems, but infection in the airways is the most important clinical problem and the primary cause of death. *P. aeruginosa* can coordinate its activities and behave as a group through a population-dependent process termed quorum sensing. *P. aeruginosa* uses N-acyl homoserine lactone (AHL) quorum-sensing molecules to regulate the expression of genes implicated in bacterial virulence and biofilm formation. AHLs produced and recognized by *P. aeruginosa* include N-3-oxododecanoyl homoserine lactone (3OC12-HSL) and N-butanoyl homoserine lactone (C4-HSL). Certain molecules and bacterial lactonases have been identified that degrade or compete with quorum-sensing signals and alter the pathogenesis of Gram-negative organisms, including *P. aeruginosa*. We have found that paraoxonases (PONs) present in airway epithelial cells and serum can inactivate 3OC12-HSL's biological activity by hydrolyzing its lactone ring (lactonase activity). Of clinical interest, we have also shown that PON1 can inhibit *P. aeruginosa* biofilm formation. Finally, based upon studies using airway cells from PON2-deficient mice, we have demonstrated that PON2 is required for airway epithelia inactivation of 3OC12-HSL. These findings suggest that PONs may represent a novel component of the innate immune system designed to interfere with *P. aeruginosa* quorum-sensing control

Keywords: Paraoxonase, *Pseudomonas aeruginosa*, quorum sensing, 3OC12-HSL, biofilms, airway epithelial cells, host defense
quo·rum n. – a fixed minimum number of members of Congress who must be present for a session to legislate. From Latin quo: "where is (the)" and rum: "alcoholic beverage".

1.   INTRODUCTION

Most bacteria exist as individual microorganisms undergoing replication and growth, based upon environmental stimuli and nutrient availability. However, certain bacteria can monitor their population density and then coordinate their

307

B. Mackness et al. (eds.), The Paraoxonases: Their Role in Disease Development and Xenobiotic Metabolism, 307–319.
© 2008 *Springer.*

activities by behaving as a communal group. This phenomenon is termed autoinduction or "quorum sensing" (Juhas et al., 2005; Reading, 2006). Quorum sensing is a density-dependent process by which bacteria cooperate as a community and can induce phenotypic changes through the synthesis, secretion, and detection of small, signal molecules termed autoinducers (quorum-sensing signals). Most autoinducers diffuse freely across membranes and bind to the LuxR family of transcription factors thereby regulating their activity. Gram-positive bacteria utilize oligopeptides ranging from 5 to 17 amino acids in length as autoinducers, while Gram-negative bacteria utilize small widely conserved molecules termed acyl homoserine lactones (AHLs) (Camilli and Bassler, 2006)

Figure 1 depicts a model of *Pseudomonas aeruginosa* infection on the airway epithelia and the quorum-sensing pathway. In the early stages of an infection, the number of bacterial pathogens and hence the concentration of quorum-sensing molecules is low. However, the bacteria continue to replicate and secrete quorum-sensing molecules. Therefore, as the cell density increases so does the concentration of autoinducers. Once the bacterial population reaches a critical level, the quorum-sensing signals reach inducing concentrations and trigger a cascade of cellular events that leads to the synchronous induction or repression of quorum-sensing dependent genes (Fig. 1).

P. aeruginosa is an aerobic Gram-negative bacterium, ubiquitously found in water and soil, which rarely causes infection in normal hosts. However, it is a major cause of clinically relevant infections in hospitalized, immunocompromised, and chronic lung disease patients. *P. aeruginosa* can cause a spectrum of disease states from an acute, life-threatening infection, such as nosocomial pneumonia or

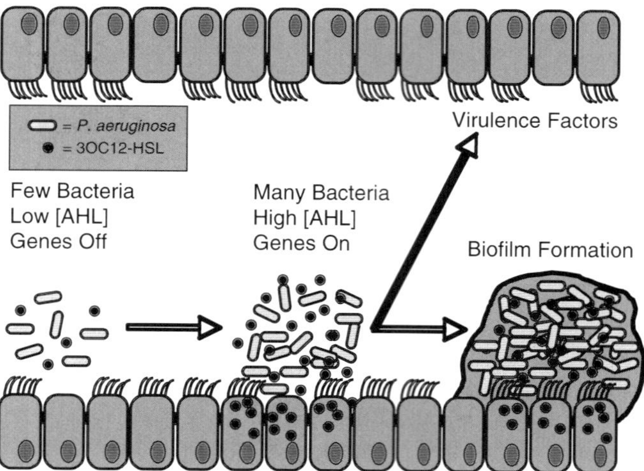

Figure 1. Overview of the *P. aeruginosa* quorum-sensing pathway. Schematic shows an airway lined with epithelial cells and depicts a simplified version of how *P. aeruginosa* might infect the airways of cystic fibrosis patients. AHL = acyl homoserine lactones

sepsis, to a more chronically destructive infection, such as that which occurs in cystic fibrosis (CF) patients. *P. aeruginosa* utilizes quorum sensing to regulate a number of its virulence factors (including exotoxin A, alkaline protease, superoxide dismutase, elastase, and pyocyanin) in addition to controlling biofilm formation (Smith and Iglewski, 2003).

The quorum-sensing signals self-produced by *P. aeruginosa* are in the form of AHLs. Two hierarchically regulated AHL systems are present in *P. aeruginosa*: the *las* and *rhl* systems. LasI codes for an enzyme that synthesizes N-3-oxododecanoyl homoserine lactone (3OC12-HSL) and lasR is the transcription factor that responds to 3OC12-HSL. RhlI codes for an enzyme that synthesizes N-butanoyl homoserine lactone (C4-HSL) and RhlR is the transcription factor that responds to C4-HSL (Juhas et al., 2005). Production of C4-HSL is positively regulated by 3OC12-HSL synthesis. The *las* and *rhl* quorum-sensing systems regulate expression of as many as 4–6% of the approximately 6,000 genes expressed by *P. aeruginosa* (Whiteley et al., 1999; Schuster et al., 2003; Wagner et al., 2003). Quorum-sensing deficient bacteria have been shown to be less virulent in a *C. elegans* infection model (Tan et al., 1999) and in animal models of burn wound infection, neonatal pneumonia, and chronic pulmonary *P. aeruginosa* infection (Tang et al., 1996; Rumbaugh et al., 1999; Pearson et al., 2000; Wu et al., 2001; Imamura et al., 2005).

Interference with quorum-sensing pathways has been studied as a potential therapeutic modality against human and plant bacterial pathogens (Gonzalez and Keshavan, 2006). An example in prokaryotes is the quorum-sensing inhibitor AiiA, a lactonase from the *Bacillus* sp. bacteria (Dong et al., 2000; Dong, et al., 2002). Transgenic expression of *aiiA* in tobacco or cabbage plants protected them from soft-rot infection induced by *Erwinia carotovora*, a quorum-sensing dependent plant pathogen (Dong et al., 2000, Dong et al., 2001). The Australian red alga, *Delisea pulchra* , synthesizes furanones that are structural analogs of AHLs (Givskov et al., 1996). Furanones bind to LuxR proteins, similar to AHLs, but they form an unstable complex and accelerate LuxR turnover (Manefield et al., 2002). Treatment of mice with a synthetic analogue of furanone improved bacterial clearance and mortality in a murine model of *P. aeruginosa* pulmonary infection (Hentzer et al., 2003; Wu et al., 2004). Similarly, mice treated with garlic extract, another recently described quorum-sensing inhibitor, showed enhanced bacterial clearance in a pneumonia model (Bjarnsholt et al., 2005). These data suggest that quorum-sensing inhibitors may represent a novel approach to treatment of *P. aeruginosa* infections.

The total surface area of the alveolar epithelium is approximately 100 m^2 and is oftentimes equated to the size of a tennis court. This extensive epithelial surface is required for gas exchange, the primary function of the lung. We breathe approximately 8,000–10,000 liters of air per day. This means that the pulmonary system is continuously exposed to a large number of foreign substances, including infectious pathogens and toxic agents. While cough and mucociliary clearance do a very good job of preventing distal passage of many of these toxins, other defense mechanisms exist to clear and/or destroy pathogens that escape the lung's initial defense systems. These include the airway surface liquid, airway epithelial cells, and resident alveolar macrophages. The airway epithelium is uniquely poised to combat invading microbes through elaboration of a number of antimicrobial peptides, including cathelicidin, defensins, lactoferrin, lysozyme, and secretory leukocyte proteinase inhibitor (Bals and Hiemstra, 2004).

Our laboratory has had a long-standing interest in studying airway epithelial cell biology and novel innate immune mechanisms important for host defense against bacterial pathogens. This chapter will summarize our recent data showing that human and murine airway epithelial cells inactivate the key *P. aeruginosa* quorum-sensing molecule, 3OC12-HSL. Furthermore, we have shown that paraoxonase (PON) 1, 2, and 3 can also degrade 3OC12-HSL and are present in airway epithelial cells. While the endogenous PON substrate(s) is/are unknown, it is now generally accepted that PON's native enzymatic activity is that of a lactonase (Draganov et al., 2005; Khersonsky and Tawfik, 2005; Ozer et al., 2005). We propose that AHLs, important in *P. aeruginosa* pathogenesis, may represent a natural substrate for members of the PON family.

2. AIRWAY EPITHELIAL CELLS AND PARAOXONASES INACTIVATE 3OC12-HSL

Our initial interest in PON resulted from earlier experiments in which we tested if transgenic expression of AiiA in human airway epithelia would inactivate 3OC12-HSL. Indeed airway epithelia, expressing AiiA, inactivated 3OC12-HSL, but to a similar degree as airway epithelia infected with the control adenovirus vector. This led to the hypothesis that airway epithelial cells possess the capacity to inactivate quorum-sensing signals (Fig. 2A). This concept was attractive from a host defense standpoint because it suggested that the airways had adapted a defense mechanism against quorum-sensing dependent pathogens. Further experiments showed that:

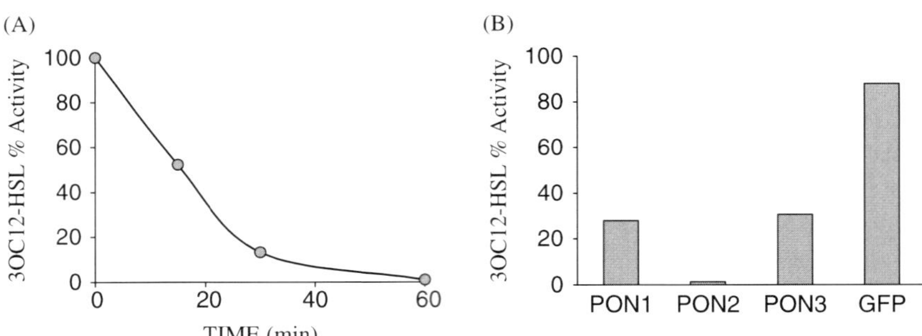

Figure 2. (A) 3OC12-HSL inactivation by primary human airway epithelial cells cultured at the air-liquid interface. (B) Human PON1, 2, and 3 inactivate 3OC12-HSL. CHO cells were transfected with an adenovirus expressing hPON1, hPON2, hPON3, or GFP. 3OC12-HSL was added (final concentration of 10 μM) to the apical and basolateral surfaces of airway cultures or to CHO medium. 3OC12-HSL biological activity was determined in a bioassay

(1) airway cells could inactivate 3OC12-HSL, but not C4-HSL, (2) boiling of airway cell lysates destroyed the 3OC12-HSL degrading activity, (3) loss of biological activity of 3OC12-HSL was dependent upon cell lysate concentration and was time dependent, and (4) the degrading activity was localized to the cell membrane-containing particulate fraction of cells and not secreted (Chun et al., 2004). These results suggested that an enzyme was responsible for the 3OC12-HSL degrading activity in airway cells. Bacteria possess enzymes able to inactivate AHLs, including AHL acylases, which cleave an amide bond resulting in the homoserine lactone and corresponding fatty acid (Lin et al., 2003), and AHL lactonases, which cleave the lactone ring and yield acyl-homoserines (Dong et al., 2001). A mammalian equivalent had yet to be described.

Although the mechanism of 3OC12-HSL inactivation by airway cells was unknown, the data suggested that this inactivating activity likely represented a not yet identified enzyme. Using NMR, MS, and HPLC techniques, we showed that exposure of 3OC12-HSL to A459 airway cells demonstrated findings consistent with lactone ring hydrolysis (Ozer et al., 2005). Similar data was obtained when 3OC12-HSL was exposed to alkaline conditions, a treatment that is known to hydrolyze AHL lactone rings (Yates et al., 2002). These data suggested that airway epithelial cells expressed an enzyme with lactonase activity. Since the human enzyme PON has lactonase activity, we hypothesized that PON may be responsible for disrupting the biological activity of 3OC12-HSL.

To test this hypothesis, 3OC12-HSL was incubated with serum, an environment rich in PON1. 3OC12-HSL was degraded in the presence of serum and this degradation was impaired by treatment with several PON inhibitors including EDTA, oxindole, and isatin (Ozer et al., 2005). Similar findings were reported by Yang et al. (2005) who demonstrated that sera from 6 mammalian species could all degrade 3OC12-HSL. When serum from PON1-KO mice was incubated with 3OC12-HSL, the quorum-sensing molecule retained its biological activity (Table 1). Adding back purified PON1 to serum from PON1-KO mice restored the inactivating capacity of PON1-KO serum. These data show that in mouse serum PON1 is both required and sufficient to degrade 3OC12-HSL. PON2 and PON3 also have lactonase activity and we next tested their individual ability to degrade 3OC12-HSL. CHO cells were infected with adenoviruses expressing hPON1, 2, or 3 and 3OC12-HSL was incubated with the cells. 3OC12-HSL biological activity was decreased in the presence of all three PON family members, with PON2 having the greatest lactonase activity (Fig. 2B) (Ozer et al., 2005). Based upon structure-function studies, PON1's native enzyme activity is predicted to be a lactonase.

Table 1. PON1-KO Serum is deficient in 3OC12-HSL Degrading activity

	Control	Wildtype serum	PON1-KO Serum
3OC12-HSL% Activity	100%	0%	78.9%
Biofilm Biomass	84.1%	32.2%	69.8%

Draganov et al. (2005) and Yang et al. (2005) have also reported that members of the PON family can inactivate AHLs, and that PON2 has the greatest lactonase activity.

3. PON1 INHIBITS *P. AERUGINOSA* BIOFILM FORMATION

We were next interested in determining if PON could interfere with *P. aeruginosa* quorum-sensing mediated events. Quorum sensing has been shown to regulate the ability of these bacteria to form mature biofilms. Biofilms represent structured communities of bacteria, composed of bacterial pathogens encased in an extracellular matrix (Fig. 3). This mode of bacterial growth, which is highly resistant to antibiotics (Nickel et al., 1985), is important in many clinically relevant infections, including urinary tract infections, burn wounds, and chronic lung infections (Parsek and Singh, 2003). Chronic *P. aeruginosa* lung colonization/infection in cystic fibrosis patients represents a biofilm mode of bacterial growth (Costerton et al., 1999; Singh et al., 2000; Prince, 2002). *P. aeruginosa* deficient in 3OC12-HSL production show reduced biofilm formation (Davies et al., 1998; De Kievit et al., 2001; Kirisits and Parsek, 2006). We hypothesized that PON would inhibit 3OC12-HSL signaling activity leading to decreased biofilm growth. Using an in vitro model of *P. aeruginosa* biofilm growth, bacteria were grown in the presence of serum from wildtype or PON1-KO mice. Wildtype serum attenuated biofilm growth by approximately 70%, but PON1 deficient serum only decreased growth by 30% (Table 1). PON1 addition to PON1-KO serum restored the biofilm inhibitory effect (Ozer et al., 2005). These data support the hypothesis that PON1 inactivates 3OC12-HSL-mediated quorum sensing and can inhibit biofilm growth in vitro.

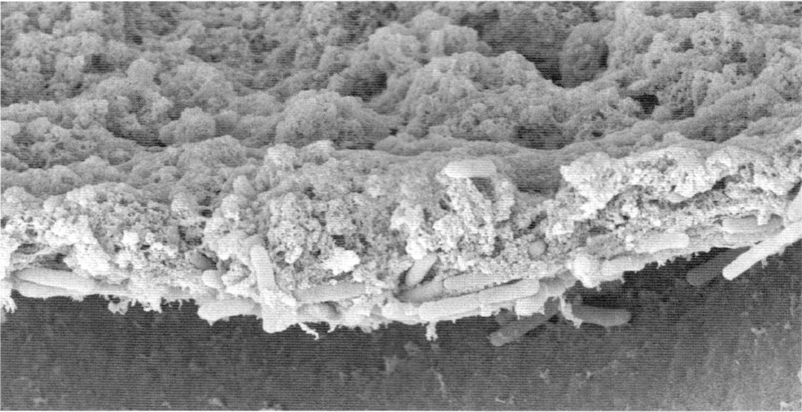

Figure 3. P. aeruginosa biofilm formation on the surface of a vascular catheter Rod-shaped bacteria are encased in a self-produced extracellular matrix

4. PON1-KO MICE ARE PARADOXICALLY PROTECTED FROM *P. AERUGINOSA* SEPSIS

Quorum-sensing mutants of *P. aeruginosa*, deficient in 3OC12-HSL and/or C4-HSL production, show reduced virulence in several animal models of infection including burn wound infection, neonatal pneumonia, and chronic pulmonary *P. aeruginosa* infection (Tang et al., 1996; Rumbaugh et al., 1999; Pearson et al., 2000; Wu et al., 2001; Imamura et al., 2005). To investigate the role of PONs in host defense, we used a murine model of *P. aeruginosa* sepsis. Control mice infected with a wildtype strain of *P. aeruginosa* (PAO1) experienced a 55% mortality compared to less than 10% mortality in mice inoculated with a 3OC12-HSL deficient *P. aeruginosa* strain. These data were the basis for our hypothesis that mice deficient in PON1 would have increased lethality when infected with PAO1. In contrast, we observed that PON1-KO mice were paradoxically protected from a lethal *P. aeruginosa* infection with 90% of PON1-KO mice surviving at 50 hours post-infection, compared to 50% of wildtype control mice (Ozer et al., 2005). Several possible explanations may account for these findings. First, PON2 and PON3 levels may have been up-regulated in PON1-KO mice as a compensatory mechanism to the loss of PON1. Second, a priming effect may have occurred in the PON1-KO mice. In the absence of PON1, key components of the inflammatory and host defense pathways may have been up-regulated leading to a more robust innate immune response against *P. aeruginosa*. Finally, absence of PON1 activity may have led to increased quorum-sensing signals and enhanced biofilm formation. This mode of bacterial growth may be less inflammatory in nature (more representative of a chronic infection) and may have protected the mice from the septic response that was responsible for *P. aeruginosa*-induced lethality.

5. PON2-KO MURINE AIRWAY CELLS HAVE IMPAIRED 3OC12-HSL INACTIVATION

The loss of 3OC12-HSL's biological activity following exposure to human airway epithelial cells is consistent with an enzyme with lactonase activity. Using microarray analysis for human airway epithelial cultures and RT-PCR for murine airway epithelial cultures, we determined if PON1, 2, or 3 were present in these cells. Transcripts for all three PONs were present in both human and murine airway cultures. However, the relative levels and distributions differed between species. PON2 was present in the greatest concentration in human cells, while PON2 and 3 transcripts were higher than PON1 in murine airway cells. To test the importance of PON for 3OC12-HSL inactivation by airway cells, we prepared polarized murine tracheal epithelial cell cultures grown at the air-liquid interface from mice deficient in either PON1, 2 or 3 and tested 3OC12-HSL inactivation. 3OC12-HSL degradation was similar in all three groups when the quorum-sensing molecule was applied to the apical and basolateral surfaces of intact airway epithelia. However, when 3OC12-HSL was incubated in the presence of cell lysates from

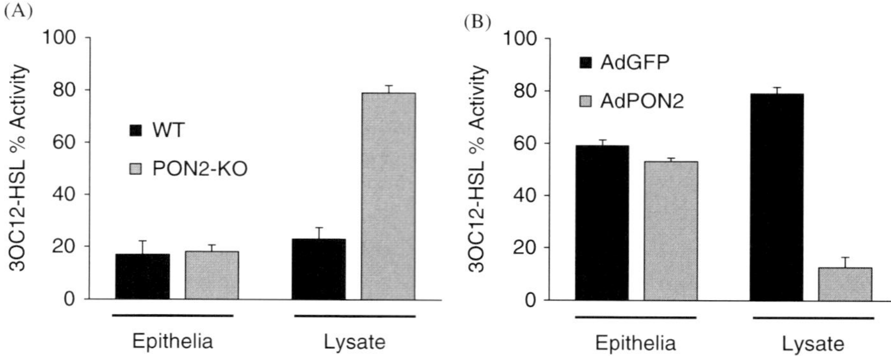

Figure 4. (A) 3OC12-HSL inactivation by intact murine airway epithelia and cell lysates from wildtype and PON2-KO mice. (B) 3OC12-HSL inactivation by intact human airway epithelia and cell lysates following transfection with an adenovirus expressing hPON2 or GFP

these airway cultures, PON2-KO lysates demonstrated impaired 3OC12-HSL inactivation (Fig. 4A) (Stoltz et al., 2006). This suggested that diffusion of 3OC12-HSL across the cell membrane might be the rate-limiting step for 3OC12-HSL inactivation.

If diffusion of 3OC12-HSL across the cell membrane is the rate-limiting step for lactone ring hydrolysis, we hypothesized that over-expression of PON2 in airway cells would not enhance 3OC12-HSL inactivation when tested on intact epithelia. However, we should observe increased inactivation with cell lysate preparations. Human airway epithelial cells were infected with an adenovirus expressing PON2 and 48 hours later 3OC12-HSL inactivation by intact epithelia and cell lysates was quantified. Over-expression of PON2 had no effect on 3OC12-HSL inactivation by intact epithelia, but inactivation was more than doubled in lysate preparations (Fig. 4B) (Stoltz et al., 2006). These data support the concept that PONs are responsible for airway cell 3OC12-HSL inactivation and at the concentration of 3OC12-HSL (10 μM) used in these studies, diffusion across the cell membrane is the rate-limiting step for inactivation.

6. LOSS OF PON ENHANCES *P. AERUGINOSA* QUORUM-SENSING ACTIVITY

To more closely examine the role of airway cells in the regulation of *P. aeruginosa* quorum-sensing, we performed experiments with a quorum-sensing reporter strain termed PAO1-qsc-102-lacZ that produces β-galactosidase (β-gal) in response to increased quorum-sensing signals. We used PON2-KO intact airway epithelia since PON2 has the greatest lactonase activity and is present in high levels in airway cells. A concentrated aliquot of *P. aeruginosa* (10^5 cfu in 2 μl) was applied to the apical surface of wildtype and PON2-KO airway cells. Six hours later bacteria were harvested and β-gal levels were determined. β-gal levels were 2–2.5 fold higher in bacteria harvested from PON2-KO airway epithelia, compared to controls (Stoltz et al., 2006). These results demonstrate that PON2 is important for airway epithelial cell control of *P. aeruginosa* quorum sensing.

7.   CONCLUDING REMARKS

The data suggest that the PONs may interfere with quorum-sensing regulation by *P. aeruginosa*. In addition to 3OC12-HSL's effects on *P. aeruginosa* virulence and biofilm formation, multiple studies have shown that 3OC12-HSL can regulate several cellular processes in eukaryotic cells, including components of the host defense system. Recently, Kravchenko et al. demonstrated that 3OC12-HSL treatment of alveolar macrophages leads to phosphorylation of both p38 and eukaryotic translation initiation factor 2a (Kravchenko et al., 2006). 3OC12-HSL treatment of bone marrow-derived macrophages, neutrophils, and monocyte cell lines has been shown to accelerate apoptosis in a time- and concentration-dependent manner (Tateda et al., 2003). Several investigators have examined 3OC12-HSL's effects on cytokine production and regulation. Tumor necrosis factor-alpha (TNF-α) production is a critical component of the early inflammatory response to invading bacterial pathogens. Lipopolysaccharide-induced TNF-α production was inhibited in peritoneal macrophages following treatment with 3OC12-HSL (Telford et al., 1998). In contrast, 3OC12-HSL has also been shown to up-regulate other proinflammatory molecules including interleukin-1α, interleukin-6, interleukin-8, cyclooxygenase-2, and prostaglandin E2 (Smith et al., 2002).

So what might be the role of the PONs in host defense? This will be a difficult question to answer until a PON triple-knockout mouse becomes available. Results from bacterial infection studies using mice deficient in either PON1, 2, or 3 are difficult to interpret due to the redundancy of the PON family members. However, we speculate that PONs may play a key role in *P. aeruginosa* quorum-sensing control. Figure 5 depicts a model of how PONs may regulate *P. aeruginosa* quorum sensing and components of the host defense system that respond to 3OC12-HSL. As *P. aeruginosa* replicates, the concentration of 3OC12-HSL increases over time (Fig. 5A). High extracellular concentrations of 3OC12-HSL will promote the molecule's diffusion into airway, inflammatory, and bacterial cells. This will lead to activation of quorum-sensing dependent genes in *P. aeruginosa*. However, PON2 may modulate this effect by degrading 3OC12-HSL (Fig. 5B). While we have been unable to detect lactonase activity in the airway surface liquid or bronchoalveolar lavage fluid (De Kievit et al., 2001; Chun et al., 2004; Stoltz et al., 2006), suggesting that PON2 is not secreted, Shamir et al. reported that PON2 was preferentially secreted to the apical compartment in cultures of polarized Caco-2 cells (Shamir et al., 2005). Therefore, it is conceivable that under certain conditions (i.e., during a bacterial infection) PONs may be secreted into the infected area and further degrade quorum-sensing signals (Fig. 5C). Alternatively, PON2 may be important for control of intracellular bacterial pathogens. A recent report by

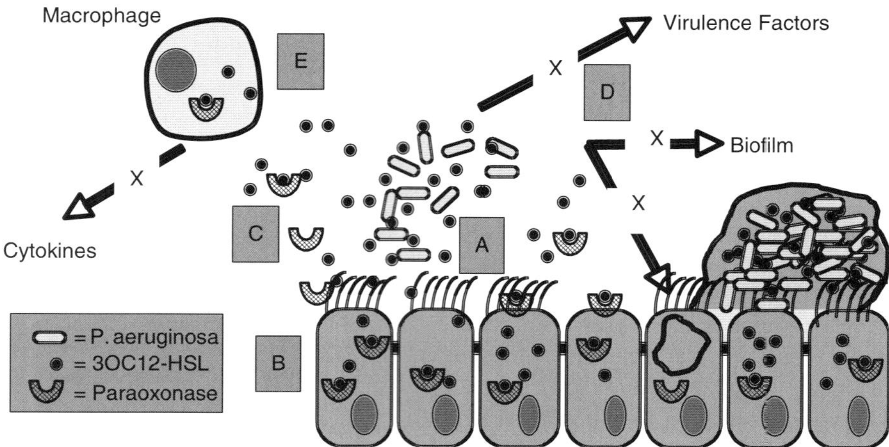

Figure 5. Proposed model of paraoxonase's role in *P. aeruginosa* quorum sensing and host defense

Garcia-Medina et al. (2005) showed that *P. aeruginosa* can form biofilm-like structures in the cytosolic compartment of airway epithelial cells. One could hypothesize that PON may regulate this intracellular process (Fig. 5D). PON1 may function similarly to PON2, by defending against quorum-sensing dependent bacterial infections in the bloodstream, such as endocarditis, septicemia, or intravascular-catheter related infections. Finally, members of the PON family may regulate 3OC12-HSL-induced effects on cells and pathways involved in the host defense system (Fig. 5E). In conclusion, the exact role and importance of PON in the host defense system remains to be fully determined, but we propose that members of the PON family may represent a novel component of innate immunity.

ACKNOWLEDGEMENT

We would like to thank Tom Moninger and Phil Karp for exceptional technical assistance in electron microscopy and cell culture. This work was funded by NHLBI 61234-09.

REFERENCES

Bals, R., and P. S. Hiemstra. 2004. Innate immunity in the lung: how epithelial cells fight against respiratory pathogens. Eur Respir J **23**:327–333.

Bjarnsholt, T., P. O. Jensen, T. B. Rasmussen, L. Christophersen, H. Calum, M. Hentzer, H. P. Hougen, J. Rygaard, C. Moser, L. Eberl, N. Hoiby, and M. Givskov. 2005. Garlic blocks quorum sensing and promotes rapid clearing of pulmonary *Pseudomonas aeruginosa* infections. Microbiology **151**:3873–3880.

Camilli, A., and B. L. Bassler. 2006. Bacterial small-molecule signaling pathways. Science **311**: 1113–1116.

Chun, C. K., E. A. Ozer, M. J. Welsh, J. Zabner, and E. P. Greenberg. 2004. Inactivation of a *Pseudomonas aeruginosa* quorum-sensing signal by human airway epithelia. Proc Natl Acad Sci U S A **101**:3587–3590.

Costerton, J. W., P. S. Stewart, and E. P. Greenberg. 1999. Bacterial biofilms: a common cause of persistent infection. Science **284**:1318–1322.

Davies, D. G., M. R. Parsek, J. P. Pearson, B. H. Iglewski, J. W. Costerton, and E. P. Greenberg. 1998. The involvement of cell-to-cell signals in the development of a bacterial biofilm. Science **280**:295–298.

De Kievit, T. R., R. Gillis, S. Marx, C. Brown, and B. H. Iglewski. 2001. Quorum-sensing genes in *Pseudomonas aeruginosa* biofilms: their role and expression patterns. Appl Environ Microbiol **67**:1865–1873.

Dong, Y. H., A. R. Gusti, Q. Zhang, J. L. Xu, and L. H. Zhang. 2002. Identification of quorum-quenching N-acyl homoserine lactonases from Bacillus species. Appl Environ Microbiol **68**:1754–1759.

Dong, Y. H., L. H. Wang, J. L. Xu, H. B. Zhang, X. F. Zhang, and L. H. Zhang. 2001. Quenching quorum-sensing-dependent bacterial infection by an N-acyl homoserine lactonase. Nature **411**:813–817.

Dong, Y. H., J. L. Xu, X. Z. Li, and L. H. Zhang. 2000. AiiA, an enzyme that inactivates the acylhomoserine lactone quorum-sensing signal and attenuates the virulence of Erwinia carotovora. Proc Natl Acad Sci U S A **97**:3526–3531.

Draganov, D. I., J. F. Teiber, A. Speelman, Y. Osawa, R. Sunahara, and B. N. La Du. 2005. Human paraoxonases (PON1, PON2, and PON3) are lactonases with overlapping and distinct substrate specificities. J Lipid Res **46**:1239–1247.

Garcia-Medina, R., W. M. Dunne, P. K. Singh, and S. L. Brody. 2005. *Pseudomonas aeruginosa* acquires biofilm-like properties within airway epithelial cells. Infect Immun **73**:8298–8305.

Givskov, M., R. de Nys, M. Manefield, L. Gram, R. Maximilien, L. Eberl, S. Molin, P. D. Steinberg, and S. Kjelleberg. 1996. Eukaryotic interference with homoserine lactone-mediated prokaryotic signalling. J Bacteriol **178**:6618–6622.

Gonzalez, J. E., and N. D. Keshavan. 2006. Messing with bacterial quorum sensing. Microbiol Mol Biol Rev **70**:859–875.

Hentzer, M., H. Wu, J. B. Andersen, K. Riedel, T. B. Rasmussen, N. Bagge, N. Kumar, M. A. Schembri, Z. Song, P. Kristoffersen, M. Manefield, J. W. Costerton, S. Molin, L. Eberl, P. Steinberg, S. Kjelleberg, N. Hoiby, and M. Givskov. 2003. Attenuation of *Pseudomonas aeruginosa* virulence by quorum sensing inhibitors. Embo J **22**:3803–3815.

Imamura, Y., K. Yanagihara, K. Tomono, H. Ohno, Y. Higashiyama, Y. Miyazaki, Y. Hirakata, Y. Mizuta, J. Kadota, K. Tsukamoto, and S. Kohno. 2005. Role of *Pseudomonas aeruginosa* quorum-sensing systems in a mouse model of chronic respiratory infection. J Med Microbiol **54**:515–518.

Juhas, M., L. Eberl, and B. Tummler. 2005. Quorum sensing: the power of cooperation in the world of Pseudomonas. Environ Microbiol **7**:459–471.

Khersonsky, O., and D. S. Tawfik. 2005. Structure-reactivity studies of serum paraoxonase PON1 suggest that its native activity is lactonase. Biochemistry **44**:6371–6382.

Kirisits, M. J., and M. R. Parsek. 2006. Does *Pseudomonas aeruginosa* use intercellular signalling to build biofilm communities? Cell Microbiol **8**:1841–1849.

Kravchenko, V. V., G. F. Kaufmann, J. C. Mathison, D. A. Scott, A. Z. Katz, M. R. Wood, A. B. Brogan, M. Lehmann, J. M. Mee, K. Iwata, Q. Pan, C. Fearns, U. G. Knaus, M. M. Meijler, K. D. Janda, and R. J. Ulevitch. 2006. N-(3-OXO-ACYL) homoserine lactones signal cell activation through a mechanism distinct from the canonical pathogen-associated molecular pattern recognition receptor pathways. J Biol Chem **281**:28822–28830.

Lin, Y. H., J. L. Xu, J. Hu, L. H. Wang, S. L. Ong, J. R. Leadbetter, and L. H. Zhang. 2003. Acyl-homoserine lactone acylase from Ralstonia strain XJ12B represents a novel and potent class of quorum-quenching enzymes. Mol Microbiol **47**:849–860.

Manefield, M., T. B. Rasmussen, M. Henzter, J. B. Andersen, P. Steinberg, S. Kjelleberg, and M. Givskov. 2002. Halogenated furanones inhibit quorum sensing through accelerated LuxR turnover. Microbiology **148**:1119–1127.

Nickel, J. C., I. Ruseska, J. B. Wright, and J. W. Costerton. 1985. Tobramycin resistance cf *Pseudomonas aeruginosa* cells growing as a biofilm on urinary catheter material. Antimicrob Agents Chemother **27**:619–624.

Ozer, E. A., A. Pezzulo, D. M. Shih, C. Chun, C. Furlong, A. J. Lusis, E. P. Greenberg, and J. Zabner. 2005. Human and murine paraoxonase 1 are host modulators of *Pseudomonas aeruginosa* quorum-sensing. FEMS Microbiol Lett **253**:29–37.

Parsek, M. R., and P. K. Singh. 2003. Bacterial biofilms: an emerging link to disease pathogenesis. Annu Rev Microbiol **57**:677–701.

Pearson, J. P., M. Feldman, B. H. Iglewski, and A. Prince. 2000. *Pseudomonas aeruginosa* cell-to-cell signaling is required for virulence in a model of acute pulmonary infection. Infect Immun **68**:4331–4334.

Prince, A. S. 2002. Biofilms, antimicrobial resistance, and airway infection. N Engl J Med **347**:1110–1111.

Reading, N. C. 2006. Quorum sensing: the many languages of bacteria. FEMS Microbiol Lett **251**:1–11.

Rumbaugh, K. P., J. A. Griswold, B. H. Iglewski, and A. N. Hamood. 1999. Contribution of quorum sensing to the virulence of *Pseudomonas aeruginosa* in burn wound infections. Infect Immun **67**:5854–5862.

Schuster, M., C. P. Lostroh, T. Ogi, and E. P. Greenberg. 2003. Identification, timing, and signal specificity of *Pseudomonas aeruginosa* quorum-controlled genes: a transcriptome analysis. J Bacteriol **185**:2066–2079.

Shamir, R., C. Hartman, R. Karry, E. Pavlotzky, R. Eliakim, J. Lachter, A. Suissa, and M. Aviram. 2005. Paraoxonases (PONs) 1, 2, and 3 are expressed in human and mouse gastrointestinal tract and in Caco-2 cell line: selective secretion of PON1 and PON2. Free Radic Biol Med **39**:336–344.

Singh, P. K., A. L. Schaefer, M. R. Parsek, T. O. Moninger, M. J. Welsh, and E. P. Greenberg. 2000. Quorum-sensing signals indicate that cystic fibrosis lungs are infected with bacterial biofilms. Nature **407**:762–764.

Smith, R. S., S. G. Harris, R. Phipps, and B. Iglewski. 2002. The *Pseudomonas aeruginosa* quorum-sensing molecule N-(3-oxododecanoyl)homoserine lactone contributes to virulence and induces inflammation in vivo. J Bacteriol **184**:1132–1139.

Smith, R. S., and B. H. Iglewski. 2003. *P. aeruginosa* quorum-sensiing systems and virulence. Curr Opin Microbiol **6**:56–60.

Stoltz, D. A., E. A. Ozer, C. J. Ng, J. Yu, S. T. Reddy, A. J. Lusis, N. Bourquard, M. R. Parsek, J. Zabner, and D. M. Shih. 2006. Paraoxonase-2 Deficiency Enhances *Pseudomonas aeruginosa* Quorum Sensing in Murine Tracheal Epithelia. Am J Physiol Lung Cell Mol Physiol.

Tan, M. W., S. Mahajan-Miklos, and F. M. Ausubel. 1999. Killing of Caenorhabditis elegans by *Pseudomonas aeruginosa* used to model mammalian bacterial pathogenesis. Proc Natl Acad Sci U S A **96**:715–720.

Tang, H. B., E. DiMango, R. Bryan, M. Gambello, B. H. Iglewski, J. B. Goldberg, and A. Prince. 1996. Contribution of specific *Pseudomonas aeruginosa* virulence factors to pathogenesis of pneumonia in a neonatal mouse model of infection. Infect Immun **64**:37–43.

Tateda, K., Y. Ishii, M. Horikawa, T. Matsumoto, S. Miyairi, J. C. Pechere, T. J. Standiford, M. Ishiguro, and K. Yamaguchi. 2003. The *Pseudomonas aeruginosa* autoinducer N-3-oxododecanoyl homoserine lactone accelerates apoptosis in macrophages and neutrophils. Infect Immun **71**:5785–5793.

Telford, G., D. Wheeler, P. Williams, P. T. Tomkins, P. Appleby, H. Sewell, G. S. Stewart, B. W. Bycroft, and D. I. Pritchard. 1998. The *Pseudomonas aeruginosa* quorum-sensing signal molecule N-(3-oxododecanoyl)-L-homoserine lactone has immunomodulatory activity. Infect Immun **66**:36–42.

Wagner, V. E., D. Bushnell, L. Passador, A. I. Brooks, and B. H. Iglewski. 2003. Microarray analysis of *Pseudomonas aeruginosa* quorum-sensing regulons: effects of growth phase and environment. J Bacteriol **185**:2080–2095.

Whiteley, M., K. M. Lee, and E. P. Greenberg. 1999. Identification of genes controlled by quorum sensing in *Pseudomonas aeruginosa*. Proc Natl Acad Sci U S A **96**:13904–13909.

Wu, H., Z. Song, M. Givskov, G. Doring, D. Worlitzsch, K. Mathee, J. Rygaard, and N. Hoiby. 2001. *Pseudomonas aeruginosa* mutations in lasI and rhlI quorum sensing systems result in milder chronic lung infection. Microbiology **147**:1105–1113.

Wu, H., Z. Song, M. Hentzer, J. B. Andersen, S. Molin, M. Givskov, and N. Hoiby. 2004. Synthetic furanones inhibit quorum-sensing and enhance bacterial clearance in *Pseudomonas aeruginosa* lung infection in mice. J Antimicrob Chemother **53**:1054–1061.

Yang, F., L. H. Wang, J. Wang, Y. H. Dong, J. Y. Hu, and L. H. Zhang. 2005. Quorum quenching enzyme activity is widely conserved in the sera of mammalian species. FEBS Lett **579**:3713–3717.

Yates, E. A., B. Philipp, C. Buckley, S. Atkinson, S. R. Chhabra, R. E. Sockett, M. Goldner, Y. Dessaux, M. Camara, H. Smith, and P. Williams. 2002. N-acylhomoserine lactones undergo lactonolysis in a pH-, temperature-, and acyl chain length-dependent manner during growth of a Yersinia pseudotuberculosis and *Pseudomonas aeruginosa*. Infect Immun **70**:5635–5646.

INDEX

Proteins and Cell Regulation

1. R.A. Kahn (ed.): *ARF Family GTPases*. 2004   ISBN 1-4020-1719-7
2. J. Frampton (ed.): *Myb Transcription Factors: Their Role in Growth, Differentiation and Disease*. 2004.   ISBN 1-4020-2779-6
3. E. Manser (ed.): *RHO Family GTPases*. 2005   ISBN 1-4020-3461-X
4. C. Der (ed.): *RAS Family GTPases*. 2006   ISBN 1-4020-4328-7
5. P. ten Dijke and C.-H. Heldin (eds.): *Smad Signal Transduction: Smads in Proliferation, Differentiation and Disease*. 2006   ISBN 1-4020-4542-5
6. B. Mackness, M. Mackness, M. Aviram and G. Paragh (eds.): *The Paraoxonases: Their Role in Disease Development and Xenobiotic Metabolism*. 2008   ISBN 978-1-4020-6560-6